Handbook of Mobile Data Privacy

Aris Gkoulalas-Divanis • Claudio Bettini

Editors

Handbook of Mobile Data Privacy

 Springer

Editors
Aris Gkoulalas-Divanis
IBM Watson Health Headquarters
Cambridge, MA, USA

Claudio Bettini
Dipartimento di Informatica
University of Milan
Milan, Italy

ISBN 978-3-030-07459-3   ISBN 978-3-319-98161-1   (eBook)
https://doi.org/10.1007/978-3-319-98161-1

This Springer imprint is published by the registered company Springer Nature Switzerland AG.
The registered company address is: Gewerbestrasse 11, 6330 Cham, Switzerland

Dedicated to my parents, Aspa and Dimitris, and to my lovely wife Elena for all their support.

–Aris Gkoulalas-Divanis

Dedicated to my parents, Anna Maria and Giovanni, and to my son Matteo.

–Claudio Bettini

Preface

The advances in mobile devices and positioning technologies, together with the progress in spatiotemporal database research, have made possible the tracking of mobile devices (and their human companions) at a very high accuracy while supporting the efficient storage of mobility data in data warehouses. This has provided the means to collect, store, and process mobility data of an unprecedented quantity, quality, and timeliness. As ubiquitous computing pervades our society, user mobility data represents a very useful, but also extremely sensitive, source of information. On the one hand, the movement traces that are left behind by the mobile devices of the users can be very useful in a wide spectrum of applications such as urban planning, traffic engineering, and environmental pollution management. On the other hand, the disclosure of mobility data to third parties may severely jeopardize the privacy of the users whose movement is recorded, leading to abuse scenarios such as user tailing and profiling.

A significant body of research work has been conducted in the last 15 years in the area of mobility data privacy, along a number of important research directions such as privacy-preserving mobility data management, privacy in location sensing technologies and location-based services, privacy in vehicular communication networks, privacy in location-based social networks, and privacy in participatory sensing systems. This work has identified important privacy gaps in the use of human mobility data and has resulted to the adoption of international laws for location privacy protection (e.g., in EU, the USA, Canada, Australia, New Zealand, Japan, Singapore), as well as to a large number of interesting technologies for privacy-protecting mobility data, some of which have been made available through open-source systems and featured in real-world applications.

The overarching aim of this book is to survey the field of mobility data privacy and to present the fundamental principles and theory, as well as the state-of-the-art research, systems, and applications, to a wide audience including non-experts. Emphasis in the book is given toward coverage of the most recent directions in mobility data privacy. The structure of the book closely follows the main categories of research works that have been recently undertaken to protect user privacy in the context of mobility data and applications. After the first introductory chapters, which cover the fundamentals around the offering of privacy in mobility data, each subsequent chapter of the book surveys an important research problem related to mobility data privacy and discusses the corresponding privacy threats that have been identified and the solutions that have been proposed. The last part of the book is devoted to state-of-the-art systems for mobility data privacy, as well as to real-world applications where privacy-protection techniques have been applied.

We would like to note that this book is primarily addressed to computer science and statistics researchers and educators, who are interested in topics related to mobility privacy. We expect that the book will be also valuable to industry developers, as it covers the state-of-the-art algorithms for offering privacy. To ease understanding by nonexperts, the chapters contain a lot of background material, as well as many examples and citations to related literature. By discussing a wide range of privacy techniques, providing in-depth coverage of the most important ones, and highlighting promising avenues for future research, this book also aims at attracting computer science and statistics students to this fascinating field of research.

Boston, MA, USA   Aris Gkoulalas-Divanis
Milan, Italy   Claudio Bettini
June 2018

Acknowledgments

We would like to thank all authors who contributed chapters to this handbook, for their valuable contributions. This work would not have been possible without their efforts. A total of 31 authors who hold positions in leading academic institutions and industry, in Italy, Hungary, Austria, Switzerland, Greece, France, Germany, the USA, and the UK, have contributed 15 chapters to this book. We sincerely thank them for their hard work and the time they devoted to this effort.

In addition, we would like to express our deep gratitude to all the expert reviewers of the chapters for their constructive comments, which significantly helped toward improving the organization, readability, and overall quality of the book.

Last but not least, we are indebted to Susan Lagerstrom-Fife and Caroline Flanagan from Springer, for their great guidance and advice in the preparation and completion of this handbook, as well as to the typesetting and publication team at Springer for their valuable assistance in the editing process.

Contents

1   **Introduction to Mobility Data Privacy** 1
    Aris Gkoulalas-Divanis and Claudio Bettini

Part I   Fundamentals for Privacy in Mobility Data

2   **Modeling and Understanding Intrinsic Characteristics
    of Human Mobility** 13
    Jameson L. Toole, Yves-Alexandre de Montjoye, Marta C. González,
    and Alex (Sandy) Pentland

3   **Privacy in Location-Sensing Technologies** 35
    Andreas Solti, Sushant Agarwal, and Sarah Spiekermann-Hoff

Part II   Main Research Directions in Mobility Data Privacy

4   **Privacy Protection in Location-Based Services: A Survey** 73
    Claudio Bettini

5   **Analyzing Your Location Data with Provable
    Privacy Guarantees** 97
    Ashwin Machanavajjhala and Xi He

6   **Opportunities and Risks of Delegating Sensing Tasks
    to the Crowd** 129
    Delphine Reinhardt and Frank Dürr

7   **Location Privacy in Spatial Crowdsourcing** 167
    Hien To and Cyrus Shahabi

8   **Privacy in Geospatial Applications and Location-Based
    Social Networks** 195
    Igor Bilogrevic

9   Privacy of Connected Vehicles 229
      Jonathan Petit, Stefan Dietzel, and Frank Kargl

10   Privacy by Design for Mobility Data Analytics 253
      Francesca Pratesi, Anna Monreale, and Dino Pedreschi

Part III   Usability, Systems and Applications

11   Systems for Privacy-Preserving Mobility Data Management 281
      Despina Kopanaki, Nikos Pelekis, and Yannis Theodoridis

12   Privacy-Preserving Release of Spatio-Temporal Density.............. 307
      Gergely Acs, Gergely Biczók, and Claude Castelluccia

13   Context-Adaptive Privacy Mechanisms................................. 337
      Florian Schaub

14   Location Privacy-Preserving Applications and Services 373
      Ioannis Boutsis and Vana Kalogeraki

Index... 399

About the Authors

Gergely Acs CrySyS Lab, Budapest University of Technology and Economics, Budapest, Hungary (acs@crysys.hu).

Gergely Acs is an Assistant Professor at the Budapest University of Technology and Economics (BUTE), Hungary, and a member of the Laboratory of Cryptography and System Security (CrySyS). Before joining BUTE, Gergely was a research scholar and engineer at INRIA, France. His research focuses on different aspects of data privacy and security, including privacy-preserving machine learning, data anonymization, and privacy risk analysis. He received his MS and PhD degrees from BUTE.

Sushant Agarwal Institute for Management Information Systems, Vienna University of Economics and Business, Vienna, Austria (sushant.agarwal@wu.ac.at).

Sushant works on the European Union project SERAMIS as a PhD researcher at WU, where he explores questions related to privacy in pervasive computing. He received an MTech degree in Aerospace Engineering from the Indian Institute of Technology (IIT), Bombay, in 2014. During his studies, he interned at the Institute for Manufacturing (IfM) in the University of Cambridge and was involved in projects dealing with data quality and failure diagnosis for Boeing and Hitachi, respectively.

Claudio Bettini EveryWare Lab, Department of Computer Science, University of Milan, Italy (claudio.bettini@unimi.it).

Claudio Bettini is Professor of Computer Science at the University of Milan, where he leads the EveryWare laboratory at the Computer Science Department. Claudio received his PhD in Computer Science from the University of Milan in 1993. He has been a postdoc at IBM Research, NY, and, for more than a decade, an Affiliate Research Professor at the Center for Secure Information Systems at George Mason University, VA. His research interests cover the areas of mobile and pervasive computing, data privacy, temporal and spatial data management, and intelligent systems.

In 2009, he co-edited the Springer book *Privacy in Location-Based Applications*, offering a first overview of the main research results in the field. Among his many organizational activities, he acted as General Chair of IEEE PerCom 2017 and IEEE MDM 2013 and as TPC Chair of IEEE PerCom 2013. He is a member of the steering committee of IEEE PerCom and Associate Editor of the *Pervasive and Mobile Computing* journal and previously Editor of *The VLDB Journal* and of the *IEEE Transactions on Knowledge and Data Engineering*. In 2011, he founded EveryWare Technologies, a startup developing innovative mobile apps and services for privacy and assistive technologies. He is a senior member of the IEEE Computer Society.

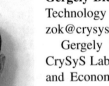

Gergely Biczók CrySyS Lab, Budapest University of Technology and Economics, Budapest, Hungary (biczok@crysys.hu).

Gergely Biczók is an Assistant Professor at the CrySyS Lab at the Budapest University of Technology and Economics (BME). He received his PhD (2010) and MSc (2003) degrees in Computer Science from BME. He was a Postdoctoral Fellow at the Norwegian University of Science and Technology (2011–2014) and at the Future Internet Research Group of the Hungarian Academy of Sciences (2014–2016). Previously, he was a Fulbright Visiting Researcher to Northwestern University (2007–2008) and held a researcher position at Ericsson Research (2003–2007). His research interests center around the economics of networked systems including privacy, incentives, and game theory.

Igor Bilogrevic Senior Research Scientist, Google Inc., Zurich, Switzerland (ibilogrevic@google.com).

Igor Bilogrevic is a Senior Research Scientist at Google, where he works on machine learning and data mining techniques in order to build novel privacy and security features in products. He obtained his PhD on the privacy of context-aware mobile networks from EPFL (Switzerland), where he investigated privacy issues at different layers of the network stack and proposed privacy-enhancing mechanisms for information sharing that work across different layers. He has previously worked in collaboration with the Nokia Research Center and PARC (a Xerox Company) on privacy challenges in pervasive mobile networks, encompassing data privacy, social community privacy, location privacy, information sharing, and private data analytics. His main areas of interest include applications of machine learning for privacy, private data analytics, contextual intelligence, applied cryptography, and user experience.

Ioannis Boutsis Athens University of Economics and Business, Athens, Greece (mpoutsis@aueb.gr).

Ioannis Boutsis received BSc (2009) and MSc degrees in Computer Science (2011) and PhD degree in Distributed Systems (2017), all from the Athens University of Economics and Business (AUEB). During his PhD, Ioannis worked in several EU projects including Insight-FP7 and NGHCS-ERC. Since 2017, he works for Amazon Video at the Amazon Development Centre, London, as a Software Development Engineer.

Claude Castelluccia Saint Ismier Cedex, France (claude.castelluccia@inria.fr).

Claude Castelluccia is a Research Director at Inria, in France, where he leads the PRIVATICS team (Privacy Models, Architectures and Tools for the Information Society). Claude has held visiting research positions at UC Irvine and Stanford University, USA. His past research interests include networking, Internet protocols, network security, and applied cryptography. His current research is on Internet privacy and security with a focus on anonymized analytics, data anonymization, data transparency, and Internet-based surveillance analysis. He is also interested in the economical and legal aspects of data privacy. He has chaired and participated in many Program Committees (ACM CCS, PETS, WiSec, etc.), co-founded the ACM WiSec conference, and advised more that 10 PhD students.

Yves-Alexandre de Montjoye Department of Computing, Imperial College, London, UK (demontjoye@imperial.ac.uk).

Yves-Alexandre de Montjoye is an Assistant Professor (Lecturer) at Imperial College London where he leads the Computational Privacy Group, and a Special Adviser to EC Commissioner Vestager. He is affiliated with the Data Science Institute and Department of Computing. Previously, he was Postdoctoral Researcher at Harvard University, working with Latanya Sweeney and Gary King, and he received his PhD from MIT, under the supervision of Alex "Sandy" Pentland. His research aims at understanding how the unicity of human behavior impacts the privacy of individuals in large-scale metadata datasets, and his work has been covered in *The New York Times*, BBC News, CNN, *Wall Street Journal*, *Harvard Business Review*, *Le Monde*, *Die Spiegel*, *Die Zeit*, and *El Pais*, and in reports of the World Economic Forum, United Nations, OECD, FTC, and the European Commission, as well as in talks at TEDxLLN and TEDxULg. Yves-Alexandre was recently named an Innovator under 35 for Belgium (TR35). He is a Fellow of the ID^3 Foundation and the B.A.E.F. Foundation and a Research Associate at Data-Pop.

Stefan Dietzel Humboldt-Universität zu Berlin, Department of Computer Science, 10099 Berlin, Germany (stefan.dietzel@hu-berlin.de).

Dr. Stefan Dietzel is a Postdoctoral Researcher with the Chair of Computer Engineering at Humboldt-Universität zu Berlin. Stefan's research interests are in security, scalability, and privacy aspects of wireless distributed computing. In several projects, he is investigating how to create communication protocols that are scalable and efficient while, at the same time, protecting the users' privacy and being secure against insider and outsider attackers.

Stefan holds a doctoral degree in Computer Science from the University of Twente, The Netherlands. During his time as a doctoral candidate, Stefan worked at the Institute of Media Informatics and the Institute of Distributed Systems at Ulm University and at the Distributed and Embedded Systems group at the University of Twente; before, he studied media informatics at Ulm University in Germany.

Frank Dürr Institute of Parallel and Distributed Systems, University of Stuttgart, Stuttgart, Germany (frank.duerr@ipvs.uni-stuttgart.de).

Frank Dürr is a Senior Researcher and Lecturer at the Institute of Parallel and Distributed Systems (IPVS) at the University of Stuttgart in Germany. He received both his doctoral degree and diploma in Computer Science from the University of Stuttgart. Frank Dürr is currently leading the mobile and context-aware systems and software-defined networking (SDN) groups of the Department of Distributed Systems at IPVS. He was technical coordinator of the Collaborative Research Center (SFB) 627 "Spatial World Models for Mobile Context-Aware Applications (Nexus)," funded by the German Research Foundation (DFG). His research interests include mobile and pervasive computing—in particular, location privacy and mobile sensing—as well as software-defined networking (SDN) and time-sensitive networking (TSN).

Aris   Gkoulalas-Divanis IBM   Watson   Health Headquarters, Cambridge, MA 02142-1123, USA (gkoulala@us.ibm.com).

Aris Gkoulalas-Divanis is the Technical Lead on Data Protection and Privacy for IBM Watson Health. He received his PhD in Computer Science from the University of Thessaly. His PhD dissertation was awarded the Certificate of Recognition and Honorable Mention in the 2009 ACM SIGKDD Dissertation Awards. Aris has been a Postdoctoral Research Fellow in the Department of Biomedical Informatics of Vanderbilt University (2009–2010) and a Research Scientist in IBM Research-Zurich (2010–2012) and in IBM Research-Ireland (2012–2016). His research interests are in the areas of privacy-preserving data mining, privacy in trajectories and location-based services, privacy in medical data, and knowledge hiding. In these areas, he has published more than 80 research works, including four Springer books, and he has applied for or being granted more than 25 patents. He is an Associate Editor of the *Knowledge and Information Systems* (KAIS) journal, the *IEEE Transactions on Information Forensics and Security* (T-IFS), the *International Journal of Research & Development Innovation Strategy* (IJRDIS), and the *International Journal of Knowledge-Based Organizations* (IJKBO). Since 2014 he is an Area Editor for *ACM Computing Reviews* (CR). Aris is a senior member of IEEE; a professional member of ACM, SIAM, and AAAS.

Marta C. González University of California, Berkeley, CA, USA (martag@berkeley.edu).

Marta C. González is an Associate Professor of City and Regional Planning at the University of California, Berkeley, and a Physics Research faculty in the Energy Technology Area (ETA) at the Lawrence Berkeley National Laboratory (Berkeley Lab). With the support of several companies, cities, and foundations, her research team develops computer models to analyze digital traces of information mediated by devices. They process this information to manage the demand in urban infrastructures in relation to energy and mobility. Her recent research uses billions of mobile phone records to understand the appearance of traffic jams and the integration of electric vehicles into the grid, smart meter data records to compare the policy of solar energy adoption, and card transactions to identify habits in spending behavior. Prior to joining Berkeley, Marta worked as an Associate Professor of Civil and Environmental Engineering at MIT, and a member of the Operations Research Center and the Center for Advanced Urbanism. She is a member of the scientific council of technology companies such as Gran Data, PTV, and the Pecan Street Project consortium.

Xi He Duke University, Durham, NC, USA (hexi88@cs.duke.edu).

Xi He is a PhD student at the Computer Science Department of Duke University. Her research interests lie in privacy-preserving data analysis and security. She has an MS degree from Duke University and a double degree in Applied Mathematics and Computer Science from the University of Singapore. Xi has been working with Prof. Machanavajjhala on privacy since 2012. She has published in SIGMOD, VLDB, and CCS and has given tutorials on privacy at VLDB 2016 and SIGMOD 2017. She received best demo award on differential privacy at VLDB 2016 and was awarded a 2017 Google PhD Fellowship in Privacy and Security.

Vana Kalogeraki Athens University of Economics and Business, Athens, Greece (vana@aueb.gr).

Vana Kalogeraki is an Associate Professor at Athens University of Economics and Business leading the Distributed and Real-Time Systems research. Previously, she has held positions as an Associate and Assistant Professor at the Department of Computer Science at the University of California, Riverside, and as a Research Scientist at Hewlett-Packard Labs in Palo Alto, CA. She received her PhD from the University of California, Santa Barbara, in 2000. Prof. Vana Kalogeraki has been working in the field of distributed and real-time systems, participatory sensing systems, mobility, and crowdsourcing for over 20 years and has co-authored over 170 papers in journals and conferences proceedings, including co-authoring the OMG CORBA Dynamic Scheduling Standard. She was awarded a Marie Curie Fellowship; three best paper awards at ACM DEBS 2017, IEEE IPDPS 2009, SAINT 2008; two Best Student Paper Awards at PETRA 2016 and SAINT 2011; a UC Regents Fellowship Award, UC Academic Senate Research Awards; and a research award from HP Labs. Her research has been supported by an ERC Starting Independent Researcher Grant, the European Union, joint EU/Greek "Aristeia" grants, joint EU/Greek "Thalis" grant, NSF, and gifts from SUN and Nokia.

Frank Kargl Ulm University, 89069 Ulm, Germany (frank.kargl@uni-ulm.de).

Prof. Dr. Frank Kargl is the Director of the Institute of Distributed Systems at Ulm University, where a main focus of his research is on security and privacy protection in distributed systems in domains like automotive and industrial control systems. He holds a doctoral degree in Computer Science from the Ulm University and previously held a Professor position at the University of Twente.

Prof. Dr. Fank Kargl was involved in a number of national and international projects related to automotive security and privacy like SeVeCom, PRECIOSA, CONVERGE, and PRESERVE, that all worked toward a holistic security and privacy solution for connected cars and Intelligent Transport Systems (ITS). He has co-authored more than 100 peer-reviewed publications in this area and actively contributed to standardization of security and privacy in ITS and beyond, through participation in bodies like C2C-CC or ETSI, and as co-chair of major conferences and workshops like ACM WiSec, ACM Vanet, IEEE WiVeC, and IEEE VNC.

Despina Kopanaki Department of Informatics, University of Piraeus, Athens, Greece (dkopanak@unipi.gr).

Despina Kopanaki is a PhD candidate in Informatics at the Department of Informatics, University of Piraeus, Greece, under the supervision of Professor Yannis Theodoridis. She got her bachelor degree from the Department of Statistics, Athens University of Economics and Business in Greece (2000–2004). She holds an MSc in Applied Economics and Finance from the Department of Economics, Athens University of Economics and Business (2004–2006). Her research interests include privacy preservation and data mining in moving object databases.

Ashwin Machanavajjhala Duke University, Durham, NC, USA (ashwin@cs.duke.edu).

Ashwin Machanavajjhala is an Assistant Professor in the Department of Computer Science, Duke University, and an Associate Director at the Information Initiative at Duke (iiD). Previously, he was a Senior Research Scientist in the Knowledge Management group at Yahoo! Research. His primary research interests lie in algorithms for ensuring privacy in statistical databases and augmented reality applications. His work on ℓ-Diversity that appeared in IEEE ICDE 2006 received the Influential Paper Award in 2017. He is also a recipient of the National Science Foundation Faculty Early CAREER Award in 2013, and the 2008 ACM SIGMOD Jim Gray Dissertation Award Honorable Mention. Ashwin graduated with a PhD from the Department of Computer Science, Cornell University, and a BTech in Computer Science and Engineering from the Indian Institute of Technology, Madras.

Anna Monreale Computer Science Department of University of Pisa, Pisa, Italy (annam@di.unipi.it).

Anna Monreale is a post-doc at the Computer Science Department of the Pisa University and a member of the KDD-LAB, a joint research group with the Information Science and Technology Institute of the National Research Council in Pisa. She received her MS and PhD degrees in Computer Science from the University of Pisa in 2007 and 2011, respectively. Her research is in privacy-aware data mining and data publishing, and in privacy-preserving outsourcing of analytical tasks.

Dino Pedreschi Computer Science Department of University of Pisa, Pisa, Italy (pedre@di.unipi.it).

Dino Pedreschi is a Full Professor of Computer Science at the University of Pisa. His research interests are in data mining and in privacy-preserving data mining. He is a member of the program committee of the main international conferences on data mining and knowledge discovery. He has been granted a Google Research Award (2009) for his research on privacy-preserving data mining and anonymity-preserving data publishing.

Nikos Pelekis Department of Statistics & Insurance Science, University of Piraeus, Athens, Greece (pedre@di.unipi.it).

Nikos Pelekis is an Assistant Professor at the Department of Statistics and Insurance Science, University of Piraeus, Greece. His research interests include all topics of data science. He has been particularly working for almost 20 years in the field of Mobility Data Management and Mining. Nikos has co-authored one monograph and more than 60 refereed articles in scientific journals and conferences, receiving more than 700 citations, while he has received three best paper awards, won the SemEval'17 competition, and was ranked third in the ACM SIGSPATIAL'16 data challenge. He has offered several invited lectures in Greece and abroad (including PhD/MSc/summer courses at Rhodes, Milano, KAUST, Aalborg, Trento, Ghent, JRC Ispra) on mobility data management and data mining topics. He has been actively involved in more than ten European and National R&D projects. Among them he is or was principal researcher in GeoPKDD, MODAP, MOVE, DATASIM, SEEK, DART and datAcron, and Track & Know.

Alex (Sandy) Pentland MIT Media Lab, Cambridge, MA 02139, USA (sandy@media.mit.edu).

Professor Alex "Sandy" Pentland directs the MIT Connection Science and Human Dynamics labs and previously helped create and direct the MIT Media Lab and the Media Lab Asia in India. He is one of the most-cited scientists in the world, and *Forbes* recently declared him one of the "seven most powerful data scientists in the world," along with Google founders and the Chief Technical Officer of the United States. He has received numerous awards and prizes such as the McKinsey Award from *Harvard Business Review*, the 40th Anniversary of the Internet from DARPA, and the Brandeis Award for work in privacy.

Jonathan Petit OnBoard Security, Wilmington, MA 01887, USA (jpetit@onboardsecurity.com).

Dr. Petit is the Senior Director of Research for Security Innovation, Inc. He is in charge of leading projects in security and privacy of automated and connected vehicles. He has conducted extensive research in detecting security vulnerabilities in automotive systems. He published the first work on potential cyber attacks on automated vehicles and remote attacks on automated vehicle sensors. He has supported communications security design and automotive cybersecurity analysis through OEM and NHTSA-sponsored projects. Previously, he was a Research Fellow in the Computer Security Group at University College Cork, Ireland. From 2011 to 2014, he was a Postdoctoral Researcher at the University of Twente, The Netherlands, where he co-coordinated the European-Union funded. FP7 PRESERVE project. He received his PhD in 2011 from Paul Sabatier University, Toulouse, France.

Francesca Pratesi Information Science and Technology Institute of the National Research Council, Pisa, Italy (francesca.pratesi@isti.cnr.it).

Francesca Pratesi is a Postdoctoral Researcher at the Information Science and Technology Institute of the Italian National Research Council, and a member of the Knowledge Discovery and Data Mining Lab, a joint research group with the Italian National Research Council in Pisa. She received her MS degree and PhD degree in Computer Science from the University of Pisa in 2013 and 2017, respectively. Her research interests include data mining, data privacy, and spatiotemporal data analysis.

Delphine Reinhardt University of Göttingen, Göttingen, Germany (reinhardt@cs.uni-goettingen.de).

Delphine Reinhardt is a full professor for Computer Security and Privacy at University of GÃŰttingen, Germany. Before 2018, she was an assistant professor at the University of Bonn and also associated to the Fraunhofer Institute for Communication, Information Processing and Ergonomics (FKIE). She completed her doctoral degree in computer science with distinction in 2013 at Technische Universität Darmstadt, Germany. Her doctoral thesis received awards by the Communication and Distributed Systems Group (KuVS) of the German Informatics Society (GI) and Information Technology Society of the Association for Electrical, Electronic & Information Technologies (VDE-ITG) as well as the "Vereinigung von Freunden der Technische Universität zu Darmstadt e.V." for outstanding academic achievements. Since 2009, she holds a double-degree in electrical engineering from TU Darmstadt and Ecole Nationale Supérieure de l'Electronique et ses Applications (ENSEA), France. She has served as program committee member and reviewer for more than 60 international conferences and journals. Her research interests include privacy and security, trust and reputation, and usability in ubiquitous computing and beyond.

Florian Schaub University of Michigan, Ann Arbor, MI, USA (fschaub@umich.edu).

Florian Schaub is an Assistant Professor in the School of Information at the University of Michigan. His research focuses on empowering users to effectively manage their privacy in complex socio-technological systems. His research interests span privacy, human-computer interaction, mobile and ubiquitous computing, and the Internet of Things. Before joining UMSI, he was a Postdoctoral Fellow in the School of Computer Science at Carnegie Mellon University. He holds a doctoral degree and Diploma in Computer Science from the University of Ulm, Germany, and a Bachelor in Information Technology from Deakin University, Australia.

Cyrus Shahabi University of Southern California, Los Angeles, CA, USA (shahabi@usc.edu).

Cyrus Shahabi is a Professor of Computer Science, Electrical Engineering and Spatial Sciences at the University of Southern California (USC). He is also the Director of the NSF's Integrated Media Systems Center (IMSC), the Informatics Program and the Information Laboratory (InfoLAB) at USC's Viterbi School of Engineering. He received his BS in Computer Engineering from Sharif University in 1989 and then his MS and PhD degrees in Computer Science from USC in May 1993 and August 1996, respectively. He authored two books and more than 200 research papers in the areas of databases, GIS, and multimedia. Dr. Shahabi is a fellow of IEEE and a recipient of the ACM Distinguished Scientist Award in 2009, the 2003 US Presidential Early Career Awards for Scientists and Engineers (PECASE), and the NSF CAREER Award in 2002.

Andreas Solti Institute of Information System, Vienna University of Economics and Business, Vienna, Austria (solti@ai.wu.ac.at).

Andreas Solti is a Postdoctoral Researcher at the Institute for Information Business at the WU Vienna, Austria. He completed his PhD studies entitled "Probabilistic Estimation of Unobserved Process Events" in the business process technology group of the Hasso Plattner Institute of the University of Potsdam. Currently, he is working on the European Union FP7 project SERAMIS on topics related to sensor data analysis. Andreas has published more than 25 international research papers in journals and highly competitive conferences, including Information Systems, BPM, CAiSE, ICSOC, EDOC. He has served in several program committees of conferences, including BPI, BPIC, and EDOC.

Sarah Spiekermann-Hoff Institute for Management Information Systems, Vienna University of Economics and Business, Vienna, Austria (spiek@wu.ac.at).

Sarah chairs the Institute for Management Information Systems at the Vienna University of Economics and Business (WU Vienna). She has published more than 80 articles on the social and ethical implications of computer systems, and given more than a hundred presentations and talks about her work throughout the world. Her main expertise is electronic privacy, disclosure behavior, and ethical computing. Sarah has co-authored US/EU privacy regulation for RFID technology and has regularly worked as an expert and advisor to companies and governmental institutions, including the EU Commission and the OECD. She was on the advisory board of the Foundation of Data Protection of the German Parliament. She maintains a blog on *The Ethical Machine* at Austria's leading daily newspaper *Standard.at* and is on the board of the Austrian art and science think-tank GlobArt. Before being tenured in Vienna in 2009, Sarah was Assistant Professor at the Institute of Information Systems at Humboldt University Berlin (Germany), where she headed the Berlin Research Centre on Internet Economics (2003–2009).

Yannis Theodoridis Department of Informatics, University of Piraeus, Athens, Greece (ytheod@unipi.gr).

Professor Yannis Theodoridis is faculty member with the Department of Informatics and Director of the Data Science Lab, University of Piraeus, Greece. He serves or has served as member of the editorial boards of *ACM Computing Surveys* (2016–) and the *International Journal of Data Warehousing and Mining* (2005–), general co-chair for SSTD'03 and ECML/PKDD'11, PC vice-chair for IEEE ICDM'08, and PC member for numerous conferences in the fields of data management and data mining. His research interests include data science (big data management and analytics) for mobility information. He has co-authored three monographs and more than 100 refereed articles in scientific journals and conferences, with over 10,000 citations according to Google Scholar. He holds a Diploma (1990) and PhD (1996) in Computer Engineering, both from the National Technical University of Athens, Greece.

Hien To University of Southern California, Los Angeles, USA (hto@usc.edu).

Hien To is a Software Engineer at Amazon Mechanical Turk. He is a PhD candidate in Computer Science, University of Southern California since 2017. He received his Bachelor of Engineering degree in Computer Science from Hanoi University of Science and Technology in 2010. He worked as a Software Engineer at the sVNG Corporation from 2010 to 2011. Hien's research interest is on data privacy and crowdsourcing. He developed new privacy-preserving techniques for efficient task assignment in spatial crowdsourcing. He contributed to various projects funded by NSF, Google, Microsoft, Oracle, and Northrop Grumman. He authored more than ten research papers in top conferences and journals. Hien serves as the publicity Chair of GeoRich 2017 and local Chair of SIGSPATIAL 2018. He is a member of ACM and IEEE.

Jameson L. Toole Fritz Inc., Boston, MA. (jltoole@mit.edu).

Jameson Toole is the co-founder and CEO of Fritz, a company helping developers to optimize, deploy, and manage machine learning models on every mobile device and platform, and of Wherehouse Inc., which aims to build efficient, robust, and useful technologies for mobile network operators. Previously, he ran the Data Engineering team at Jana Mobile and spent summers as a Software Engineering Intern at GoogleX, working with Project Wing to change the way we transport things with drones. Dr. Toole holds a PhD degree in Engineering Systems from the Massachusetts Institute of Technology, where he used data collected from mobile phones to understand urban mobility, social behavior, and economic outcomes. In addition to the PhD, he holds an MS and undergraduate degrees in physics, economics, and mathematics from the University of Michigan. He is a leading expert on using big data to understand transportation, human behavior, and mobility. Dr. Toole has authored multiple peer-reviewed papers and built teams of data scientists for companies like Google and Jana Mobile.

About the Editors

Aris Gkoulalas-Divanis IBM Watson Health Headquarters, Cambridge, MA 02142-1123, United States (gkoulala@us.ibm.com).

Aris Gkoulalas-Divanis is the Technical Lead on Data Protection and Privacy for IBM Watson Health. In this role, he is responsible for the design and integration of novel privacy-protection methods to the IBM Watson Platform for Health and the IBM Cloud, to support IBM customers' privacy and utility requirements. Aris received his PhD in Computer Science from the University of Thessaly in 2009. His PhD dissertation was awarded the Certificate of Recognition and Honorable Mention in the 2009 ACM SIGKDD Dissertation Awards. Aris has been a Postdoctoral Research Fellow in the Department of Biomedical Informatics of Vanderbilt University (2009–2010) and a Research Scientist in IBM Research-Zurich (2010–2012) and IBM Research-Ireland (2012–2016). His research interests are in the areas of databases, data mining, privacy-preserving data mining, privacy in trajectories and location-based services, privacy in medical data, and knowledge hiding. In these areas, he has given many seminars and two tutorials (ECML/PKDD 2011, SDM 2012), he has published more than 80 research works—including four Springer books, and he has applied for or being granted more than 25 patents (received three IBM Invention Achievement Awards). Aris is a regular reviewer in top-quality journals, as well as participates in the program committee of many prestigious conferences. He serves as an associate editor of the *Knowledge and Information Systems* (KAIS) journal, the *IEEE Trans-*

actions on Information Forensics and Security (T-IFS), the *International Journal of Research & Development Innovation Strategy* (IJRDIS), and the *International Journal of Knowledge-Based Organizations* (IJKBO). Since 2014 he is an area editor for *ACM Computing Reviews* (CR) covering the Information Systems category. Aris is a senior member of IEEE; a professional member of ACM, SIAM, and AAAS; and an at-large member of UPE and Sigma Xi.

Claudio Bettini EveryWare Lab, Department of Computer Science, University of Milan, Italy (claudio.bettini@unimi.it).

Claudio Bettini is Professor of Computer Science at the University of Milan, where he leads the EveryWare laboratory at the Computer Science Department. Claudio received his PhD in Computer Science from the University of Milan in 1993. He has been post-doc at IBM Research, NY, and, for more than a decade, an Affiliate Research Professor at the Center for Secure Information Systems at George Mason University, VA. His research interests cover the areas of mobile and pervasive computing, data privacy, temporal and spatial data management, and intelligent systems. He is the author of over 150 peer-reviewed publications on these topics. He acted as co-PI of several international projects on data privacy and contributed as project reviewer in the EU Horizon H2020 ERC and FET initiatives. Among his many organizational activities, he acted as General Chair of IEEE PerCom 2017 and IEEE MDM 2013 and as TPC Chair of IEEE PerCom 2013. He is a member of the steering committee of IEEE PerCom and Associate Editor of the *Pervasive and Mobile Computing* journal and previously Editor of *The VLDB Journal* and of the *IEEE Transactions on Knowledge and Data Engineering*. In 2011, he founded EveryWare Technologies, a startup developing innovative mobile apps and services for privacy and assistive technologies. He is a senior member of the IEEE Computer Society.

List of Figures

Fig. 2.1   Mobile phones are increasingly being used to collect
high-resolution mobility data. This figure from de
Montjoye et al. [15] depicts (**a**) a sequence of calling events
made by a user at different locations. (**b**) These events are
localized to the area served by the closest mobile phone
tower to the use and (**c**) can be aggregated into individual
specific neighborhoods where a user is likely to be found
at different times of the day or week   16

Fig. 2.2   (**a**) Individual mobility trajectories are passively collected
from mobile devices [23]. (**b**) Measuring the distribution of
radius of gyrations, r_g within a population of 100,000 users
in a European country reveals considerable heterogeneity
in typical travel distance of individuals. Moreover, this
distribution cannot be explained by modeling each
individual's movement as realizations of a single Levy
flight process [23]. (**c** and **d**) Show the slower than linear
growth in new locations visited over time $S(t)$ and that
the probability a location is visited next is inversely
proportional to the frequency it has been visited in the
past [54]. (**e**) This preferential return contributes to
strikingly high predictability $R(t)$ over time while (**f**) the
number of unique locations visited in any given hour is
highly periodic and corresponds to the sleep-wake cycles
of individuals [55]   19

Fig. 2.3   (a) Removing geographic coordinates from locations and
             only focusing on a set of unique places and the directed
             travel between them, mobility motifs reveal that the
             daily routines of people are remarkably similar. Despite
             over 1 million unique ways to travel between 6 or fewer
             points, just 17 motifs are used by 90% of the population.
             Moreover, the frequency of their appearance in CDR
             data matches very closely with more traditional survey
             methods [52]. (b) Despite this similarity and predictability,
             our movement displays a high degree of unicity. Just four
             spatiotemporal points is enough to differentiate a user from
             95% of all others individuals [15]   23
Fig. 2.4   (a) The radiation model accounts for intervening
             opportunities, producing more accurate estimates of
             flows between two places than more traditional gravity
             models [53]. (b) Routing millions of trips measured
             from CDR data to real road networks makes it possible
             to measure the importance of a road based on how
             many different locations contribute traffic to it, K_{road}.
             Understanding how transportation systems perform
             under different loads presents new opportunities to solve
             problems related to congestion and make infrastructure
             more efficient [63]   26
Fig. 2.5   (a) Global air travel has dramatically increased the speed at
             which diseases can spread from city to city and continent
             to continent [44]. (b) Mobility also adds context to social
             networks. When two individuals visit the same locations
             can suggest the nature of a social relationship [60]. (c)
             Mobility and the access it provides has strong correlations
             with economic outcomes. Children have dramatically
             different chances at upward economic mobility in certain
             places of the United States than others [10].......................   29

Fig. 3.1   Illustration for trilateration   41
Fig. 3.2   Illustration for 2-step fingerprinting. (a) and (b) depict the
             first step for training and (c) shows the second step for
             positioning   41
Fig. 3.3   An illustration of a shopping trip with RFID readers at a
             point of sales and an exit gate   49
Fig. 3.4   An illustration of a shopping trip with RFID readers at an
             interaction point, a POS and an exit gate   51
Fig. 3.5   An illustration of a shopping trip with RFID RTLS, a POS
             and an exit gate   52
Fig. 3.6   An illustration of a shopping trip with WiFi locating system   54

Fig. 3.7   An illustration of a shopping trip with RFID RTLS, Wi-Fi
              locating system, a POS and an exit gate 56
Fig. 3.8   Balancing act for legitimate interests as defined in the
              GDPR [81] 60

Fig. 5.1   DPT framework overview... 111

Fig. 6.1   Crowdsensing architecture.. 131

Fig. 7.1   Threat models in spatial crowdsourcing. W and R denote
              workers and requesters, respectively. The dotted circles
              surrounding them denote that they are protected from an
              untrusted entity shown in the first column. After tasking
              and reporting, the assignment and reporting links between
              W and R represent the established connections during
              each phase. The dashed links indicate connections that are
              oblivious to the corresponding malicious entity. (a) Push
              mode. (b) Pull mode 172
Fig. 7.2   Screenshots of TaskRabbit web application from worker
              Bob. (a) Task locations. (b) Task price. (c) Task status. (d)
              Performed tasks 174
Fig. 7.3   A framework for privacy protection during tasking and
              reporting in the pull mode. Dashed entities are malicious,
              while others are trusted 177
Fig. 7.4   Examples of range dependency and all-inclusivity.
              (a) Query formation. (b) Range dependency leak. (c)
              All-inclusivity leak 181
Fig. 7.5   Distance estimation methods. (a) Centroid-point method.
              (b) Expected probabilistic method 183
Fig. 7.6   Differentially private framework for spatial crowdsourcing.
              (a) System architecture. (b) Worker PSD using adaptive
              grid 186

Fig. 8.1   A thematic map (top) and a reference map (bottom). The
              thematic maps shows the poverty rate by county in the U.S.
              in 2014 [123], whereas the reference map shows the U.S.
              territory by State, together with topographic, transportation
              and demographic information (images: (top) https://www.
              census.gov/did/www/saipe/data/statecounty/maps/iy2014/
              Tot_Pct_Poor2014.pdf, (bottom) https://upload.wikimedia.
              org/wikipedia/commons/7/7d/United_states_wall_2002_
              us.jpg, last retrieved Dec. 6, 2016) 205

Fig. 8.2   Online crime map for the region of Berkeley, California,
            during a 1-week period between Oct. 18–24th, 2016. The
            map shows the different types of reports, such as thefts,
            burglaries, assaults, vandalisms, and the place where
            they were reported by the Berkeley Police (image shown
            with permission from crimemapping.com, http://www.
            crimemapping.com/, last retrieved Dec. 6, 2016) 206
Fig. 8.3   System architecture of a generic Location-Based Social
            Network. The links represent possible ways by which
            location data can be attached to the content posted by
            either people or organizations (adapted from [129]). The
            solid and dashed lines correspond to the actions that people
            and organizations can perform, respectively 211

Fig. 9.1   The connected car ecosystem 234
Fig. 9.2   Abstract pseudonym lifecycle 240
Fig. 9.3   Pay-as-You-Drive insurance models, once with the
            classical, privacy-invasive model (left) and once with the
            PriPAYD model that ensures no user data is inadvertently
            leaked (right) 244
Fig. 9.4   POPCORN protocol for privacy-preserving charging of
            electric vehicles 246

Fig. 10.1   (a) Milan GPS Trajectories, (b) characteristic points, (c)
            spatial clusters, (d) tessellation of the territory, and (e)
            generalized trajectories... 260
Fig. 10.2   10 largest clusters of the original trajectories (top) and of
            the anonymized trajectories (down) 262
Fig. 10.3   F-measure for comparison of the clusterings of the
            anonymized dataset versus the clustering of the original
            trajectories 263
Fig. 10.4   The places taxonomy... 265
Fig. 10.5   The empirical disclosure probability on Milano dataset 266
Fig. 10.6   (a) Number of patterns extracted from Milan data and (b)
            coverage of the patterns varying the support threshold........... 267
Fig. 10.7   CCDFs of Flow per Link (Left); CCDFs of Flow per Zone
            (Right) 272
Fig. 10.8   Visualization of Flow per Link (A-B) and Flow per Zone
            (C-D) 273

Fig. 11.1   A big picture of the system architecture [8] 285
Fig. 11.2   The architecture of Hermes++ [22] 287
Fig. 11.3   Generating a fake trajectory over a set of line segments [22] 288
Fig. 11.4   Sensitive location tracking and sequential tracking attacks
            to user privacy [22] 292
Fig. 11.5   Protecting sensitive locations of user trajectories [22] 293

Fig. 11.6   Selecting segments from real trajectories [22] 294
Fig. 11.7   Prohibiting sequential tracking: (a) case I, (b) case II [22] 296
Fig. 11.8   The result of a range query in its (a) original vs. (b) NWA
            anonymized version [23] 299
Fig. 11.9   T-Optics applied on (a) the original vs. (b) the anonymized
            dataset [23] 300
Fig. 11.10   Private-Hermes architecture [23] 301
Fig. 11.11   HipStream architecture [31] 303

Fig. 12.1   Never-Walk-Alone anonymization. Original dataset (city
            of Oldenburg in Germany) with 1000 trajectories (left) and
            its anonymized version (NWA from [2]) with $k = 3$ where
            the distance between any points of two trajectories within
            the same cluster is at most 2000 m (right) (image courtesy
            of Gábor György Gulyás) 320
Fig. 12.2   IRIS cells of Paris (left) and Voronoi-tesselation of tower
            cells (right).. 324
Fig. 12.3   Algorithm 12.1 before improvements ($\varepsilon = 0.3$,
            $\delta = 2 \cdot 10^{-6}, \ell = 30$). (a) Large counts. (b) Small counts
            around local minimas 328
Fig. 12.4   Algorithm 12.1 after improvements
            ($\varepsilon = 0.3, \delta = 2 \cdot 10^{-6}, \ell = 30$). (a) Scaling. (b)
            Smoothing 329
Fig. 12.5   Mean relative error and Pearson correlation of each
            IRIS cell ($\varepsilon = 0.3, \delta = 2 \cdot 10^{-6}, \ell = 30$). (a) Naive
            Gaussian Perturbation (Avg. MRE: 1.01, PC: 0.47). (b)
            Algorithm 12.1 (Avg. MRE: 0.17, PC: 0.96) 330

Fig. 13.1   Conceptual view of a context-aware system based on the
            layer models by Baldauf et al. [13] and Knappmeyer et
            al. [74] 341
Fig. 13.2   Context-adaptive privacy mechanisms align context-aware
            system capabilities with the cognitive privacy regulation
            process in order to support privacy decision making and
            regulation 349

Fig. 14.1   The PCube app. (a) Login. (b) Add friends. (c) Set
            proximity 382
Fig. 14.2   The Locaccino app. (a) Map. (b) Privacy settings. (c)
            History 383
Fig. 14.3   The CrowdAlert app. (a) Login. (b) Report. (c) Settings 392

List of Tables

Table 3.1   Comparison of different technologies for location tracking [43] 44

Table 4.1   A classification of location-based services (SP = Service Provider) 78

Table 5.1   Table of notation 100

Table 7.1   Attacks on SC users.. 169

Table 7.2   Three tasks requested by requester Alice 175

Table 7.3   Overview of problem focuses (Re: reporting, Ta: tasking); privacy techniques used (Ps: pseudonym, Cl: cloaking, Pt: perturbation, Ex: exchange-based, En: encryption-based); threats (W: worker, T: requester, S: server); trusted third party (TTP); optimization type (ST: single task, MT: multiple tasks). x and (x) represent primary and secondary aspects, respectively 176

Table 8.1   Properties of different methods for geographic content generation 201

Table 8.2   Categorization of different research works according to the adversary type (In: internal, E: external), adversarial model (M: malicious, S: semi-honest, T: trusted third-party), the goal of the adversary (L: location, Ab: Absence, C: co-location, Id: identity, Ac: activity) and the proposed or suggested privacy protection mechanism (spatial/temporal cloaking, elimination, fake data, cryptography) 216

Table 12.1   Examples for k- and k^m-anonymity, where each row
              represents a record, public and sensitive attributes
              are not distinguished, and temporal information is omitted
              for simplicity 313

Table 13.1   Overview of context-adaptive privacy mechanisms 355

Table 14.1   Location-based applications, data sharing and privacy
              mechanisms 385

Table 14.2   University initiative location based app applications, data
              sharing and privacy mechanisms 386

Chapter 1
Introduction to Mobility Data Privacy

Aris Gkoulalas-Divanis and Claudio Bettini

Abstract The recent advances in mobile computing and positioning technologies have resulted in a tremendous increase to the amount and accuracy in which human location data can be collected and processed. Human mobility traces can be used to support a number of real-world applications spanning from urban planning and traffic engineering, to studying the spread of diseases and managing environmental pollution. At the same time, research studies have shown that individual mobility is highly predictable and mostly unique, thus information about individuals' movement can be used by adversaries to re-identify them and to learn sensitive information about their whereabouts. To address such privacy concerns, a significant body of research has emerged in the last 15 years, studying privacy issues related to human mobility and location information, in a number of contexts and real-world applications. This work has led to the adoption of privacy laws worldwide, for location privacy protection, as well as to the proposal of novel privacy models and techniques for technically protecting user privacy, while maintaining data utility. This chapter provides an introduction to the field of mobility data privacy, discusses the emerging research directions, along with the real-world systems and applications that have been proposed.

1.1 Introduction

The recent developments in mobile computing and positioning technologies, along with the tremendous advances in information technology, have made possible the collection, processing, storage and analysis of very detailed human location traces. Modern mobile devices, including smartphones and wearable devices, are equipped

A. Gkoulalas-Divanis (✉)
IBM Watson Health Headquarters, Cambridge, MA, USA
e-mail: gkoulala@us.ibm.com

C. Bettini
Dipartimento di Informatica, University of Milan, Milan, Italy
e-mail: claudio.bettini@unimi.it

© Springer Nature Switzerland AG 2018
A. Gkoulalas-Divanis, C. Bettini (eds.), *Handbook of Mobile Data Privacy*,
https://doi.org/10.1007/978-3-319-98161-1_1

with a myriad of sensors that can be used to collect detailed location-specific readings. The tracking of mobile devices (and their human companions) can be used to support a large number of real-world applications, such as urban planning [2], traffic engineering [22], managing of environmental pollution [20], studying the spread of diseases [21, 23], encouraging vehicle pooling [25], understanding human purchasing behavior [26], and many more. At the same time, real-time services whose offering depends on user location, have also emerged. Location-based services and location-based social networks [13] enable users to receive high quality services based on their current location, as well as to locate nearby friends, geotag friends and pictures, and share their whereabouts with other users. On the negative side, however, studies have shown that individuals' mobility information can be highly predictable and unique [5], thereby making it highly possible to identify an individual based on some of his or her mobility traces. Such attacks have led to the adoption of international laws for location privacy protection (in the EU, US, Canada, Australia, New Zealand, Japan, and Singapore), as well as to a plethora of novel privacy models and technical approaches proposed in the research literature to protect user privacy under specific guarantees, while maintaining high data utility.

This book aims to survey the field of mobility data privacy and to present the fundamental principles and theory, as well as the state-of-the-art research, systems and applications, to a wide audience including non-experts. Emphasis in the book is given towards coverage of the most recent directions in mobility data privacy. The book consists of three parts and its structure closely follows the main categories of research works that have been recently undertaken to protect user privacy in the context of mobility data and applications. In the sections that follow, we provide an overview of the contents of each part of the book.

1.2   Part I: Fundamentals for Privacy in Mobility Data

The first part of the book aims to introduce the readers to the area of mobility data privacy. Chapter 2 starts from the basics to explain why the knowledge of an individual's mobile traces alone, without any additional information about the individual, is typically sufficient to perform successful re-identification attacks. Empirical evidence and statistical models have demonstrated the existence of important intrinsic and universal characteristics about human movement. Individuals' mobility information can be highly predictable and unique, not only due to the places that each person visits during their daily movement but also due to the time associated with these visits. The more information that an attacker has for an individual about the places he or she visited (and the respective times of visit), the higher is the probability of re-identifying the individual. Considering human mobility at an aggregate level, by modeling aggregate movement of people from one place to another, allows mining important mobility patterns that can be beneficial in a number of real-world applications. It is important, however, that the insights gained from the mining of mobility patterns, do not come at a cost to individuals' privacy.

Location information can be sensed from a multitude of devices [27], including smart devices (e.g., smart phones, smart watches, smart cars, etc.), radio-frequency identification (RFID) devices, bluetooth devices, global-positioning system (GPS) tracking devices, and many others. Depending on the particular use case, certain location sensing technologies are preferred over others. Chapter 3 sheds light on popular location sensing technologies and use cases they can support, giving more emphasis on the use of such technologies in retail environments. The authors discuss in detail the different techniques that are used for location sensing, such as trilateration and fingerprinting, satellite-based location sensing, WiFi-based location sensing, bluetooth-based location sensing, cellular tower based sensing, etc., explaining their advantages and disadvantages. Following that, they introduce readers to the different privacy threats that emerge from the use of such technologies, the legal requirements associated with their use, and the technical controls that are available to protect location data from being personally identifiable.

1.3 Part II: Main Research Directions in Mobility Data Privacy

The second part of the book is devoted to important, emerging research directions in mobility data privacy. Each chapter in this part of the book surveys an important research problem related to mobility data privacy, discusses the corresponding privacy threats that have been identified and the solutions that have been proposed.

Chapter 4 offers a survey of privacy protection in the context of location based services. Location based services (LBS) are widely spread and commonly used nowadays, especially by mobile users who need navigation instructions, discovery of resources (such as nearby gas stations, restaurants, etc.), emergency services, etc. At the same time, social network LBS are very popular among users, allowing them to geotag and post information to their user profile in a social network, to find nearby friends, or to engage in location-based games [9]. This chapter presents the main privacy threats that are associated with the use of LBS, discusses regulatory compliance and personal preferences as the main requirements for LBS privacy-protection, and proceeds to survey state-of-the-art privacy-protection techniques that have been proposed in the research literature [4]. An important message that is conveyed in this chapter is that in order to decide on the privacy protection method that should be used for a given LBS, it is important to analyze the service in terms of the information being exchanged, the service architecture and the different parties to which the information is exposed, as well as the necessary location data accuracy.

Chapter 5 presents techniques for analyzing human mobility data under provable privacy guarantees, even when adversaries have strong prior knowledge. Instead of focusing on location cloaking techniques and traditional k-anonymity based approaches, the authors rely on differential privacy [7, 8] to ensure the private information of individuals while allowing the learning of useful information about a population of users. The chapter serves as a nice introduction to the area of differential privacy for location data. The authors explain the properties of location data that constitute them vulnerable to privacy threats, discuss why cloaking techniques are inadequate to offer strong protection, and present the key variants of differential privacy in the context of location data. Following that, they cover differentially private algorithms that satisfy privacy guarantees while resulting in different utilities. Using these algorithms, one can support answering COUNT queries over a single-time snapshot of location data, continuous queries over location streams, as well as releasing synthetic location trajectory databases. Last, the authors introduce a framework for defining privacy and use it to protect privacy when adversaries may know correlations within a location stream.

Chapters 6 and 7 of the book are devoted to participatory sensing applications [3, 14, 15], where users—equipped with personal mobile devices that may have a myriad of embedded sensors—collect sensor readings from different locations in order to support a common objective. The delegation of these sensing tasks to volunteers opens new opportunities for the timely support of large-scale crowdsourcing tasks, but, at the same time, comes with serious risks as volunteers have to disclose their location information to other (potentially untrustworthy) entities.

Chapter 6 provides a nice introduction to crowdsourcing applications by covering the different classes of these applications as well as the opportunities and benefits they offer. Following that, the authors describe the risks in which crowdsourcing participants are exposed by participating to crowdsensing applications, and the risks for crowdsensing applications to rely on volunteers to fulfill the sensing tasks.

Chapter 7 delves into more detail on the various attacks in the context of spatial crowdsourcing applications, which include location-based attacks during tasking in the push mode and collusion attacks during reporting in the pull mode. Following that, the authors survey the state-of-the-art solutions for addressing privacy issues in spatial crowdsourcing, covering techniques for privacy protection in the pull mode and techniques for protection in the push mode. These techniques range from pseudonym, clocking, and perturbation, to exchange-based and encryption-based privacy-protection approaches.

The focus of Chap. 8 is on geospatial applications, such as thematic maps and crowdsourced geo-information, and on location-based social networks. Geospatial applications enable users to both consume and contribute geographic information to the online community. The first part of this chapter is devoted to privacy challenges in the context of geospatial applications from the crowdsourcing aspect and from aspects related to surveillance. Following that, the second part of the chapter focuses on the subcategory of geospatial applications that involves location-based social networks. It covers privacy threats and protection mechanisms that have been proposed for users of location-based social networks. The chapter provides a

large number of references to state-of-the-art works that will be helpful for readers
interested to learn more in these areas.

Chapter 9 surveys the emerging area of privacy in the context of the connected car
ecosystem [12]. By enabling vehicles to exchange information with an underlying
infrastructure, as well as with other vehicles in their vicinity, a large number of
safety applications and services can be provisioned. At the same time, the connected
car ecosystem comes with significant privacy challenges [6], as broadcasting the
location of a vehicle discloses also the precise location of its passengers. To prohibit
such disclosures while allowing the offering of different safety applications and
services, several privacy protection approaches have been recently proposed. This
chapter discusses these approaches, giving particular emphasis to the approach of
using short-term identifiers—or pseudonyms [10]—in vehicle-to-X communications. A number of important research challenges that will pave the way for future
work in this area, are outlined in the end of the chapter.

Chapter 10 is the last chapter of this part of the book and covers privacy-by-
design[1] in the context of Big mobility data analytics. In this chapter, the authors
propose the use of the privacy-by-design paradigm when developing technological
frameworks to offer sufficient privacy, without obstructing knowledge discovery.
To achieve this, they propose inscribing privacy protection into the knowledge
discovery technology by design, so that the analysis incorporates the relevant
privacy requirements from the start. The authors illustrate this idea in three
different scenarios of mobility data analytics: (a) privacy-preserving mobility data
publication to support clustering analysis, (b) privacy-preserving publication of
semantic trajectories to extract frequent sequential patterns, and (c) protection of
movement data (collected in a distributed fashion from individual vehicles) using
differential privacy, in order to support their subsequent analysis by a central station.
The presented scenarios illustrate that under suitable conditions it is feasible to reach
a good balance between data privacy and utility.

1.4   Part III: Usability, Systems and Applications

The last part of the book is devoted to state-of-the-art systems for mobility data
privacy, as well as to real-world applications where privacy-protection techniques
have been applied.

Chapter 11 discusses real-world systems and research prototypes that have
been developed for privacy-preserving mobility data management. With respect
to privacy-protection, the focus of the chapter is on systems that maintain human
mobility data in-house to a hosting organization, enabling external users to query
these data while ensuring that the returned results protect privacy [11]. Along
these lines, they present Hermes++ [16], a query engine for sensitive trajectory

---

[1]https://www.ryerson.ca/pbdce/about/.

data that allows subscribed end-users to gain restricted access to the database to accomplish various analytic tasks. Following that, they present Private-Hermes [17], a benchmark framework for privacy-preserving mobility data querying and mining methods. Private-Hermes supports privacy-preserving publishing of user mobility data by implementing state-of-the-art algorithms for trajectory anonymization. Last, the authors cover HipStream [24], a privacy-preserving system for managing mobility data streams. HipStream enforces the fundamental Hippocratic principles of limited use, limited disclosure, and limited collection of data during stream management.

Chapter 12 presents a real-world application where the goal is to release spatiotemporal density information, i.e., the number of individuals visiting a given set of locations as a function of time, in a privacy-preserving way [1]. The authors survey some fundamental approaches that have been proposed for anonymizing and releasing spatiotemporal density, which follow different privacy models and come with different privacy and accuracy offerings. They then demonstrate some anonymization techniques with provable guarantees by releasing the spatiotemporal density of Paris, France. An important finding of this work is that in order to achieve sufficient accuracy in the release of spatiotemporal density information, the anonymization process has to be first carefully customized to the public characteristics of the spatiotemporal data.

Chapter 13 is devoted to context-aware computing [19] and specifically to context-adaptive privacy mechanisms and systems [18]. Leveraging context in privacy management is becoming increasingly important. While context-aware systems allow for building smarter and more adaptive applications, at the same time they pose serious challenges for personal privacy due to their extensive collection of detailed information about individuals as well as the usual inability of individuals to properly manage their information flows. This chapter surveys context-adaptive privacy mechanisms that have been proposed and discusses issues around privacy management. Context-adaptive privacy mechanisms can leverage contextual information to determine privacy-relevant context changes in a user's environment and either provide context-specific privacy recommendations, or automatically adapt the user's privacy configuration to meet the new needs.

Chapter 14 is the last chapter of Part III and focuses on privacy threats and solutions for location sharing applications. The chapter provides a survey of the most popular location-based applications, describes the important privacy implications that arise from contributing information in such applications and the existing privacy mechanisms for protecting user privacy. A number of popular location-based applications are considered by the authors, including social applications, transportation and travel applications, fitness applications, image sharing and location sharing applications. Existing approaches for privacy preservation of location sharing applications include path confusion, mix zones, fake data injection, data perturbation, data generalization and suppression, k-anonymity, and encryption.

1.5 Conclusion

Privacy in the context of human mobility is a very popular and widely researched area with a broad spectrum of applications, ranging from urban planning and traffic engineering, to studying the spread of diseases and managing environmental pollution. The recent developments in mobile computing and positioning technologies, along with the advances in information technology, have led to a number of new location sharing applications and systems. This chapter provided an introduction to the field of mobility data privacy, and an overview of the main research directions, real-world systems and applications that are covered in the remainder of this book. These include privacy in the context of location based services, privacy in location based social networks and geo-social applications, privacy in vehicular communication networks, privacy in participatory sensing systems, privacy in RFID applications, privacy-by-design to support mobility data analytics, and many more. At the same time, real-world systems and research prototypes that have been developed for privacy-preserving mobility data management, along with real-world applications where privacy methods need to be selected while accounting for the level of accuracy that is required by the application, were discussed.

Since mobility data privacy is a very active research topic, new techniques, new results and new systems have been probably proposed and published while this book was in production and by the time you are reading it. It is also inevitable that some existing approaches did not find appropriate space for their presentation. However, by providing a solid base to allow readers understand the many facets of the privacy problems involved in mobility data management as well as the wide spectrum of technical solutions, we believe that the material offered in this book will guide researchers and software engineers in better understanding other solutions and possibly proposing new ones.

References

1. G. Acs and C. Castelluccia. A case study: Privacy preserving release of spatio-temporal density in Paris. In *Proceedings of the 20th ACM SIGKDD International Conference on Knowledge Discovery and Data Mining*, KDD '14, pages 1679–1688. ACM, 2014.
2. R. A. Becker, R. Caceres, K. Hanson, J. M. Loh, S. Urbanek, A. Varshavsky, and C. Volinsky. A tale of one city: Using cellular network data for urban planning. *IEEE Pervasive Computing*, 10(4):18–26, 2011.
3. A. T. Campbell, S. B. Eisenman, N. D. Lane, E. Miluzzo, R. A. Peterson, H. Lu, X. Zheng, M. Musolesi, K. Fodor, and G. S. Ahn. The rise of people-centric sensing. *IEEE Internet Computing*, 12(4):12–21, 2008.
4. L. Chen, S. Thombre, K. Järvinen, E. S. Lohan, A. AlÖén-Savikko, H. Leppäkoski, M. Z. H. Bhuiyan, S. Bu-Pasha, G. N. Ferrara, S. Honkala, J. Lindqvist, L. Ruotsalainen, P. Korpisaari, and H. Kuusniemi. Robustness, security and privacy in location-based services for future IoT: A survey. *IEEE Access*, 5:8956–8977, 2017.
5. Y.-A. de Montjoye, C. A. Hidalgo, M. Verleysen, and V. D. Blondel. Unique in the crowd: The privacy bounds of human mobility. *Nature Scientific Reports*, 3, 2013.

6. M. Douriez, H. Doraiswamy, J. Freire, and C. T. Silva. Anonymizing nyc taxi data: Does it matter? In *2016 IEEE International Conference on Data Science and Advanced Analytics (DSAA)*, pages 140–148, 2016.
7. C. Dwork. Differential privacy. In *Proceedings of the 33rd International Conference on Automata, Languages and Programming - Volume Part II*, ICALP'06, pages 1–12. Springer-Verlag, 2006.
8. C. Dwork, F. McSherry, K. Nissim, and A. Smith. Calibrating noise to sensitivity in private data analysis. In S. Halevi and T. Rabin, editors, *Theory of Cryptography*, pages 265–284. Springer Berlin Heidelberg, 2006.
9. S. Ejsing-Duun. *Location-based games: From screen to street*. PhD thesis, Department of Communication and Psychology, Aalborg, University in Copenhagen, 2011.
10. M. Gerlach and F. Guttler. Privacy in vanets using changing pseudonyms - ideal and real. In *2007 IEEE 65th Vehicular Technology Conference - VTC2007-Spring*, pages 2521–2525, 2007.
11. A. Gkoulalas-Divanis and V. S. Verykios. A privacy-aware trajectory tracking query engine. *SIGKDD Explor. Newsl.*, 10(1):40–49, 2008.
12. H. Hartenstein and L. P. Laberteaux. A tutorial survey on vehicular ad hoc networks. *IEEE Communications Magazine*, 46(6):164–171, 2008.
13. X. Hu, T. H. S. Chu, V. C. M. Leung, E. C. H. Ngai, P. Kruchten, and H. C. B. Chan. A survey on mobile social networks: Applications, platforms, system architectures, and future research directions. *IEEE Communications Surveys Tutorials*, 17(3):1557–1581, 2015.
14. L. Kazemi and C. Shahabi. Geocrowd: Enabling query answering with spatial crowdsourcing. In *Proceedings of the 20th International Conference on Advances in Geographic Information Systems*, SIGSPATIAL '12, pages 189–198. ACM, 2012.
15. F. Mahmud and H. Aris. State of mobile crowdsourcing applications: A review. In *2015 4th International Conference on Software Engineering and Computer Systems (ICSECS)*, pages 27–32, 2015.
16. N. Pelekis, A. Gkoulalas-Divanis, M. Vodas, D. Kopanaki, and Y. Theodoridis. Privacy-aware querying over sensitive trajectory data. In *Proceedings of the 20th ACM International Conference on Information and Knowledge Management*, CIKM '11, pages 895–904. ACM, 2011.
17. N. Pelekis, A. Gkoulalas-Divanis, M. Vodas, A. Plemenos, D. Kopanaki, and Y. Theodoridis. Private-HERMES: A benchmark framework for privacy-preserving mobility data querying and mining methods. In *Proceedings of the 15th International Conference on Extending Database Technology*, EDBT '12, pages 598–601. ACM, 2012.
18. F. Schaub, B. Könings, and M. Weber. Context-adaptive privacy: Leveraging context awareness to support privacy decision making. *IEEE Pervasive Computing*, 14(1):34–43, 2015.
19. B. Schilit, N. Adams, and R. Want. Context-aware computing applications. In *1994 First Workshop on Mobile Computing Systems and Applications*, pages 85–90, 1994.
20. V. Sivaraman, J. Carrapetta, K. Hu, and B. G. Luxan. Hazewatch: A participatory sensor system for monitoring air pollution in sydney. In *38th Annual IEEE Conference on Local Computer Networks - Workshops*, pages 56–64, 2013.
21. C. E. Walters, M. M. Meslé, and I. M. Hall. Modelling the global spread of diseases: A review of current practice and capability. *Epidemics*, 2018 (in press). https://www.sciencedirect.com/science/article/pii/S1755436517301135?via%3Dihub.
22. T. D. Wemegah and S. Zhu. Big data challenges in transportation: A case study of traffic volume count from massive radio frequency identification (RFID) data. In *2017 IEEE International Conference on the Frontiers and Advances in Data Science (FADS)*, pages 58–63, 2017.
23. A. Wesolowski, N. Eagle, A. J. Tatem, D. L. Smith, A. M. Noor, R. W. Snow, and C. O. Buckee. Quantifying the impact of human mobility on malaria. *Science (New York, N.Y.)*, 338(6104):267–70, 2012.
24. H. Wu, S. Xiang, W. S. Ng, W. Wu, and M. Xue. HipStream: A privacy-preserving system for managing mobility data streams. In *2014 IEEE 15th International Conference on Mobile Data Management*, volume 1, pages 360–363, 2014.

25. J. Xia, K. M. Curtin, W. Li, and Y. Zhao. A new model for a carpool matching service. *PLoS One*, 10(6), 2015.
26. A. Yaeli, P. Bak, G. Feigenblat, S. Nadler, H. Roitman, G. Saadoun, H. J. Ship, D. Cohen, O. Fuchs, S. Ofek-Koifman, and T. Sandbank. Understanding customer behavior using indoor location analysis and visualization. *IBM Journal of Research and Development*, 58(5/6):3: 1–3:12, 2014.
27. A. Yassin, Y. Nasser, M. Awad, A. Al-Dubai, R. Liu, C. Yuen, R. Raulefs, and E. Aboutanios. Recent advances in indoor localization: A survey on theoretical approaches and applications. *IEEE Communications Surveys Tutorials*, 19(2):1327–1346, 2017.

Part I
Fundamentals for Privacy in Mobility Data

Chapter 2
Modeling and Understanding Intrinsic Characteristics of Human Mobility

Jameson L. Toole, Yves-Alexandre de Montjoye, Marta C. González, and Alex (Sandy) Pentland

Abstract Humans are intrinsically social creatures and our mobility is central to understanding how our societies grow and function. Movement allows us to congregate with our peers, access things we need, and exchange information. Human mobility has huge impacts on topics like urban and transportation planning, social and biologic spreading, and economic outcomes. Modeling these processes has however been hindered so far by a lack of data. This is radically changing with the rise of ubiquitous devices. In this chapter, we discuss recent progress deriving insights from the massive, high resolution data sets collected from mobile phone and other devices. We begin with individual mobility, where empirical evidence and statistical models have shown important intrinsic and universal characteristics about our movement: we as human are fundamentally slow to explore new places, relatively predictable, and mostly unique. We then explore methods of modeling aggregate movement of people from place to place and discuss how these estimates can be used to understand and optimize transportation infrastructure. Finally, we highlight applications of these findings to the dynamics of disease spread, social networks, and economic outcomes.

J. L. Toole
Engineering Systems Division, MIT, Cambridge, MA, USA
e-mail: jltoole@alum.mit.edu

Y.-A. de Montjoye (✉)
Imperial College London, Dept. of Computing, London, UK

Imperial College London, Data Science Institute, London, UK
e-mail: deMontjoye@imperial.ac.uk

A. (Sandy) Pentland
Media Lab, MIT, Cambridge, MA, USA
e-mail: pentland@mit.edu

M. C. González
Department of Civil and Environmental Engineering, MIT, Cambridge, MA, USA
e-mail: martag@berkeley.edu

© Springer Nature Switzerland AG 2018
A. Gkoulalas-Divanis, C. Bettini (eds.), *Handbook of Mobile Data Privacy*,
https://doi.org/10.1007/978-3-319-98161-1_2

2.1 Introduction

Mobility has been a steering force for much of human history. The movement of peoples has determined the dynamics of numerous social and biological processes from tribal mixing and population genetics to the creation of nation-states and the very definition of our living areas and identities. Urban and transportation planners, for example, have long been interested in the flow of vehicles, pedestrians, or goods from place to place.

With more than half of the world's population is now living in urban areas,[1] understanding how these systems work and how we can improve the lives of people using them is more important than ever. Insights from models informed by novel data sources can identify critical points in road infrastructure, optimize public services such as busses or subways, or study how urban form influences its function. Epidemiologists are also relying heavily on models of human movement to predict and prevent disease outbreaks [13, 66] as global air travel makes it possible for viruses to quickly jump continents and dense urban spaces facilitate human-to-human contagion. This has made understanding human movement a crucial part of controlling recent disease outbreaks.[2] Finally, social scientists are increasingly interested in understanding how mobility impacts a number of social processes such as how information spreads from person to person in offices and cafes across the world. These interactions have been theorized to impacts crime rates, social mobility, and economic growth [6, 46] and understanding their dynamics may improve how we live, work, and play.

The growing need to understand and model human mobility has driven a large body of research seeking to answer basic questions. However, the lack of reliable and accessible data sources of individual mobility has greatly slowed progress testing and verifying these theories and models. Data on human mobility has thus far been collected through pen and paper surveys that are prohibitively expensive to administer and are plagued by small and potentially biased sample sizes. Digital surveys, though more convenient still require active participation and often rely on self-reporting [14]. Despite the development of statistical methods to carefully treat this data [5, 26, 45] new, cheaper, and larger data sources are needed to push our understanding of human mobility efforts further.

The evolution of technology over the past decade has given rise to ubiquitous mobile computing, a revolution that allows billions of individuals to access people, goods, and services through 'smart' devices such as cellular phones. The penetration of these devices is astounding. The six billion mobile phones currently in use triples the number of internet users and boast penetration rates above 100% in the developed word, e.g. 104% in the United States and 128% in Europe.[3] Even

---

[1]United Nations Department of Economic and Social Affairs—World Urbanization Prospects—2014 Update. http://esa.un.org/unpd/wup/Highlights/WUP2014-Highlights.pdf.

[2]http://www.worldpop.org.uk/ebola/.

[3]GSMA European Mobile Industry Observatory 2011. http://www.gsma.com/publicpolicy/wp-content/uploads/2012/04/emofullwebfinal.pdf.

in developing countries, penetration rates are of 89%[4] and growing fast. These devices and the applications that run on them passively record the actions of their users including social behavior and information on location[5] with high spatial and temporal resolution. Cellular antennas, wifi access points, and GPS receivers are used to measure the geographic position of users to within a few hundred meters or less. While the collection, storage, and analysis of this data presents very real and important privacy concerns [15, 16], it also offers an unprecedented opportunity for researchers to quantify human behavior at large-scale. With billions of data points captured on millions of users each day, new research into computational social science [37] has begun to augment and sometimes replace sparse, traditional data sources, helping to answer old questions and raise new.

In this chapter, we present an overview of mobility research in the current data rich environment. We describe a variety of new data sources and detail the new models and analytic techniques they have inspired. We start by exploring research on individuals that emphasizes important intrinsic and universal characteristics about our movement: we are slow to explore, we are relatively predictable, and we are mostly unique. We then discuss efforts to add context and semantic meaning to these movements. Finally, we review research that models aggregates of human movements such as the flow of people from place to place. Throughout and at the end of this chapter, we point out applications of this research to areas such as congestion management, economic growth, or the spreading of both information and disease.

2.2 New Data Sources

Traditional data sources for human mobility range from census estimates of daily commutes to travel diaries filled out by individuals. These surveys are generally expensive to administer and participate in as they require intensive manual data encoding. To extract high-resolution data, individuals are often asked to recall large amounts of information on when, where, and how they have traveled making them prone to mistakes and biases. These challenges make it hard for surveys to cover more than a day or week at a time or to include more than a small portion of the population (typically less than 1%).

Mobile phones, however, with their high penetration rates, represent a fantastic sensor for human behavior. A large fraction of location data from mobile phones are currently in the form of call detail records (CDRs) collected by carriers when users perform actions on their devices that make use of the telecommunications

---

[4]ITU (2013). ICT facts and figures. http://www.itu.int/en/ITU-D/Statistics/Documents/facts/ICTFactsFigures2013-e.pdf.

[5]Lookout (2010). Introducing the app genome project. https://blog.lookout.com/blog/2010/07/27/introducing-the-app-genome-project/.

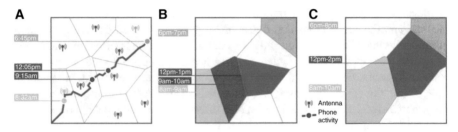

Fig. 2.1 Mobile phones are increasingly being used to collect high-resolution mobility data. This figure from de Montjoye et al. [15] depicts (**a**) a sequence of calling events made by a user at different locations. (**b**) These events are localized to the area served by the closest mobile phone tower to the use and (**c**) can be aggregated into individual specific neighborhoods where a user is likely to be found at different times of the day or week

network. The location of each device at the time a call, text, or data request is registered (Fig. 2.1) is recorded by carriers for billing, network performance, and legal purposes. Locations are inferred either by observing the tower through which the phone is connected or by triangulation with nearby towers. With the increasing use of mobile phones, each individual generates tens to hundreds of these digital breadcrumbs on a daily basis and this number is only increasing. Through specific agreements or through open-data challenges [17], location data on millions of users is readily available to researchers and has been used extensively to augment and sometimes replace traditional travel surveys. This data now forms the core of numerous new mobility studies and models some of which we describe below.

Though generally less common than CDRs, applications running on smartphones may access even more precise estimates of a user's position. A variety of these sensors, from GPS to wifi, can pinpoint the location of a device to within just a few meters and can record data every few minutes [1]. Similarly, protocols such bluetooth and NFC allow devices to discover and connect to one another within a few meter radius, creating ad hoc sensor and social proximity networks [21]. Some of these applications and underlying social-networks explicitly add crucial context to mobility data. Foursquare invites users to "check-in" at specific places and establishments, Twitter will automatically geotag tweets with precise coordinates from where they were sent, and the Future Mobility survey app passively maintains an activity diary [14] requiring little input from users.

Infrastructure and public services have also become much smarter and now collect data on their usage to improve and help plan operations. Toll booths automatically count and track cars and this data has helped create accurate and real-time traffic estimates used by mapping and navigation services to provide better routing information. Subways, streetcars, and busses use electronic fare systems that record when millions of users enter and exit transportation systems to help better predict demand. In addition to smarter public infrastructure, the ecosystem created by digital devices has given birth to entirely new transportation services such as Hubway, the Boston bike rental service, that collects data on every bike ride and has even released some publicly[6] or Uber, an on-demand car service, that uses historical usage data to balance the time a user has to wait for a car to arrive and the time drivers spend without clients. Finally, on-board devices and real-time data feeds from automatic vehicle location (AVL) systems power applications such as NextBus to track the location of thousands of busses and subways across the world to display and predict when the next bus will arrive. While smart infrastructure comes with its own privacy challenges [35],[7] vehicle and public transport data offer additional information to urban planners and mobility modelers to better understand these systems.

Finally, most practical mobility models need to properly account for geography such as mountains and rivers, transportation infrastructure such as bridges and highways, differences in density between urban and rural areas, and numerous other factors. Thankfully, the digitization of maps has led to an explosion of geographic data layers. Geographic information systems (GIS) have improved dramatically while falling data storage prices have made it possible for small and large cities to offer their public mapping data to citizens in an online, machine readable format. The U.S. Census Bureau's TIGERline program, San Francisco's OpenSF, and New York City's PLUTO data warehouse are just a few sources that offer huge repositories of publicly accessible geographic data on everything from building footprints and the location of individual trees in a city. Open- and crowd-sourced initiatives like OpenStreetMap allow anyone in the world to contribute and download high-resolution digital maps of roads, buildings, subways, and more, even in developing areas that may not have institutional resources to create them. Private efforts such as Google Maps and MapBox offer high-resolution satellite imagery, route planning, or point of interest information through free or low cost APIs. Put together, these resources provide a digital map of the world that serves as a rich backdrop on which to study human mobility and the infrastructure built to facilitate it.

Put together, new sources from CDRs to public transport data, from mobile phone applications to AVLs generate a dataset with size and richness prohibitively expensive to match via traditional methods. Collected passively and without any effort from the user, this data is often more robust to manipulation by conscious or unconscious biases and provide a signal that is difficult to fake. While we are convinced of the potential of this data, it is always important to remember that it is not without pitfalls. It would be illusory to think that all of the old biases or hidden variables would simply disappear because the data is large. In some cases, data is only recorded when an individual interacts with a device which may bias when samples are taken [47]. Similarly it is important to keep in mind that even if it covers a significant fraction of the population this data might not be representative. Finally, these data generally come stripped of context. We do not know why an individual

---

[6]Hubway Data Visualization Challenge (2012). http://hubwaydatachallenge.org/.
[7]New York taxi details can be extracted from anonymized data, researchers say (2014). http://www.theguardian.com/technology/2014/jun/27/new-york-taxi-details-anonymised-data-researchers-warn.

has chosen to move or what they will be doing there. For these reasons, sampling and robust statistical methods are still—maybe more than ever[8]—needed to use this data to augment our current understanding of human mobility while still providing robust conclusions. We now discuss a number of studies that aim to do just this.

2.3   Individual Mobility Models

Understanding mobility at an individual level entails collecting and analyzing sets of times, places, and semantic attributes about how and why users travel between them. For example, on a typical morning one may wake up at home, walk to a local coffee shop on the way to the bus that takes them to work. After work they may go to the grocery store or meet a friend for dinner before returning home only to repeat the process the next day. The goal modeling this mobility is to understand the underlying patterns of individuals using new high resolution data. While models have been used to plan infrastructure or public transport, they have also uncovered insights into the underlying nature of human behavior: we are slow to explore, relatively predictable, and mostly unique.

Early modeling work draws a great amount of inspiration from statistical physics, with numerous efforts making parallels with human mobility and random walk or diffusion processes. One of the used data from the crowdsourced "Where's George" project. Named after George Washington, whose head appears on the $1 bill, the project stamped bills asking volunteers to enter the geographic location and serial number of the bills in order to build a travel history of various banknotes. As bills are primarily carried by people when traveling from store to store, a note's movement serves as a proxy for human movement. Modeling the bills trajectories as continuous random walks, Brockmann et al. found that their movement appears to follow a Levy flight process [8]. This process is characterized by subsequent steps whose angular direction is uniformly distributed, but whose step-lengths follow a fat-tailed distribution. While small jumps are most probable, bills have a significant probability of making long jumps from time to time. These findings are aligned with observations that humans tend to make many short trips in a familiar area, but also take longer journey's now and then.

In 2008, Gonzalez et al. [23] showed that the movement of these bills does not tell the whole story. Using a CDRs dataset of more than 100,000 users over a 6 month period in a European country (Fig. 2.2a), they showed that the step-length distribution for the entire population was better approximated by a truncated power-law $P(\Delta r) = (\Delta r + \Delta r_0)^{-\beta} \exp(-\Delta r/\kappa)$ with exponent $\beta = 1.79$ and cutoff distances between 80 and 400 km. This suggests that Levy flights are only a good approximation of individual's mobility for short distances. To understand the

---

[8]Flowing data—where people run in major cities. http://flowingdata.com/2014/02/05/where-people-run/.

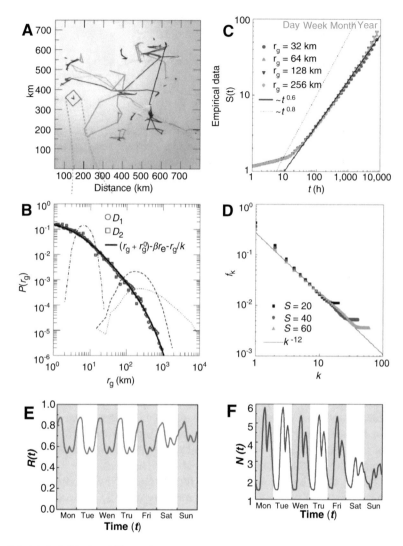

Fig. 2.2 (**a**) Individual mobility trajectories are passively collected from mobile devices [23]. (**b**) Measuring the distribution of radius of gyrations, r_g within a population of 100,000 users in a European country reveals considerable heterogeneity in typical travel distance of individuals. Moreover, this distribution cannot be explained by modeling each individual's movement as realizations of a single Levy flight process [23]. (**c** and **d**) Show the slower than linear growth in new locations visited over time $S(t)$ and that the probability a location is visited next is inversely proportional to the frequency it has been visited in the past [54]. (**e**) This preferential return contributes to strikingly high predictability $R(t)$ over time while (**f**) the number of unique locations visited in any given hour is highly periodic and corresponds to the sleep-wake cycles of individuals [55]

mechanism that gives rise to this distribution, the authors borrowed a quantity from polymer physics known as the "radius of gyration" r_g:

$$r_g(t) = \sqrt{\frac{1}{N(t)} \sum_{i=1}^{N(t)} (\mathbf{r} - \mathbf{r_{cm}})^2}, \qquad (2.1)$$

where N(t) are the number of observed locations and r_{cm} is the mean location of the user during the observation period. In essence, the radius of gyration is a measurement of the characteristic distance an individual travels during an observation period t. The authors then showed that the distribution of r_g in the population is itself well approximated by a truncated power-law with $r_g^0 = 5.8\,\text{km}$, $\beta_{r_g} = 1.65$, and a cutoff of $\kappa = 350\,\text{km}$ (Fig. 2.2b). Simulations suggest that the step-length distribution of the entire population is produced by the convolution of heterogeneous Levy flight processes, each with a different characteristic jump size determined by an individual's radius of gyration. Put differently, each person's mobility can be approximated by a Levy flight process up to trips of some individual characteristic distance r_g. After this distance, however, the probability of long trips drops far faster than would be expected from a traditional Levy flight.

Further investigation by the authors revealed the source of this behavior: the idiosyncrasy of human movements. Unlike random processes, humans are creature of habits and tend to returns to previously visited locations such as home or work. The nature of these returns was also found to follow a very particular pattern. An individual returns to a previously visited location with a probability proportional to that location's rank $P(L)$ $1/L$ amongst all the places he or she visits. These non-random, predictable return visits are unaccounted for in random walk and Levy flight models and have been shown to be at the heart of deviations of observed behavior from random processes. Additional studies [9] have found similar patterns in both other CDRs datasets and Foursquare or Twitter check-ins.

Subsequent work by Song et al. [54] further studied how individual-specific locations need to be taken into account in mobility models. Using a similar CDR dataset, the authors showed three important characteristics of human behavior. First, the number of unique locations visited by individuals $S(t)$ scales sub-linearly with time $S(t)$ t^μ where $\mu = 0.6$ (Fig. 2.2c). Second, the probability an individual returning to a previously visited locations scales with the inverse of the rank of that location $P(L)$ $L^{-\zeta}$ where $\zeta = 1.2$ (Fig. 2.2d), a phenomena labeled as 'preferential return'. And third, the mean displacement (Δr) of an individual from a given starting point shows slower than logarithmic growth, demonstrating the extremely slow diffusion of humans in space. In essence, these finding pinpoint the dampening of explorative human movement overtime. Long jumps are observed so infrequently that they do not affect the average displacement of individuals. The authors then propose a new model of human mobility to capture these three characteristics. The model is as follows: starting at time t, an individual will make a trip at some future time Δt drawn from a fat-tailed probability distribution measured from CDRs. With probability $\rho S^{-\gamma}$, the individual travels to a new, never-before visited location some distance Δr away, where Δr is drawn from the fat-tailed distribution characterized in the previous model. With probability $1 - \rho S^{-\gamma}$ an individual returns to a previously visited location according to the inverse rank equation.

These early models do not attempt to recover periodic aspects of movement (e.g. daily commuting) or semantic meaning of visits (e.g. to visit a friend or go shopping), or attempt to do so. They do, however, emphasize important statistical and scaling properties of human mobility and often successfully reproduce them. Taken together, these models show how we slow we human are in our exploration, returning more often than not to known places and with less long steps than predicted by a power-law distribution.

Approaching the problem from the perspective of machine and statistical learning, another set of models has uncovered and explored another facet of human mobility: how predictable we are. In [55], Song et al. used information theory metrics on CDRs to show the theoretical upper-bound on predictability using three entropy measures the entropy S, the random entropy S^{rand}, and the uncorrelated entropy S^{unc}. They then use their empirical distributions to derived an upper bound on a user's predictability ($\prod^{max}$, $\prod^{rand}$, and $\prod^{unc}$). On average, the potential predictability of an individual's movement is an astounding 93% and no user displayed a potential predictability of less than 80%. To further quantify predictability, the author introduced two new metrics. They defined regularity $R(t)$ as the probability a user is found at their most visited location during a given hour t, along with the number of unique locations visited during a typical hour of the week $N(t)$ (Fig. 2.2e and f). Both show strong periodicity and regularity. These quantities have since been measured in different data sets in different cities and countries and have been shown to be consistent among them [9].

While the previous study provided a theoretic upper bound on the predictability of an individual, a number of statistical learning techniques have been developed to make predictions of where an individual will be at a given time. Early work in the area, predating even analytic computations, used Markov models and information on underlying transportation networks to predict transitions between mobile phone towers within cities. These models have been used to improve quality of service of wireless networks through proper resource allocation [33, 36, 40, 58]. Later work incorporated various trajectory estimation and Kalman filtering algorithms to predict movements in small spaces such as college campuses [38, 43].

Temporal periodicity was used by Cho et al. [12] in their Periodic Mobility Model and social behavior incorporated in the Period Social and Mobility Model. At their core, these models are mixture models in two-dimensional space that learn the probability distribution of a user to be at any given location at a given time from previous location data. The latter also account for the location history of social contacts. The authors used these models to estimate that as much as 30% of our trips may be taken for social purposes. Multivariate nonlinear time series forecasting produced similar results [19, 51] predicting where an individual will be either in the next few hours or at a given time of a typical day. These models, however, are all focused on predicting the geographic position of individuals at different times and do not attempt to understand what individuals may be doing there or any other semantics of place.

Though acquiring semantic information about mobility is more difficult than simply measuring geographic coordinates, it provides a much richer abstraction to study behavior. In one of the first studies to mine the behavior of college students using mobile phones, Eagle and Pentland [20] gave a few hundred students smart phones that recorded not only locations, but asked users to label each place with its function such as home or work. Applying principal component analysis to these abstract movements from semantic place to semantic place (as opposed to geographic movements alone), the authors found that an individual's behavior could be represented as a linear combination of just a few 'eigenbehaviors'. These eigenbehaviors are temporal vectors whose components represent activities such as being at home or being at work. They can be used to predict future behaviors, perform long range forecasts of mobility, and label social interactions [21, 48]. The price paid for such detailed predictions, however, is the need for semantic information about locations. Geographic positions need to be tagged with attributes such as home or work in order for them to be grouped and compared across individuals.

Another approach to studying more abstract measurements of individual location information comes from recent work by Schneider et al. [52]. The authors introduced *mobility motifs* by examining abstract trip chains over the course of a day. A daily mobility motif is defined a set of locations and a particular order that a person visits them over the course of a day. More formally, these motifs constitute directed networks where nodes are locations and edges are trips from one location to another. For example, the motif of an individual whose only trips in a day are to and from work will consist of two nodes with a two directed edges (one in both directions). Counting motifs in mobility data from both CDRs and traditional travel surveys, they find on average individuals visit three different places in a given day. They then construct all possible daily motifs for a given number of locations n and compute the frequencies that those motifs appear in human mobility data. Shockingly, while there exist over 1 million ways for a user to travel between 6 or fewer locations, 90% of people use one of just 17 motifs and nearly a quarter follow the simple two location commute motif introduced earlier (Fig. 2.3a). The authors found similar results in travel survey data and introduced a simple Markov model for daily mobility patterns which reproduces empirical results.

It is tempting to hypothesize that high theoretical and practical predictability results from high levels of similarity between individuals in a region. Perhaps the pace of life, pull of mono-centric downtowns, or the structure of transportation systems funnel users to the same places and route choices. de Montjoye et al. [15] explored this hypothesis and found that, while predictable, an individual's movement patterns are also unique. The authors introduced *unicity*, $\mathscr{E}_p$, as the fraction of traces uniquely defined by a random set of p spatiotemporal points where a trace T is a set of spatiotemporal points, each containing a location and a timestamp. A trace is said to be uniquely defined by a set of points I_p if it is the only trace that matches I_p in the entire dataset. Applying this measure to a CDR

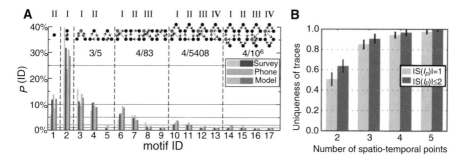

Fig. 2.3 (**a**) Removing geographic coordinates from locations and only focusing on a set of unique places and the directed travel between them, mobility motifs reveal that the daily routines of people are remarkably similar. Despite over 1 million unique ways to travel between 6 or fewer points, just 17 motifs are used by 90% of the population. Moreover, the frequency of their appearance in CDR data matches very closely with more traditional survey methods [52]. (**b**) Despite this similarity and predictability, our movement displays a high degree of unicity. Just four spatiotemporal points is enough to differentiate a user from 95% of all others individuals [15]

dataset on 1.5 million users, the authors found that just four spatiotemporal points is enough to uniquely identify 95% of all users (Fig. 2.3b). The authors further study unicity when the data is coarsened spatially or temporally. They found $\mathcal{E} \sim (v*h)^{\beta}$ unicity decrease as a power function with the spatial (v) and temporal resolution of the data (h) and that $\beta \sim -p/100$. Taken together, these equations show that unicity decreases slowly with the spatial and temporal resolution of the data and that this decrease is easily compensated by the number of points p. High uniqueness in human mobility traces exists across many spatiotemporal scales. These results raise many questions about the privacy of massive, passively collected metadata datasets, but also highlight an interesting nuance of human mobility: though individuals are predictable, they are also unique.

Merging concepts of predictability and unicity, work by Sun et al. [57] used temporal encounter networks to study repeated co-locations between passengers using data from bus passengers in Singapore. Temporal encounter networks were constructed by connecting individuals if they rode the same bus at the same time. An average individual encountered roughly 50 people per trip and these trips were highly periodic, occurring at intervals associated with working hours as well as daily and weekly trips. A pair of individuals who encountered each other tended to meet an average of 2.5 times over the course of a week. The distribution of time between encounters reveals strong periodicity, with passengers riding the same bus to work in the morning riding the same home, or riding the same bus at the same time each morning. This finding illustrates the idiosyncrasies of human mobility. We not only visit just a few places during the day, we do so at the same times and by the same routes. Though both of these results suggest that our unicity should be low, the previous work shows us that this is not the case.

In summary, new data sources have allowed researchers to show that, over weeks and months, human movement is characterized by slow exploration, preferential return to previous visited places, exploration of daily motifs, and predictable unique-ness. These regularities have been used to develop algorithms capable of predicting movement with high degrees of accuracy and have been shown to mediate other important processes such as social behavior and disease spread. Individual mobility patterns, however, are not the only level of granularity of interest to researchers, city planners, or epidemiologist. Aggregate movement can be either derived from individual level model or modeled as an emergent, personified phenomena. In the next section, we discuss works and models which aim at describing and modeling aggregate movement and flows of many individuals from place to place.

2.4 Aggregate Mobility

Aggregated mobility is used for planning urban spaces, optimizing transportation networks, studying the spread of ideas or disease, and much more. Perhaps the largest component in these models are origin-destination matrices that store the number of people traveling from any location to any other at different times or by different means. Like many complex systems, aggregate behavior is often more than the sum of individual parts and can be modeled separately. Additional layers of complexity are also needed to account for and sometimes explain individual choice of mode of transportation or route as described by the "four step model" [41, 45].

Like their individual-focused counterparts, many of these aggregate models are inspired by physical processes. Some of the earliest techniques for estimating origin-destination matrices are gravity models which have been used to model flows on multiple scales, from intra-city to international [27, 45]. Borrowed directly from Newton's law of gravitation, the number of trips T_{ij} taken from place i to place j is modeled as a function of the population of each place m_i and m_j and some function of the distance between them $f(r_{ij})$. The intuition is that the population of a place, it's mass, is responsible for generating and attracting trips and thus the total flux between the two places should be proportional to the product of the two masses while the distance between them mitigates the strength of this connection. In the fully parameterized version of this model, an exponent is applied to the population at the origin and destination $T_{ij} = a \frac{m_i^\alpha m_j^\beta}{f(r_{ij})}$ to account for hidden variables that may be specific to local regions or populations. While the classical gravity model from physics is recovered by setting $\alpha = \beta = 1$, and $f(r_{ij}) = r_{ij}^2$, these parameters are generally calibrated for specific application using survey data.

Gravity models, however, are not without limitation. First, they rely on a large number of parameters to be estimated from sparse survey data which often leads to overfitting and, second, they fail to account for opportunities that exist between the two masses of people. The latter fault results in the same flow of people being estimated between two locations whether there is an entire city or an empty desert between them. Intuitively, one would expect that trips between places would be affected by the intervening opportunities to complete a journey. These shortcomings led Simini et al. to develop the radiation model [53]. Again borrowing from physics (this time radiation and absorption), they imagined individuals being emitted from a place at a rate proportional to its population and absorbed by other locations at a rate proportional to the population there. In this model, the probability that an emitted person arrives at any particular place is a function of their probability of not being absorbed before getting there. The model is as follows: $T_{ij} = T_i \frac{m_i m_j}{(m_i + s_{ij})(m_i + m_j + s_{ij})}$, where T_i is total number of trips originating from location i and s_{ij} is the population within a disc centered on location i with a radius equal to the distance between i and j. The radiation model does not directly depend on the distance between the two places, taking instead into account the opportunities in-between them (Fig. 2.4a). Unlike the gravity model, the radiation model is parameterless and requires only data on populations to estimate flow. The authors showed that despite its lack of parameters, the radiation model provides better estimates of origin-destination flows than the gravity model for areas the size of counties or larger.

Yang et al. adapted Simini's radiation model to correct for distortions caused at different scales [67]. They showed the original radiation model's lower accuracy in urban environment is due to the relatively uniform density and small distances that characterize cities. In dense urban areas, distances are all relatively short and an individual may choose to visit a particular location due to hedonic attributes regardless of whether it is convenient to get to or not. Yang et al. subsequently introduced a scaling parameter α in the function describing the conditional probability an individual is absorbed at a location. This single parameter was enough to correct for these distortions and to provide a model that works on any length scale. Moreover, the authors suggested that for urban areas, the density of points of interest (POIs) such as restaurants and businesses is a better predictor of the absorption of a place than its population. Iqbal et al. [31] have demonstrated an improved way to extract valid, empirical OD matrices from call detail records (CDRs) data to validate the model.

Finally, activity-based models [5] model user intent more explicitly. They hypothesize that all trips are made to fulfill certain needs or desires of an individual. Travel and survey diaries are used to identify those needs for different segments of the population and how they are typically fulfilled. This knowledge can then be used by the model given the demographics of individuals and environmental factors. These models are closely related to agent-based models simulating the behavior of city residents and rely heavily on the idea of economic utility.

From a practical perspective, city planners need to know not only how many people will go from point A to point B at a certain time of the day but also the mode of transportation and route choice of these individuals. For example, we would like to predict which route they will take so that we can properly estimate the stress placed on transportation systems and potentially optimize performance. Models of route choice typically assume that individual rationally chose the path from A to B that minimize some cost function such as total travel time or distance. Paths can be computed on a road network using shortest path algorithms such as the traditional Dijkstra algorithm or A-Star, an extension that enjoy better performance thanks to heuristics. Other information such as speed limits can also be taken into account to estimate free flow travel times.

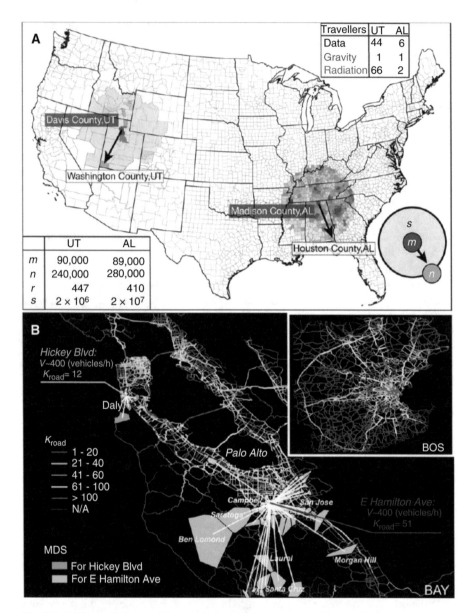

Fig. 2.4 (**a**) The radiation model accounts for intervening opportunities, producing more accurate estimates of flows between two places than more traditional gravity models [53]. (**b**) Routing millions of trips measured from CDR data to real road networks makes it possible to measure the importance of a road based on how many different locations contribute traffic to it, K_{road}. Understanding how transportation systems perform under different loads presents new opportunities to solve problems related to congestion and make infrastructure more efficient [63]

More advanced models are needed to account for the impact of congestion as drivers rarely encounter completely empty freeways. Incremental traffic assignment algorithms model congestion endogenously [56]. Trips are first split into increments containing only a fraction of total flow between two points. Trips in each increment are then routed along shortest paths independently of all other trips in that increment keeping counts of how many trips were assigned to each road. The travel times are then adjusted according to a volume delay function that accounts for the current congestion on a road where congestion is computed as the ratio between the volume of traffic assigned to the segment and the capacity of the road (referred to as volume-over-capacity). Trips in the next increment are then routed using updated costs until all flow has been accounted for. In this way, as roads become more congested and the travel time increases, drivers in later iterations are assigned to different, less congested routes. Values of total volume on each road, congestion, and travel times can then be validated against traffic counters, speed sensors, or data from vehicle fleets like taxis and busses but also smartphones such as in the Mobile Millennium project [28, 30, 32, 49].

Wang et al. [63] further explored the use of CDRs as input for these incremental algorithms to estimate traffic volume and congestion. After correcting for differences in market share and vehicle usage rates, they measure trips by counting consecutive phone calls of individuals as they move through the city to generate flow estimates that were then routed. Using this approach, Wang et al. show the distribution of traffic volume and congestion to be well approximated by an exponential mixture model. This model depends on the number of major and minor roadways in a cities network. Using the same approach, the authors describe the usage patterns of drivers by a bipartite usage graph connecting locations in the city to roads used by those travelers (Fig. 2.4b). Roads can be defined by the number of locations that contribute traffic them and places can be described by the roads used to visit. The "function" of a road can then be classified by comparing its topological to its behavioral importance. For example, a bridge may be topologically important because it is the only way to cross a river, but a main street may be behaviorally important because it attracts motorists from many different neighborhoods. Using these measures, researchers were able to devise congestion reduction strategies that target the 2% of neighborhoods where trip reduction will have the largest network wide effect. They found this smart reduction strategy is three to six times as effective as a random trip reduction strategy. Further work used this analysis to predict traffic jams [62, 64].

Private cars, however, are not the only mode of transportation studied. Using smartphones and AVL data, researchers have been mapping the routes followed by public transport and even privately owned mini-buses in the developing countries [11, 18, 50]. Similarly, data on air travel has been increasingly available to study aggregated mobilities between cities for applications in epidemiology (see below).

2.5   Human Behavior and Mobility

While of obvious interest to travelers, urban planners and transportation engineers, people's movement strongly impacts other areas. Though by no means an exhaustive list, we highlight three areas here: social behavior, disease and information spread, and economic outcomes. Many of these dynamics are discussed in greater detail in further sections of this volume.

2.5.1   Mobility and Disease Spread

Human movement via cars, trains, or planes has always been a major vector in the propagation of diseases. Consequently, the human mobility data and models discussed so far have increasingly been used to study the propagation of diseases. For example, CDR data has been used to map mobility patterns in Kenya helping researchers in their fight against Malaria [65, 66]. More recently, CDR and other data from West-Africa has been used to model regional transportation patterns to help control the spread of Ebola.[9] Finally, air travel data has become central to the study of global epidemics when planes allow an individual to travel between nearly any two points on the globe in a matter of hours. The global airline network therefore often determines how potent an epidemic could be and its likely path across the globe [3, 4, 13, 42, 44] (Fig. 2.5a).

2.5.2   Mobility and Social Behavior

Intent is a crucial element of human mobility and movement is often a means to a social end. Despite new communications technologies making it easier than ever to connect across vast distances, face to face interactions still play an important role in social behavior whether it is the employees of a company commuting to a central workplaces or friends meeting at a restaurant on a weekend. The link between social contacts and mobility has becoming increasingly prominent in research as mobility data is often collected through mobile phones or location-based social networks.

Using data from an online social-network, Liben-Nowell showed the probability of being friends with another individual to decrease at a rate inversely proportional to the distance between them suggesting a gravity model of the form discussed above [39]. Subsequent work verified Liben-Nowell findings in other social networks [2, 24] while Toole et al. [59] showed the importance of taking into account geography when studying social-networks and how information spreads through

---

[9]Cell-Phone Data Might Help Predict Ebola's Spread (2014). http://www.technologyreview.com/news/530296/cell-phone-data-might-help-predict-ebolas-spread/.

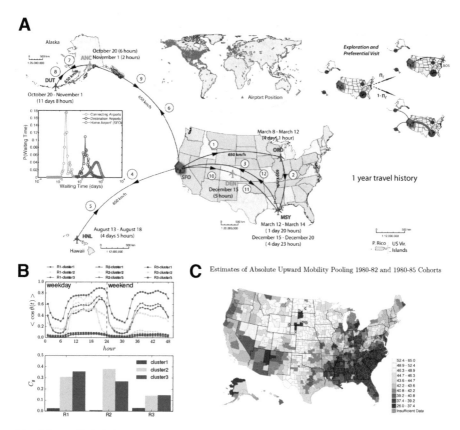

Fig. 2.5 (**a**) Global air travel has dramatically increased the speed at which diseases can spread from city to city and continent to continent [44]. (**b**) Mobility also adds context to social networks. When two individuals visit the same locations can suggest the nature of a social relationship [60]. (**c**) Mobility and the access it provides has strong correlations with economic outcomes. Children have dramatically different chances at upward economic mobility in certain places of the United States than others [10]

them. Moreover, geographic characteristics can be used to predict the social fluxes between places [29]. Conversely, social contacts are very useful in predicting where an individual would travel next [12, 19, 61] and Cho et al. find that while 50–70% of mobility can be explained as periodic behavior, another 10–30% are related to social interactions.

Models such as the one proposed by Grabowicz et al. [24] or Toole et al. [60] have subsequently been developed to incorporate this dynamic and evolve both social networks and mobility simultaneously. For example, Grabowicz et al. incorporate social interactions by having individuals travel in a continuous 2D space where an individual travel's is determined by the location of their contacts and use location as a determinant of new social tie creation. The model is as follows: with probability p_v, an individual moves to the location of a friend, and, with probability $1 - p_v$, they choose a random point to visited some distance Δr away. But, while social ties impact mobility, mobility can also impact social ties. Upon arriving at a new location, the individual can thus choose to form social ties with other individuals within a radius with probability p or random individuals anywhere in the space with probability p_c, a free parameter. A simple model is here also able to reproduce many empirical relationships found in social and mobility data.

2.5.3 Mobility and Economic Outcomes

Mobility not only provides people with social opportunities, it also provides economic ones. Economists and other social scientists have developed numerous theories on the role of face to face interactions in socio-economic outcomes and economic growth. In-person meetings are thought to unlock human capital, making us productive [22, 34]. For example, jobs in dense cities tend to pay higher wages than the same jobs in more rural areas even after controlling for factors such as age and education [68] in part due to productivity and creativity gains made possible by the rich face to face interactions that close spatial proximity facilitates. Universal urban scaling laws have been repeatedly found showing that societal attributes from the number of patents to average walking speed scales with population and theoretic models have been proposed that suggest density is at the heart of these relationships [6, 7, 46]. While density is one way to propagate these benefits, increased mobility is another. Poorer residents of cities have for example been shown to have better job prospects and higher chances of retaining jobs when given a personal car instead of being constrained by public transit [25]. Finally, Chetty et al. [10] found strong correlations between intergenerational economic mobility and variables related to the commuting times and spatial segregation of people (Fig. 2.5c). While we are only beginning to explore these relationships, early returns suggest that mobility is a critical component of many economic systems.

2.6 Conclusion

In this chapter, we reviewed a number of ways new data sources are expanding our understanding of human mobility. Applying methods from statistical physics, machine learning, and traditional transportation modeling, reproducible characteristics of human movement become visible. We explore slowly [23, 54], we are highly predictability [19, 55], and we are mostly unique [15]. Models of aggregate flows of people from place to place have also found success with analogies to statistical physics validated by new data sources [53]. More accurate measurements of city-wide traffic has made it easier than ever to assess the performance of transportation systems and devise strategies to improve them [63]. Valuable in their own rights, these insights have informed our understanding of other social phenomena as well, leading to more accurate models of disease spread, social interactions, and economic outcomes. As cities become home to millions for people each year, the insights gained from these new data are critical for making them more sustainable, safer, and better places to live.

References

1. Nadav Aharony, Wei Pan, Cory Ip, Inas Khayal, and Alex Pentland. Social fMRI: Investigating and shaping social mechanisms in the real world. In *Pervasive and Mobile Computing*, volume 7, pages 643–659, 2011.
2. Lars Backstrom, Eric Sun, and Cameron Marlow. Find me if you can: improving geographical prediction with social and spatial proximity. In *Proceedings of the 19th international conference on World wide web*, pages 61–70, 2010.
3. Duygu Balcan, Vittoria Colizza, Bruno Gonçalves, Hao Hu, José J Ramasco, and Alessandro Vespignani. Multiscale mobility networks and the spatial spreading of infectious diseases. *Proceedings of the National Academy of Sciences of the United States of America*, 106(51):21484–9, December 2009.
4. Duygu Balcan, Vittoria Colizza, Bruno Gonçalves, Hao Hu, José J Ramasco, and Alessandro Vespignani. Multiscale mobility networks and the spatial spreading of infectious diseases. *Proceedings of the National Academy of Sciences of the United States of America*, 106(51):21484–9, December 2009.
5. Moshe E. Ben-Akiva and Steven R. Lerman. *Discrete Choice Analysis: Theory and Application to Travel Demand*. MIT Press, 1985.
6. Luís M a Bettencourt. The origins of scaling in cities. *Science*, 340:1438–41, 2013.
7. Luís M A Bettencourt, José Lobo, Dirk Helbing, Christian Kühnert, and Geoffrey B West. Growth, innovation, scaling, and the pace of life in cities. *Proceedings of the National Academy of Sciences of the United States of America*, 104(17):7301–6, April 2007.
8. D Brockmann, L Hufnagel, and T Geisel. The scaling laws of human travel. *Nature*, 439:462–465, 2006.
9. Zhiyuan Cheng, James Caverlee, Kyumin Lee, and Daniel Z. Sui. Exploring Millions of Footprints in Location Sharing Services. In *ICWSM*, pages 81–88, 2011.
10. Raj Chetty, Nathaniel Hendren, Patrick Kline, and Emmanuel Saez. Where is the Land of Opportunity? The Geography of Intergenerational Mobility in the United States. January 2014.
11. Albert M. L. (Albert Man Loon) Ching. A user-flocksourced bus intelligence system for Dhaka, 2012.
12. Eunjoon Cho, Seth A Myers, and Jure Leskovec. Friendship and mobility. In *Proceedings of the 17th ACM SIGKDD international conference on Knowledge discovery and data mining KDD 11*, KDD '11, page 1082. ACM Press, 2011.
13. Vittoria Colizza, Alain Barrat, Marc Barthélemy, and Alessandro Vespignani. The role of the airline transportation network in the prediction and predictability of global epidemics. *Proceedings of the National Academy of Sciences of the United States of America*, 103(7):2015–20, February 2006.
14. C.D.a Cottrill, F.C.a Pereira, F.a Zhao, I.F.b Dias, H.B.c Lim, M.E.d Ben-Akiva, and P.C.d Zegras. Future mobility survey. *Transportation Research Record*, (2354):59–67, 2013.
15. Yves-Alexandre de Montjoye, César A Hidalgo, Michel Verleysen, and Vincent D Blondel. Unique in the Crowd: The privacy bounds of human mobility. *Nature Scientific Reports*, 3:1376, 2013.
16. Yves Alexandre De Montjoye, Erez Shmueli, Samuel S. Wang, and Alex Sandy Pentland. OpenPDS: Protecting the privacy of metadata through SafeAnswers. *PLoS One*, 9, 2014.

17. Yves-Alexandre de Montjoye, Zbigniew Smoreda, Romain Trinquart, Cezary Ziemlicki, and Vincent D. Blondel. D4D-Senegal: The Second Mobile Phone Data for Development Challenge. July 2014.
18. Giusy Di lorenzo, Marco Luca Sbodio, Francesco Calabrese, Michele Berlingerio, Rahul Nair, and Fabio Pinelli. AllAboard. In *Proceedings of the 19th international conference on Intelligent User Interfaces - IUI '14*, pages 335–340, New York, New York, USA, February 2014. ACM Press.
19. Manlio De Domenico. Interdependence and Predictability of Human Mobility and Social Interactions. *csbhamacuk*, 2012, 2012.
20. Nathan Eagle and Alex Pentland. Reality mining: Sensing complex social systems. *Personal and Ubiquitous Computing*, 10:255–268, 2006.
21. Nathan Eagle and Alex Sandy Pentland. Eigenbehaviors: Identifying structure in routine. *Behavioral Ecology and Sociobiology*, 63:1057–1066, 2009.
22. Matthew L. Freedman. Job hopping, earnings dynamics, and industrial agglomeration in the software publishing industry. *Journal of Urban Economics*, 64:590–600, 2008.
23. Marta C González, César A Hidalgo, and Albert-László Barabási. Understanding individual human mobility patterns. *Nature*, 453(7196):779–782, 2008.
24. Przemyslaw A. Grabowicz, Jose J. Ramasco, Bruno Goncalves, and Victor M. Eguiluz. Entangling mobility and interactions in social media. page 16, July 2013.
25. Tami Gurley and Donald Bruce. The effects of car access on employment outcomes for welfare recipients. *Journal of Urban Economics*, 58:250–272, 2005.
26. Randolph W. Hall, editor. *Handbook of Transportation Science*, volume 23 of *International Series in Operations Research & Management Science*. Springer US, Boston, MA, 1999.
27. Walter G. Hansen. How Accessibility Shapes Land Use. *Journal of the American Institute of Planners*, 25(2):73–76, May 1959.
28. Juan C. Herrera, Daniel B. Work, Ryan Herring, X. Ban, Quinn Jacobson, and Alexandre M. Bayen. Evaluation of traffic data obtained via GPS-enabled mobile phones: The Mobile Century field experiment. *Transportation Research Part C: Emerging Technologies*, 18:568–583, 2010.
29. C Herrera-Yagüe, C M Schneider, Z Smoreda, T Couronné, P J Zufiria, and M C González. The elliptic model for communication fluxes. *Journal of Statistical Mechanics: Theory and Experiment*, 2014(4):P04022, April 2014.
30. Ryan Herring, Tania Abou Nasr, Amin Abdel Khalek, and Alexandre Bayen. Using Mobile Phones to Forecast Arterial Traffic through Statistical Learning. *Electrical Engineering*, 59:1–22, 2010.
31. Md. Shahadat Iqbal, Charisma F. Choudhury, Pu Wang, and Marta C. González. Development of origin destination matrices using mobile phone call data. *Transportation Research Part C: Emerging Technologies*, 40:63–74, March 2014.
32. Jerald Jariyasunant. *Improving Traveler Information and Collecting Behavior Data with Smartphones*. PhD thesis, 2012.
33. H.S. Kim. QoS provisioning in cellular networks based on mobility prediction techniques. *IEEE Communications Magazine*, 41(1):86–92, January 2003.
34. Sunwoong Kim. Labor Specialization and the Extent of the Market, 1989.
35. Eleni Kosta, Hans Graux, and Jos Dumortier. Collection and Storage of Personal Data: A Critical View on Current Practices in the Transportation Sector. In *Privacy Technologies and Policy SE - 10*, volume 8319, pages 157–176. 2014.
36. John Krumm, Eric Horvitz, Paul Dourish, and Adrian Friday. Predestination: Inferring Destinations from Partial Trajectories. *UbiComp 2006: Ubiquitous Computing*, 4206:243–260, 2006.
37. David Lazer, Alex Pentland, Lada Adamic, Sinan Aral, Albert-Laszlo Barabasi, Devon Brewer, Nicholas Christakis, Noshir Contractor, James Fowler, Myron Gutmann, Tony Jebara, Gary King, Michael Macy, Deb Roy, and Marshall Van Alstyne. Computational Social Science. *Science*, 323(5915):721–723, 2009.

38. Kyunghan Lee Kyunghan Lee, Seongik Hong Seongik Hong, Seong Joon Kim Seong Joon Kim, Injong Rhee Injong Rhee, and Song Chong Song Chong. SLAW: A New Mobility Model for Human Walks. *IEEE INFOCOM 2009*, 2009.
39. David Liben-Nowell, Jasmine Novak, Ravi Kumar, Prabhakar Raghavan, and Andrew Tomkins. Geographic routing in social networks. *Proceedings of the National Academy of Sciences of the United States of America*, 102(33):11623–11628, 2005.
40. Tong Liu, Paramvir Bahl, and Imrich Chlamtac. Mobility modeling, location tracking, and trajectory prediction in wireless ATM networks. *IEEE Journal on Selected Areas in Communications*, 16:922–935, 1998.
41. Michael G. McNally. The Four Step Model. *Center for Activity Systems Analysis*, November 2008.
42. Sandro Meloni, Nicola Perra, Alex Arenas, Sergio Gómez, Yamir Moreno, and Alessandro Vespignani. Modeling human mobility responses to the large-scale spreading of infectious diseases. *Scientific reports*, 1:62, January 2011.
43. Kim Minkyong, David Kotz, and Kim Songkuk. Extracting a mobility model from real user traces. In *Proceedings - IEEE INFOCOM*, 2006.
44. Christos Nicolaides, Luis Cueto-Felgueroso, Marta C. González, and Ruben Juanes. A metric of influential spreading during contagion dynamics through the air transportation network. *PLoS One*, 7, 2012.
45. Juan de Dios Ortúzar and Luis G. Willumsen. *Modelling Transport*. 2011.
46. Wei Pan, Gourab Ghoshal, Coco Krumme, Manuel Cebrian, and Alex Pentland. Urban characteristics attributable to density-driven tie formation. *Nature communications*, 4:1961, 2013.
47. Gyan Ranjan, Hui Zang, Zhi-Li Zhang, and Jean Bolot. Are call detail records biased for sampling human mobility? *ACM SIGMOBILE Mobile Computing and Communications Review*, 16(3):33–44, 2012.
48. Adam Sadilek and John Krumm. Far Out: Predicting Long-Term Human Mobility. *AAAI*, pages 814–820, 2012.
49. Samitha Samaranayake, Sebastien Blandin, and Alexandre Bayen. Learning the dependency structure of highway networks for traffic forecast. In *Proceedings of the IEEE Conference on Decision and Control*, pages 5983–5988, 2011.
50. Paolo Santi, Giovanni Resta, Michael Szell, Stanislav Sobolevsky, Steven Strogatz, and Carlo Ratti. Taxi pooling in New York City: a network-based approach to social sharing problems. page 12, October 2013.
51. Salvatore Scellato, Mirco Musolesi, Cecilia Mascolo, Vito Latora, and Andrew T. Campbell. NextPlace: A spatio-temporal prediction framework for pervasive systems. In *Lecture Notes in Computer Science (including subseries Lecture Notes in Artificial Intelligence and Lecture Notes in Bioinformatics)*, volume 6696 LNCS, pages 152–169, 2011.
52. Christian M Schneider, Vitaly Belik, Thomas Couronné, Zbigniew Smoreda, and Marta C González. Unravelling daily human mobility motifs. *Journal of the Royal Society, Interface / the Royal Society*, 10(84):20130246, 2013.
53. Filippo Simini, Marta C González, Amos Maritan, and Albert-László Barabási. A universal model for mobility and migration patterns. *Nature*, 484(7392):8–12, 2012.
54. Chaoming Song, Tal Koren, Pu Wang, and Albert-László Barabási. Modelling the scaling properties of human mobility. *Nature Physics*, 6(10):818–823, September 2010.
55. Chaoming Song, Zehui Qu, Nicholas Blumm, and Albert-László Barabási. Limits of predictability in human mobility. *Science*, 327(5968):1018–1021, 2010.
56. Heinz Spiess. Technical Note—Conical Volume-Delay Functions. *Transportation Science*, 24(2):153–158, May 1990.
57. L. Sun, K. W. Axhausen, D.-H. Lee, and X. Huang. Understanding metropolitan patterns of daily encounters. *Proceedings of the National Academy of Sciences*, 110(34):13774–13779, August 2013.

58. Arvind Thiagarajan, Lenin Ravindranath, Katrina LaCurts, Samuel Madden, Hari Balakrishnan, Sivan Toledo, and Jakob Eriksson. VTrack: accurate, energy-aware road traffic delay estimation using mobile phones. In *Proceedings of the 7th ACM Conference on Embedded Networked Sensor Systems - SenSys '09*, pages 85–98, 2009.
59. Jameson L Toole, Meeyoung Cha, and Marta C González. Modeling the adoption of innovations in the presence of geographic and media influences. *PLoS One*, 7(1):e29528, 2012.
60. Jameson L. Toole, Carlos Herrera-Yaqüe, Christian M. Schneider, and Marta C. González. Coupling human mobility and social ties. *Journal of The Royal Society Interface*, 12(105), 2015.
61. Pauline van den Berg, Theo A. Arentze, and Harry J. P. Timmermans. Size and Composition of Ego-Centered Social Networks and Their Effect on Geographic Distance and Contact Frequency, 2010.
62. Jingyuan Wang, Yu Mao, Jing Li, Chao Li, Zhang Xiong, and Wen-Xu Wang. Predictability of road traffic and congestion in urban areas. July 2014.
63. Pu Wang, Timothy Hunter, Alexandre M Bayen, Katja Schechtner, and Marta C González. Understanding road usage patterns in urban areas. *Scientific reports*, 2:1001, January 2012.
64. Pu Wang, Like Liu, Xiamiao Li, Guanliang Li, and Marta C González. Empirical study of long-range connections in a road network offers new ingredient for navigation optimization models. *New Journal of Physics*, 16(1):013012, January 2014.
65. Amy Wesolowski, Nathan Eagle, Abdisalan M Noor, Robert W Snow, and Caroline O Buckee. The impact of biases in mobile phone ownership on estimates of human mobility. *Journal of the Royal Society, Interface / the Royal Society*, 10(81):20120986, April 2013.
66. Amy Wesolowski, Nathan Eagle, Andrew J Tatem, David L Smith, Abdisalan M Noor, Robert W Snow, and Caroline O Buckee. Quantifying the impact of human mobility on malaria. *Science (New York, N.Y.)*, 338(6104):267–70, October 2012.
67. Yingxiang Yang, Carlos Herrera, Nathan Eagle, and Marta C González. Limits of predictability in commuting flows in the absence of data for calibration. *Scientific reports*, 4:5662, January 2014.
68. Jeffrey J. Yankow. Why do cities pay more? An empirical examination of some competing theories of the urban wage premium. *Journal of Urban Economics*, 60:139–161, 2006.

Chapter 3
Privacy in Location-Sensing Technologies

Andreas Solti, Sushant Agarwal, and Sarah Spiekermann-Hoff

Abstract Data analysis is becoming a popular tool to gain marketing insights from heterogeneous and often unstructured sensor data. Online stores make use of click stream analysis to understand customer intentions. Meanwhile, retail companies transition to locating technologies like RFID to gain better control and visibility of the inventory in a store. To further exploit the potential of these technologies, retail companies invest in novel services for their customers, such as smart fitting rooms or location of items in real time. In such a setting, a company can not only get insights similar to online stores, but can potentially also monitor customers. In this chapter, we discuss various location-sensing technologies used in retail and identify possible direct and indirect privacy threats that arise with their use. Subsequently, we present technological and organizational privacy controls that can help to minimize the identified privacy threats without losing on relevant marketing insights.

3.1 Introduction

The era of sensing technologies has already begun. We use smart devices (e.g., smart phones, smart watches, smart cars, smart clothes) in our daily lives and we often cannot imagine life without internet and being online every day. The acceleration of technological progress offers ever new use cases of sensing technologies. Organizations that heedlessly implement novel use cases as they become technically feasible without considering the privacy implications to users or employees risk losses in reputation and trust [6]. Therefore, it is crucial to be aware of the privacy implications of the used technologies. Privacy risks caused by the use of information technology are rooted in operators' ability to permanently save and link information about sensed individuals [76, 83]. The anxiety of the general public with

A. Solti · S. Agarwal (✉) · S. Spiekermann-Hoff
Vienna University of Economics and Business, Vienna, Austria
e-mail: solti@ai.wu.ac.at; sagarwal@wu.ac.at; spiek@wu.ac.at

© Springer Nature Switzerland AG 2018
A. Gkoulalas-Divanis, C. Bettini (eds.), *Handbook of Mobile Data Privacy*,
https://doi.org/10.1007/978-3-319-98161-1_3

these technologies is exemplified in newspaper articles that use the term "privacy snatchers" [16] to refer to organisations monitoring workers or customers.

Throughout this chapter, we will look at the everyday example of how brick-and-mortar retailers use locating technologies to gain a better understanding of their customers. In contrast to online retailers that can tap into a rich source of information in terms of browsing behaviour of customers through click stream analysis, brick-and-mortar retailers are only recently investigating location-sensing technologies for gaining similar insights about the physical movements and behaviours of customers in shop floors. For example, radio-frequency identification (RFID) technology can detect with which items customers interact on their shopping trip. An information that is very interesting for retailers from a marketing perspective. Additionally, location-sensing technologies enable novel services for customers (e.g., locating an item, automating checkout). While such services may be interesting for the company, they often come at the risk of compromising privacy of users. This is especially the case, as the locating-technologies used are pervasive and do not generally alert the users when information is collected from them. From the legal perspective, companies need to avoid unlawful handling of privacy sensitive data. Otherwise, they risk not only a loss in reputation and trust, but also substantial fines. In the EU, for example, the new regulation extends the upper limit of fines for privacy infringements to 20 million EUR, or 4% of the annual world-wide turnover of an organisation (whichever is greater) [81]. For both organisations and users it is, however, crucial to be aware of the interplay between different technologies, the use cases supported, and the direct and indirect privacy threats entailed by using these novel technologies. In this chapter, we investigate this interplay to offer an overview and also present privacy controls that can help to minimise the identified privacy threats without losing on relevant information. Organisations that use these technologies have the responsibility to make customers aware about the technologies used and the information gathered by them. The chapter is organised as follows. Section 3.2 presents use cases that are enabled by location-sensing technology in the retail sector. In Sect. 3.3, we introduce various location-sensing technologies and compare them. We also list the privacy threats that are associated to automated location-sensing. Section 3.4 exemplifies the interplay of technology, usecases and associated privacy threats in popular scenarios in retail. Specifically, we focus on RFID and WiFi and their combination. Last, in Sect. 3.5, we present privacy controls. These controls minimise the possible privacy threats using location-sensing technologies in different scenarios. We conclude this chapter in Sect. 3.6.

3.2   Use Cases of Location-Sensing Technologies in Retail

Retailers are interested in improving their service quality to increase customer satisfaction [72], and in maximising their profits. In this chapter, we focus on how location-sensing technologies enable location-based services. To clarify terminology, location-based services belong to the general class of context-based services, where context is defined by Abowd et al. [2] as:

> "any information that can be used to characterize the situation of entities (i.e., whether a person, place or object) that are considered relevant to the interaction between a user and an application, including the user and the application themselves"

We restrict our analysis to *location* as one of the most important contextual feature in this chapter, but also discuss features that can be derived from location-sensing technology. For a broader discussion on context-aware systems, we refer to the survey by Baldauf et al. about context-aware systems and their support for security and privacy [14] and the textbook in the field of ubiquitous computing edited by Krumm [50].

Location-sensing technologies create novel use cases to increase service quality, or assist existing use cases by gaining more transparency about the customer behaviour in the retail environment.

While online retailers rely on a rich information source of customer behaviour through click-stream analysis to provide recommendations (e.g., by analysing the online browsing and search histories), brick and mortar retailers are often blind to their customers' behaviour in their shops. For decades, they employed market researchers who would follow customers around in stores to better understand the needs of customers. With location-sensing technologies, brick and mortar retailers can automatically gain insights into the interests of customers, and can react to their location context. We distinguish management use cases, marketing use cases, and operational use cases.

3.2.1 Management Use Cases

Management is typically interested in the aggregate performance of a store and the trend of the performance over time. Several *performance indicators* can be supported with location-sensing technologies. For some of them, point of sales data needs to be integrated [18]. Generally, the behaviour before, during and after the consumption can be defined as product information browsing, consumption and product usage, respectively [69], and this behaviour information can be collected using location-sensing technologies. Management use cases include any type of analysis of this information. We briefly sketch the most important use cases here that can be based on location-sensing:

Conversion rates One of the simplest location-based indicators are conversion rates. For example, a measure of interest is the conversion rate of passersby into shop visitors [17]. Another important measure is the fraction of entering customers that purchases products. When location-sensing technologies track customers' behaviour in the shop, it is possible to gather more fine-grained information and these conversion rates can be partitioned into product categories [89]. One example is to measure the number of visitors to a store section (e.g., the area where jackets are on display) and relate it to the number of purchases that contained the category (e.g., jackets).

Length of stay   Other interesting insights that can be of managerial relevance are length of stay of customers. This measure positively correlates with the probability of making a purchase (e.g., through impulse buying). Being able to automatically measure the length of stay as an indicator can help for example to select background music that increases the length of stay [61].

Queuing times   Location-sensing technologies can be used to extract waiting times of the location data. Of particular interest are the queuing times at service stations like the point of sales, or also fitting rooms in the context of fashion retail. Studies show that waiting time influences perceived service quality [51]. Therefore, timely control of these measures is important to balance service quality and resource utilisation.

Store layout optimisation   The optimisation of the store layout is important to maximise profit [55]. The layout can profit from additional location information that is available, when customer movement patterns are analysed. Furthermore, novel layouts can be quickly tested for operational efficiency by analysing the changes in customer movement patterns.

3.2.2   Marketing Use Cases

For marketing purposes, we consider interactions with the customer. We exemplify a few location-based use cases here, and refer to the survey of Adomavicius and Tuzhilin [4] and the handbook by Ricci et al. [71] for more general recommendation concepts.

Geofencing   The idea of geofencing, that is triggering notifications based on entering or leaving a defined area boundary (i.e., the geofence) belongs to location based services [17]. Geofencing in retail environments resembles traditional market places, where the passing customers heard the voices of the nearby sellers advertising special deals, when getting closer the their booths. The difference is that the marketing is now automated, and the customers that are detected in the defined areas get push notifications on their smart devices.

Context-aware browsing   More subtle than getting push notifications, when entering an area, context-aware browsing changes the services offered when browsing the web based on the context of the user [20, 29, 63]. In the retail domain, this can for example be applied to smart screens in the store that react on the items carried by the customer. Also the online shop offering to customers accessing the store with their mobile devices can be adapted based on the customer's location. Here, changing the order of items or recommendations in the online shop based on the shopper's distance to the items.

3.2.3 Operational Use Cases

Location-based sensing technology enables further use cases besides management and marketing that support daily operations. We mention some of the more common use cases in the following.

Preventing theft Location sensing of items in a brick and mortar retail environment can be used to trigger an alarm, when items pass the boundaries of the defined shopping area without having checked them out before [86].

Locating products When location-sensing technology [41] is harnessed to track the whereabouts of products, the primary novel use case is locating the products in case a client is looking for it. This use case is often supported by RFID technology based on passive RFID tags, as it is affordable to equip every item with a unique identifier. In case an item is requested, the system can be asked about the assumed position of that item, to potentially avoid a time consuming search, when the item is not at its allocated position.

Replenishing products Location-sensing applied to products has another important use case, which is replenishment of items. Typically, an item needs to be replenished, when items are sold to customers and this type of replenishment does not depend on location-sensing technologies. However, there are further reasons for items disappearing from the sales floor, which is referred to as retail shrinkage (e.g., stolen items). In these cases, location-sensing technologies can help to detect shrinkage and allow more timely replenishment in that case [27]. Furthermore, if the item is only misplaced, locating technologies can prevent unnecessary orders of available items.

Path/Layout optimization When multiple tasks need to be performed at different locations (e.g., the items of an order need to be collected from different positions) workers can be assisted to save time and traveled distance by optimizing their paths through the shop or back room [85]. Location-sensing technology can assist here to adjust the proposed path to the items and the worker's current positions. Also customer paths can be analysed and taken into consideration for store layout optimization [23].

Waiting time estimation Knowing the expected waiting time at a queue has a positive influence on perceived service quality (as long as the expectations are reliable) [51]. A good waiting time estimator in stable systems is asking the last person that exited the queue about their waiting time. There are systems that measure the waiting time by requiring customers to draw a number from the system, when they enter the queue to measure the time. Location-based sensing allows us to collect the information from the location data and enables the use case of informing customers about their expected waiting time in front of fitting rooms or checkout.

Automated checkout Perhaps one of the technically more advanced use cases that use location sensing is the automated checkout of items at the point of sale. RFID technology is an enabler for this technology, and mobile applications installed on smart devices allow identification of customers. Recently, prototypes for this use case have emerged [87].

To sum up, the new use cases of location-sensing technology are manifold, and more and more of them emerge, as combinations of different technologies and information sources are explored. In the next section, we provide an intuition on the methods for location sensing and compare technologies that can be used for this purpose.

3.3   Location-Sensing Technologies and Entailed Privacy Threats

We first introduce location sensing technologies and compare them to each other. An organisation wanting to deploy location-sensing technologies has to be aware of the ensuing privacy issues and legal demands. Thus, we discuss and categorise privacy issues by focusing, in particular, on concrete privacy threats arising on the technical level.

3.3.1   Introduction to Location-Sensing Technologies

Location sensing refers to the process of obtaining location information of a mobile agent with respect to a set of reference positions in a predefined space [35, 56]. The most common techniques for location sensing are trilateration and fingerprinting. In this section, we discuss on a high level how these two techniques work and discuss different technologies with which they can be enabled.

3.3.1.1   Locating Objects by Trilateration and Fingerprinting

Trilateration is a method to determine absolute or relative location of an object based on measurement of distances from three known points [84]. Figure 3.1 illustrates the process where for an object the distance is known with reference to three points X1, X2 and X3. With respect to point X1, see Fig. 3.1a, the radial distance is r1 and based on this information the item lies somewhere on the circumference of the highlighted circle. Then if we consider the distance from the second point X2, as shown in Fig. 3.1b, the item can lie on either points of intersection of the two circumferences, marked by A and B. Finally, if distance from all the three points is considered, as shown in Fig. 3.1c, then location of the item can be concluded, marked as B. This illustration works for location-sensing in a 2D space. If the position of an object is to be estimated in 3D space, we need a fourth reference point. In practice there are usually imprecisions regarding the distance to a reference point, which affect the accuracy of the estimated position [13, 58]. Therefore, more than three reference points are often used to increase the accuracy.

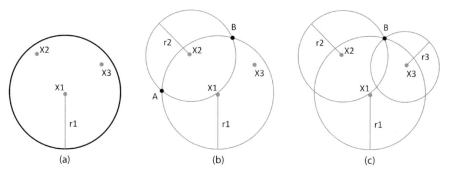

Fig. 3.1 Illustration for trilateration

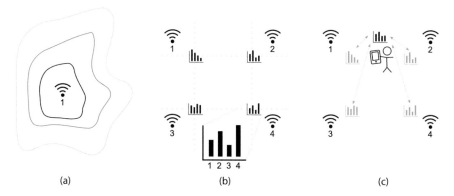

Fig. 3.2 Illustration for 2-step fingerprinting. (**a**) and (**b**) depict the first step for training and (**c**) shows the second step for positioning

The process of trilateration can be used in different ways to estimate the location: For satellite-based applications, estimation is done through measuring the time taken for a radio signal to travel from the transmitting satellites to a receiver and then multiplying it with the speed of the wave [68]. For applications like WiFi and Bluetooth it is done via measuring the received signal strength (RSS), a measurement of the power present in a received radio signal [90]. Trilateration generally works well outdoors. But due to obstacles like walls, street canyons, roofs, floors etc. the radio signals do not propagate linearly and get attenuated indoors. As a result, accuracy levels for location estimation goes down.

To attain better accuracy, a 2-step process called *fingerprinting* is used [12]. To model the attenuation, multiple reference points are considered and parameters (like signal strength) are calculated for these points. This step is called the training or calibration stage. The second step is the positioning stage where the parameters are recorded at the device's location and these parameters are then compared to the reference points to estimate the location. In other words, in the training step a fingerprint for the signals is created and in the positioning stage, the parameters are measured and compared with the fingerprints to ascertain the location. Figure 3.2

shows the two stages of fingerprinting for radio signals (such as WiFi, bluetooth or cellular signals). For the first stage, received signal strength (RSS) is measured for the radios, Fig. 3.2a shows the distorted radio signal in a field that is attenuated by obstacles. Figure 3.2b shows the training phase, where measurements at known positions (for example in a grid) are taken and recorded. Then, in the second stage, the RSS values are measured for a user (Fig. 3.2c) and compared with the data collected in the training stage. Thus, by comparing the RSS data with training data, the location is estimated.

In the following, we describe a multitude of technologies that allow location-sensing by a system with the help of trilateration and explore their feasibility of use. As outlined above, We will approach these technologies from the perspective of brick-and-mortar retail shops.

3.3.1.2 Satellite-Based Location Sensing

Smart devices capable of satellite based navigation have an integrated receiver to communicate with the satellites. The most commonly used navigation system as of 2017 is the Global Positioning System (GPS). To get an estimation of the position, the receiver needs to be in line of sight of at least satellites and solid objects like buildings, caves etc. attenuate the signals drastically. Hence, satellite navigation works well outdoors but cannot be used extensively for indoor location tracking. Also, as there is only a receiver in the devices for satellite communication, no information is directly transmitted to the satellites or any server. Thus to gain location information of such a device, a retailer has to request customers to install an app or visit their website where the device owner (customer) grants the retailer access to the device's location.

3.3.1.3 WiFi Based Location Sensing

WiFi technology enables devices to connect to a network wirelessly. Every networking chip or interface in these devices has a unique identifier called media access control (MAC address) which is broadcasted to wireless access points in range if WiFi is turned ON in the device. Uniqueness of the MAC address can be used to ascertain if a WiFi enabled device is in a proximity. For instance, it can be used to count unique customers (with a WiFi enabled device) in stores [5]. This method can further be extended by keeping two WiFi access points (A and B) and then analysing the pattern of movement of customers/devices, such whether A or B has a higher count or are customers spending more time around A or B etc. This can further be extended if an array of WiFi access points are setup. In this case, based on trilateration (similar to satellite-based location sensing systems) and the received signal strength indicator (RSSI) for each connection, location can be estimated as well as tracked [57]. Thus, by just measuring the signal strengths and the MAC addresses, the retails can track location as well as movement of customers who carry a WiFi enabled device without requesting the customer to install any extra application for such a purpose. This only works though if the WiFi is switched ON in a device. An American fashion retailer, Nordstrom, used this technology in 2012–2013 to track the customers in 17 of its stores [24] and that hampered the retailer's brand-image. Though Nordstrom ended WiFi tracking after the protests, based on the media reports, WiFi tracking is still harnessed in thousands of retail stores around the world [48].

3.3.1.4 Bluetooth Based Location Sensing

Bluetooth beacons are low-power radio transmitters which send signals in immediate vicinity using bluetooth. Martin, in his article in Harvard Business Review, refers to beacons as the *missing piece in the mobile-shopping puzzle* as they allow precise targeting of customers in a certain area [60]. Using beacons, retails can push a message, advertisement or even coupons to a customer's device. Similar to WiFi based sensing, if a cluster of these beacons are used then RSSI can be analysed for computing the location of a device [52]. In retail, beacons are currently used for pushing offers, but places like Eldheimar museum, Iceland use bluetooth beacons for indoor location sensing [77].

3.3.1.5 Cellular Tower Based

Trilateration or 2-stage fingerprinting can also be used based on the analysis of RSSI from the cellular network antennas to calculate locations of devices with cellular radios [42]. Research has shown that by just using four different location points calculated using the RSSI, more than 90% of the individuals can be uniquely identified[26]. Thus, just by sensing the location of a device, a few times in a day, there is a potential to differentiate or uniquely identity that device in a database of thousands of other devices. This method is not popular in the field of retail, however, emergency services generally use this information to estimate location of devices [34].

3.3.1.6 Ultrasonic Waves Based

Interestingly, even speakers/microphones present in smart devices can be used to track the location. Using ultrasonic sounds (inaudible to humans), it is possible to estimate distances based on the sound volume of the received signal. Thus, unlike other technologies they use sound instead of radio signals. Based on the arrangement, a customer's device can either act as a transmitter if it generate the sound signal or as a receiver if it listens to such sounds through the microphone. However, for sensing location using this technology, retailers need to convince their customers to install app with privileges to access the microphone and or speakers. In the recent past, this technology was exploited to provide analytics for

TV based advertisements [8]. In the advertisements, firstly some unique ultrasonic sound signals were attached. Secondly, malicious apps were pushed on devices like computers, tablets and cellphones which were listening and analysing ultrasonic sound signals 24 hours a day. Based on the analysis of received sound signals, the company provided rich insights like % of people watching the advertisements etc. [8].

3.3.1.7 RFID Based Location Sensing

By analysing the RSSI of the RFID tags, it is possible to estimate the rough distance of the tag from a reader [22]. If an array of RFID readers are used then through trilateration, location of tags can be estimated. Such systems, tracking location in real-time are referred to as RFID enabled Real-Time Locating Systems (RTLS) [75]. In retail, RFID tags can be added to loyalty cards, shopping baskets or even with the items on sale (either as price tags or integrated in the items) [33].Thus, movement of RFID tags could relate to the movement of people and provide additional insights to the retailers. In addition to retail, RFID tags have been used in hospitals to track movement of customers [46] and in schools to monitor the students [49].

3.3.1.8 Comparison of Technologies' Sensing Accuracy and Prerequisites

Following Table 3.1, based on Hazas et al. paper [42], compares the discussed technologies based on the requirements, accuracy and ability to track indoors. For a

Table 3.1 Comparison of different technologies for location tracking [43]

Tech	Accuracy	Indoors	Whats tracked	Prerequisites for customers to be tracked
Satellite	5–10 m	No	Devices	• Device with appropriate receiver • Application with location access internet access • Internet to share location information
WiFi	10–50 m	Yes	Devices	• Device with WiFi capability • WiFi turned on
Bluetooth	5–10 m	Yes	Devices	• Bluetooth enabled device • Some application with internet access to communicate with beacons • Internet to share beacon information
Cell tower	50–100 m	Yes	Devices	• Device with cellular capability • Some application with access to internet and network information • Internet to share network information
Ultrasonic	1–10 m	Yes	Devices	• App with permission to send/listen ultrasonic signals • Internet to share beacon information
RFID	1–10 m	Yes	Tagged items	None

more exhaustive analysis, survey by Liu et al. can be referred where they discuss 20 different solutions for indoor location sensing [56].

For the discussed use cases, Satellite based tracking is not the preferred option for the retail sector as it does not performs well in indoor conditions. Similarly, cell tower based tracking has low accuracy which makes it unattractive for the retail purpose. For Bluetooth and ultrasonic based tracking, the customers, currently need to install an extra application on their smart devices to enable the retailers for tracking them. On the other hand, WiFi and RFID based tracking are two options which do not require any extra application on customers' smart devices for retailers to track them. For WiFi, the main limitation is that the customers must carry a WiFi enabled smart device with WiFi radio turned on. This is increasingly common as many users want to connect to their home WiFis automatically and therefore keep their WiFi access enabled on their phones. For RFID, there is finally no user access activity required. The infrastructure can be set up and used by the retailers without customer knowledge and involvement. For instance, attaching RFID tags to shopping carts or simply tagging the shelved products, retailers can locate and track the movement of customers. Thus, for this chapter, we focus on WiFi and RFID based tracking as these technologies can potentially be used without any active consent of the customer.

3.3.2 Associated Privacy Threats

In IT, a threat is commonly defined as a *potential cause of an incident that may result in harm of systems* [44]. A privacy threat can therefore be understood as a potential cause of an *incident*, which may again cause *harm* to an individual's privacy. We focus on the technical causes or activities resulting in privacy harms without considering the impact of harm. Impact is subjective and varies on a case by case basis, depending on the type of data involved, privacy expectations etc. Thus, first, we discuss the different activities which lead to privacy harms. Second, we present a general overview about how these activities materialise for RFID and WiFi based location sensing. For further reading, we refer to extensive survey works on the topic of privacy threats with RFID and wireless technologies [38, 45, 53, 88].

3.3.2.1 Classification of Privacy Harms

Privacy is defined as an elusive concept [74] and there is a little agreement on how to define it [62]. As such, this makes it difficult to base the threats to privacy on its definition. Solove [73] instead of defining privacy, discusses different activities which lead to privacy harm and classifies them. He categorises the activities that cause privacy harm in four high level groups which are then further classified into 16 different forms [73].

Information collection   relates to the process of data gathering
Surveillance—watching, sensing or recording an individual's activities or
Interrogation—questioning or probing the individual for information
Information processing   relates to the activities involving storage of the collected
information, its manipulation and the use
Aggregation—combining different pieces of information about an individual
Identification—linking the information to the identity of individuals
Insecurity—carelessness in protecting the collected information
Secondary Use—using the collected information for a different purpose
Exclusion—keeping individuals unaware about their collected information
Information dissemination   relates to the activities involving revealing, sharing
or spreading information about the individuals
Breach of confidentiality—breaking a promise to keep individual's information
confidential
Disclosure—revealing true information about an individual
Exposure—revealing intimate information such as nudity, grief etc.
Increased accessibility—easing the accessibility of information by third parties
Blackmail—threatening to disclose the information
Appropriation—faking the individual's identity for mala fide interests
Distortion—disseminating false and misleading information about individuals
Invasion   impinging privacy by other means, not necessarily with the use of
information
Intrusion—disturbing an individual's solitude
Decisional interference—government's unwanted incursion into an individual's
decisions about their private life

As such, for the chapter we focus on a retail scenario and so we discuss only the threats which are directly related to the technology. Thus, we rule out interrogation and blackmailing as they are not directly related to RFID or WiFi. Also intrusion and decisional interference are not considered as these are not dependent on the use of information. We use this classification in the next section to understand the associated privacy threats. After discussing what can cause privacy harm, let us now focus on specific technologies, WiFi and RFID to understand how these harms are materialised.

3.3.2.2   Realisation of Privacy Harms Using RFID and WiFi

For RFID based tracking, companies track the items with RFID tags and then later associate with the customers. For WiFi, a customer's device is tracked to estimate the location. Privacy of customers is compromised not only by sensing the exact location but also due to processing which combines other data sources as well for rich insights about customers, for e.g. inferring preferences of customers based on the time spent in different areas of the store. Thus, we discuss the activities described by Garfinkel et al. [38] through which privacy can be comprised with the use of RFID in retail through:

Action   movement of an item triggers an action, for e.g. the disappearance of items from sensors could yield in an action of a photograph taken

Association   individuals are correlated with the RFID tags they interact with, e.g. customers are associated with items they pick in a store

Location   individual's position is tracked, e.g. in a retail store if an individual picks an item then movement of the item can give information about location of the individual

Inferred preferences   individual's preferences are estimated by associating the carried RFID tags, e.g. if an individual picks up sports garments then RFID tag of the garment can provide information about the possible preferences of that individual

Estimation of constellation   a combination of several tags used lead to a unique digital fingerprint [91] e.g. combination of different items can lead to the uniqueness of the shopping basket, creating unique constellation or group of items which can differentiate individuals

Transaction   transactions or relations can be inferred based on the movement of tags from one constellation to the other, e.g. individuals shopping together can be identified if they exchange some products during a shopping trip

"Breadcrumbs"   wrong association or association of discarded items can lead to false inference, e.g. if an individual picks up an item and later discards it then in case another individual picks up that same item, latter can be wrongly associated with the identity of the first individual

For WiFi based location sensing, only the customers' devices can be tracked. As customers are already uniquely identified based on the devices, *estimation of constellation* and *associations* are not applicable. Similarly, *transactions* cannot be identified as device exchange during a trip cannot be analysed and *actions* affecting privacy are difficult to trigger. On the other hand, *location* of the devices can be tracked, time spent in different sections can lead to *inferred preferences*. *Breadcrumb* threat is still valid if different customers carry the same device at different times (e.g. families or friends sharing a device, customers selling devices to others etc.). The activities harming privacy by using WiFi are hence a subset of the list discussed for RFID. Thus, the classification by Garfinkel et al. provides an exhaustive list for the ways through which privacy can be possibly compromised by using RFID as well as WiFi based tracking. Let us now consider some specific scenarios where RFID and WiFi based tracking is used for the use cases discussed in Sect. 3.2.

3.4   Analysis of Popular Location-Sensing Scenarios in Retail

We consider different scenarios that are relevant for brick-and-mortar retailers that plan to use, or already use location-sensing technologies, and are also relevant to customers confronted with these technologies in their daily life. We introduce the scenarios in order of increasing amount of accumulated location information about customers. Therefore, we first investigate RFID-based scenarios, then turn to WiFi-based scenarios and last consider the combination of the two technologies. The use cases supported by these scenarios and the associated privacy threats are outlined for each scenario.

3.4.1 RFID Locating Systems

The adoption of RFID technology in retail is the subject of a major ongoing privacy debate. The reason is the combination of three of its technological traits that raise consumer fears: First, humans have always been afraid of the invisible. This invisibility is manifest in many kinds of RFID that use chips too tiny to be recognised by the human eye, and communicate information without a line of sight through fabrics and even walls. Second, RFID cannot be "switched off", as other technologies. Last but not least, RFID technology is expected to be ubiquitously deployed and present on or embedded in all products and product components carrying barcodes today. This means that the technology will most likely become omnipresent in the near future.

Here, we first consider RFID data with statically installed gates and handheld readers. On an item level, we distinguish two cases. The first case is that an RFID tag is attached to the price tag of an item, which is typically removed after purchase. The second case is that RFID tags are integrated into items, such that removal becomes impossible without damaging the items. We also look at whether additional RFID enabled interaction points are existing in the retail area. For example, interactive smart kiosks allow customers to find more information about an item by presenting it to an attached reader.[1]

3.4.1.1   RFID Without Integrated Tags and Without Interaction Points

The customer interaction with the RFID system is limited to the checkout at the point of sales and potential reading at the exit (also *electronic article surveillance (EAS)*) gate, illustrated in Fig. 3.3. Sometimes, a customer brings back an item for returns at the customer centre. In such instances, there are additionally two data reads (at the EAS gate, and at the point of sales). The data collected through the RFID system in this case does not contain identifying patterns and only shows that there were items bought and perhaps returned.

---

[1]In the context of fashion retail, these interaction points can be inside the fitting room. Typically the users can interact there with a touch screen or also a smart mirror [7].

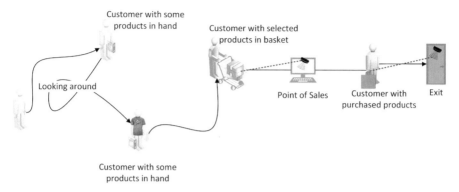

Fig. 3.3 An illustration of a shopping trip with RFID readers at a point of sales and an exit gate

Supported Use Cases

This scenario shown in Fig. 3.3 does not allow for many customer specific use cases. Nevertheless, it supports the use case of *preventing theft* (as outlines in Sect. 3.2.3). The EAS gates can automatically signal that an item has passed the gate that was not paid for to alert employees or security personnel. Additionally, as items are tagged with RFID technology, the use case of *locating products* (cf., Sect. 3.2.3) becomes possible. When the replenishment gate between back room and shop floor is RFID-enabled, the system knows whether items are on the sales floor or in the back room. Additionally, searching for misplaced items can be facilitated by handheld RFID scanners that can detect hundreds of items per second. Taking inventory with RFID technology can be sped up by handheld scanners, or fully automated (e.g., by robots). The use case of *replenishing products* (see Sect. 3.2.3) in case of retail shrinkage is supported by the updated inventory reports.

Privacy Threats

For this scenario, a major privacy threat arises from not restricting the RFID readers to only read the company's tags. This may lead to *surveillance* and *aggregation* of additional information, if there exists no mechanism to block reading of third party tags. Processing the aggregated information could be considered as a *secondary use* if the company is not transparent about it. Identifiers from the third party tags can lead to indirect *identification*. This aggregated information can further reveal more information about the individual *through association* of purchases with the unauthorised tag reads [88]. For instance, consider a customer, who carries a RFID smart card for public transportation, shops in such a store. When she leaves, not only the items bought would be recorded but also the identifier from the RFID smart card. This unauthorised read of the card is then a case of secondary use and the aggregated information leads to surveillance as retailer would gain knowledge about her shopping pattern i.e. when all does she visits the store. Additionally, the unauthorised tag can also be associated with the RFID tags of the bought items providing a personal identifier for them. Companies like *Integrity For You* have used such RFID chips in loyalty cards [19] which leads to the discussed threats if read by an authorised party.

3.4.1.2 RFID with Integrated Tags

When RFID tags are integrated into items for an increased theft protection level, they remain in the sold items without removal by the customer. In this scenario, the retail companies should take measures to deactivate (destroy, or send to sleep) the integrated tags after purchase [21], especially if the sold items are worn or carried around by customers. Otherwise the tags can be used to identify customers at later points in time or track them at other places where RFID technology is used. It is worth noting that other stores, but also any other organisation employing RFID-readers, can track people carrying or wearing items with enabled integrated tags [38].

Supported Use Cases

The supported use cases are the same as in Sect. 3.4.1.1 above (i.e., *preventing theft*, *locating* and *replenishing products*). However, there is a notable increase in protection against theft in this scenario [83]. An attacker can no longer simply remove the price tag from the product, or destroy a tag that is attached to the outside of the product. As the tags are embedded in the products, their removal becomes infeasible for most thieves.

Privacy Threats

The threats are similar to that discussed for the previous Sect. 3.4.1.1. However, in the long run for customers, all threats listed in Sect. 3.3.2.1 exist as the tags can be read by any interested party for malicious intentions. Some tags can be clipped [47] or ripped off to ensure they are not read, but such possibilities do not exist in all kinds of tagging. For instance, if tags are sewn-in with brand labels in garments then ripping them off might damage the garment. Thus, it becomes difficult to block the unwanted tag reads.

3.4.1.3 RFID with RFID-Enabled Interaction Points

When interaction points (like smart kiosks, mirrors, esp. in fitting rooms) are equipped with RFID readers, customers can benefit from more information about

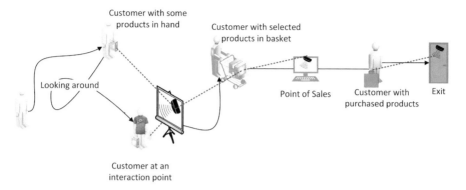

Fig. 3.4 An illustration of a shopping trip with RFID readers at an interaction point, a POS and an exit gate

the items of their choice [7]. The scenario is illustrated in Fig. 3.4. From a privacy perspective, novel information can be gathered from customers. It is possible to collect the information about items that a person was interested in, but decided to not buy. This is the case, when the sets of items that are brought to an interaction point overlap with the items that are finally bought by the customer. Additionally, the number of visits to the interaction points can be inferred to a certain degree. Prerequisite is that each subsequent visit to the interaction point has a given certain overlap in items (e.g., in the case of smart fitting rooms).

Supported Use Cases

Besides the operational use cases mentioned in Sect. 3.4.1.1 (i.e., preventing theft, locating and replenishing products), this scenario allows for additional managerial and marketing use cases. The use case of capturing *conversion rates* (as illustrated in Sect. 3.2.1) is partly supported in this scenario. In fact, the conversion rates of items that customers brought to the interaction point can be computed. In this way, it is possible to separate the items of interest that are also sold from those, that are interesting but not sold. For marketing, the use case of *context-aware browsing* (see Sect. 3.2.2) is supported. That is, the smart screens at interaction points can show the information pages according to the products detected that a customer brings there.

Privacy Threats

In addition to the threats discussed in Sect. 3.4.1.1, through *aggregation* of additional RFID data from interaction points could lead to even richer inferences about customers. The data *discloses* details of items that customers interact with which could be used for other *secondary uses*. Items brought to the interaction point reveal *inferred preferences* i.e. types of items picked up and brought there and possibly *transactions* for the customers shopping together [45]. For example, if two customers A and B come on a shopping trip then RFID tags would be read at interaction points as well as at the POS. Aggregated RFID tag reads can provide information about items which customers picked and did not buy. Also, if there exist some exchanges of items among the customers then those exchanges or transactions can also be inferred i.e. if many tags were read for customer A at an interaction point and then for B at a POS then either both have similar preference or they are shopping together.

3.4.1.4 RFID Real-Time Locating Systems

RFID real-time locating systems (RTLS) enable full visibility of inventory at all times. Usually the data is polled in periodic intervals for economic reasons. An illustration for RFID RTLS is shown in Fig. 3.5. If a retailer deployed an RTLS, it becomes possible to track customers indirectly by tracking moving items that are finally checked out [54]. It is theoretically possible to classify item movements into customer movements and employee inventory actions. For example, a large group of items moved from one area to the other on the sales floor indicates employee replenishment or store assortment activity. In contrast, smaller groups of items travelling through the shop and eventually ending at the checkout counter, could indicate customer movement. This information can be traced back to the point of the first picked up item. If customers always carried at least one item with them on their path, trading that item allows to reconstruct the customer's path. Furthermore,

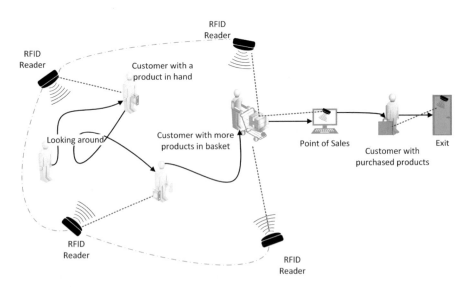

Fig. 3.5 An illustration of a shopping trip with RFID RTLS, a POS and an exit gate

if the customer carries an integrated RFID tag that is technically compatible with the RTLS, the mentioned correlation anchor is unnecessary. It can be replaced by the integrated tag. The entire path of the customer can be traced in this case. Note that even the RFID RTLS data of a store that itself does not offer any items with integrated RFID tags, could potentially track customers, as they might carry integrated tags from other organisations.

Supported Use Cases

This scenario supersedes the scenario with only limited interaction points in Sect. 3.4.1.3. That is, many additional use cases are supported here. The management use cases of *conversion rates* (see Sect. 3.2.1) extend beyond bringing an item voluntarily to an interaction point to picking up an item. The queuing times (see Sect. 3.2.1) of customers can be monitored indirectly by observing that items queue in front of the point of sale. Also the *store layout optimisation* (see Sect. 3.2.1) can benefit from the movement patterns of customers through the shop. The marketing use cases mentioned in Sect. 3.2.2 are not supported directly by this scenario. However, when interaction points exist in the store (e.g., smart kiosks) these can offer *context-aware browsing* that can react on the items present. Furthermore, items that accompanied the items on their movement paths (e.g., a second picked up item that was again dropped) can be included in the context. All operational use cases are supported, except the use case of *automated checkout* (see Sect. 3.2.3). Notably, *theft prevention* (see Sect. 3.2.3) is in place and even suspicious movement patterns can be detected. For instance, when a product suddenly disappears from the sensing infrastructure, this might indicate a destruction of a tag. Also, *locating products* (see Sect. 3.2.3) is supported to the highest degree, as the system is aware of the locations in real time.

Privacy Threats

As this scenario supersedes the previous scenario discussed in Sect. 3.4.1.3, here we can assume that a store is full of interaction points revealing the data. In this scenario, analysing the movement pattern for a group of items can correlate to the movement of customers in the store. Thus through *estimation of constellations* or groups of items, *location* information can be inferred for the customers. Thus, in addition to the threats discussed in previous Sect. 3.4.1.3, the processing of RFID data threatens *exclusion* if customers are not made aware of location sensing processing. Also, as the analysis for grouping the items or estimation of constellations is based on correlations, there exists a possibility to infer *distorted* information. For instance, wrong relations or *transactions* can be concluded between customers if a customer picks up some item left by some other customer (*"breadcrumbs"* issue) [53].

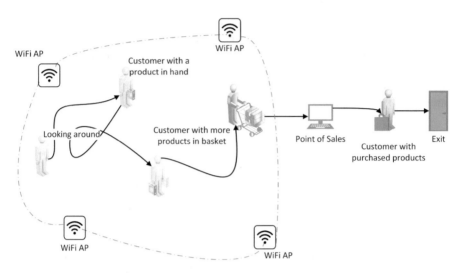

Fig. 3.6 An illustration of a shopping trip with WiFi locating system

3.4.2 Wi-Fi Locating Systems

While RFID systems focus on the identification of (passive) tags, Wi-Fi positioning systems allow identification of communication devices. These systems allow to sense the location of smart phones using a cluster of Wi-Fi access points (AP), as shown in Fig. 3.6. If customers use the wireless network of an organisation, they leave traces with the MAC address of their device, which constitutes a unique identifier to track the owner [25]. Even when they do not actively use a wireless network, the communication devices often send polling requests for currently available networks. Mostly, this happens even when Wi-Fi is set to disabled on the devices. These polling requests can be used to locate the source device by means of triangulation or fingerprinting [3]. To avoid the possibility of being tracked in this way, recent device operating systems feature a random assignment of MAC addresses for every new connection of a device to wireless networks. However, a recent study has shown that these mechanisms are not fully functional yet, and it is possible to track devices at least over the duration of a visit [80].

In Wi-Fi locating system, the frequency of gathered data points is determined by the device model and its operating system. The probe request intervals range from 10 s, when the device is active, to 500 s when the device is inactive [28]. The granularity of the data impacts the quality of conclusions that can be drawn from it. The more fine grained the resolution is, the more privacy sensitive the gathered data becomes [15].

Supported Use Cases

This scenario is the only one supporting the capturing of the *conversion rates* (see Sect. 3.2.1) of passersby into customers that enter the shop. However, there is a bias, as only the customers that have a Wi-Fi enabled device are reflected in this rate. The *length of stay* (Sect. 3.2.1) of customers can be accurately measured, as customers become visible to the system from the moment they enter a store. As in the scenario of real-time locating systems based on RFID, the queuing times can be extracted from the data. Wi-Fi technology can also be used in mobile apps to support *geofencing* (Sect. 3.2.2), although for this particular use case Bluetooth is the more common technology. The use case of context-aware browsing can be supported at interaction points, by reacting to the areas that a customer visited on their path before starting the interaction. As far as the operational use cases are concerned, this scenario only supports *path optimisation* (Sect. 3.2.3) to a limited degree, as the customer paths can be seen, but to fully understand what the customers were looking for in their paths, further information is required. The *waiting time estimation* (Sect. 3.2.3) can be supported and displayed to customers.

Privacy Threats

In general, a log for MAC addresses is maintained for technical troubleshooting. However, the *aggregation* of unique MAC addresses if used for location sensing, leads to *surveillance* as location can be tracked and sensed for every shopping trip undertaken by customers if they leave Wi-Fi turned ON. Thus, this leads to *secondary use* of the collected data. Moreover if there is lack of transparency regarding the collection and processing of MAC addresses then it also leads to *exclusion*. Through the tracked *location*, retailers can *infer preferences* of their customers based on the time spent in different sections of the store. Also, MAC addresses can provide information about the devices being used customers and could be used for marketing, for instance is a customer using a new device or a fairly old device, if it is expensive or relatively cheap etc. For an example of this scenario, consider a big mall which provides free Wi-Fi. Through the collected information, one can analyse different paths taken by customers along with the time spent in certain areas of the mall. This information *discloses* probable preferences of a customer e.g. whether she visits more of sports stores, fashion retail store etc. Additionally, if WiFi location sensing is used by retailers then they can also understand the movement paths of the customers in their store and time spent in different sections similar to the mall example.

3.4.3   Analysis of Combinations of RFID and Wi-Fi

The combination of RFID and Wi-Fi data sources is of particular interest. The scenario is illustrated in Fig. 3.7. The reason is that RFID tags are typically attached

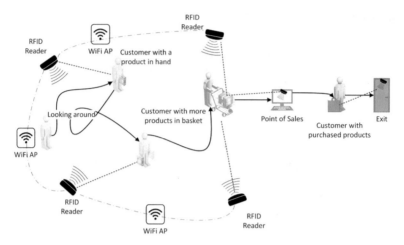

Fig. 3.7 An illustration of a shopping trip with RFID RTLS, Wi-Fi locating system, a POS and an exit gate

to items, whereas Wi-Fi is associated with actors (e.g., personnel, customers) interacting with items. In this way, when combining these two sources of information, it becomes possible to not only track a moving actor, but also track the items with which that actor is interacting. For example, the items picked up and dropped along the path are available for analysis.

Supported Use Cases

This combination of data sources enables all the use cases outlined in Sect. 3.2. Among others, specially tailored personalised marketing campaigns can take items of interest into account to tailor advertisements, and provide recommendations. Also, retailers can analyse the paths of customers for optimal shop floor layout, estimate waiting times, etc.

Privacy Threats

For this scenario, threats are simply a combination of those discussed for RTLS (Sect. 3.4.1.4) and Wi-Fi (Sect. 3.4.2). First of all, it leads to high level of *surveillance* as not only the location is sensed and continuously tracked but also information about items which are picked and carried or picked and later left are also associated with the location data. Thus, the *aggregated* information *discloses* fine-grained information about a customer's shopping trip. This scenario is comparable to the online cookies for analytics. Analytics cookies provide information about the browser used, mouse clicks, pages visited, count of visits etc. Similarly, this scenario provides information about the device used by a customer, path which they took in the store, items that were picked up and later bought, items that were picked and not bought etc. Thus, the scale or level of surveillance is much higher as compared to the previous scenarios.

3.5 Privacy Controls for Location Sensing Technologies

After having discussed the threats to privacy in different scenarios, we now turn to legal requirements and technical controls.

3.5.1 Legal Requirements

For the legal requirements, we base our analysis on the upcoming EU General Data Protection Regulation (GDPR) [81]. The GDPR supersedes the previously applied EU Data Protection Directive [82] and raises data protection standards by adapting rules in line with the recent technological developments. Based on the general principles discussed in the regulation, companies should ensure overall *lawfulness, fairness and transparency* for processing along with appropriate security measures to ensure *integrity and confidentiality*. Personal data should be only collected for specified and explicit purpose for establishing *purpose limitation*. Then, data collection should be relevant and limited to what is essential such that *data minimisation* is achieved. Next, the collected data should be kept accurate and up to date such that *accuracy* is ensured. Last, the personal data should be deleted or anonymised after the purpose has been fulfilled to assure *storage limitation*. While ensuring the basic principles for processing personal data, companies have broadly three different paths to ensure compliance: They can either (1) anonymise the data, (2) obtain informed consent for processing or (3) perform a balancing act to check if their processing can be considered as part of their legitimate interests.

3.5.1.1 Anonymisation

ISO defines anonymisation as a process by which personally identifiable information (PII) is irreversibly altered in such a way that an individual can no longer be identified directly or indirectly, either by the PII controller alone or in collaboration with any other party [1]. For anonymisation of personal data, through the techniques of randomisation and/or generalisation, personally identifiable part is removed from data sets. Article 29 in their paper on anonymisation have discussed various techniques to achieve the non-identifiability of individuals [9]. Along with the explanation of different techniques they have also provided with strengths and weaknesses as well as common mistakes and failures related to their use. Like the ISO definition, they also emphasise on the importance of irreversible non-identifiability. If an individual can directly or even indirectly be identified in a dataset (for instance, through a reference to an identifier such as a name, number, location data) then that dataset has to be considered as personal data. Hence, MAC addresses used for Wi-Fi tracking is to be considered as personal data as individuals can be indirectly identified based on the uniqueness of the devices carried by them. Similarly, if RFID tags can be associated with customers based on comparison of fields like timestamps (which are also generally associated with purchase history) then RFID data also becomes personally identifiable. To remove the personally identifiable part, in the literature, a lot of techniques have been defined for anonymisation for example noise addition [32], k-anonymity [78], l-diversity [59], differential privacy [31] etc. As individuals cannot be identified in a well anonymised dataset, it falls out of the scope of data protection and reduces the legal obligations for the companies.

However, it is quite complicated to achieve a level of anonymisation that guarantees privacy. Researchers like Ohm have discussed the failures of anonymisation in ensuring privacy [67]. There exist a number of techniques to de-anonymise data [64] i.e. reidentifying individuals from a data set which was previously believed to be anonymous. The complication arises from the tradeoff between utility and privacy [70]. Higher levels of anonymisation increase privacy but in turn decrease the utility of a dataset. Thus, it is important to attain an equilibrium where utility and privacy parameters are well balanced [39]. If anonymisation techniques are chosen intelligently based on the context (type of data involved) then adequate level of privacy can be achieved while conserving useful utility of the dataset.

Anonymisation can be used in two broad ways—(1) Collecting information which is anonymous, (2) Collecting personal information and later anonymising it for a further purpose. For the first case, the collected information should not be personally identifiable. Hence, if collected data is considered as anonymous then companies should ensure that the dataset cannot be linked to existing datasets such that it is not possible to de-anonymise the data. For the second case, if data is later anonymised then best available techniques for anonymisation have to follow suit [66], including a regular inspection of the dataset for potential re-identifiability. This becomes specially important as location traces using the discussed technologies, in general, create quite distinct traces for individuals. For instance, Montjoye et al. found that by using only four different data points for coarse location and time during a day per individual, 95% of the 1.5 Million individuals could be uniquely identified [26]. Thus, location data can easily become personally identifiable.

3.5.1.2 Obtaining Consent

If a company values the utility of a dataset with personal information or if anonymisation is not adequately maintaining privacy then it can try to obtain the consent for processing from datasubjects. For obtaining consent, a company must explain, in a transparent way, the purpose of processing along with distinguishable information on the possibility of withdrawing consent in the future. Being transparent is crucial since consent must be given in an informed manner. Friedman et al. have developed a model providing broad guidelines for informed consent provision [36, 37]. Based on their model, there are six components to be considered:

Disclosure Providing accurate information about the processing along with harms and benefits involved with the processing. For example, companies should mention details like what information is collected, who will have access to it, how long would it be stored etc.

Comprehension Understandability of the information such that individuals are able to accurately interpret the information disclosed to them. Thus, it is not only important to provide all the information about processing but to also ensure that the information is easy to understand for the data subjects.

Voluntariness Ensuring that the action of giving consent is not forced on the individuals i.e. companies should not make data processing compulsory if data is not essential for the purpose. For instance, a marketing survey collecting information on a Pizza delivery service should be voluntary and service of delivering pizzas should not be affected if a customer chooses not to take part in the survey.

Competence Mental, emotional and physical competence (capability) of the targeted data subjects should be considered to ensure that they give an informed consent. For example, if information is provided such that the font which is not readable for an average individual then the consent will not be counted as readability (vision competence) was not properly considered.

Agreement Clear options must be provided for data subjects to provide consent for the data processing. Moreover, the GDPR adds that the process of revoking consent should be as simple as giving consent.

Minimal distraction All the above criteria must be met in such a way that individuals are not unduly diverted from their task at hand. For instance, if a company asks customers to read a 50 page document before they shop in a smart store then customers would tend to ignore the information and make uninformed decisions. This becomes quite challenging to implement as all information should be provided to customers and at the same time it should not distract them from their main task. For this reason, the GDPR recommends the use of Privacy Icons that are simpler to process by users.

After a company obtains an informed consent for processing, the customers must then be given options to access their data as well as the possibility to rectify some parts if logged incorrectly. Furthermore, options to erase their personal data or withdrawing consent for further processing have to be provided, along with ensuring adequate security in order to ensure compliance with the GDPR. In the context of security of personal information, pseudonymisation is referred to as a recommended technique.

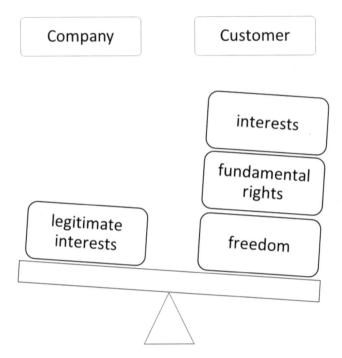

Fig. 3.8 Balancing act for legitimate interests as defined in the GDPR [81]

3.5.1.3 Legitimate Interests

Processing can also be considered lawful if done within the scope of legitimate interests of a company [10]. Legitimate interests can only be argued when the following three points are fulfilled:

1. The considered legitimate interests of the company are balanced against the interests or fundamental rights and freedoms of the data subjects, as illustrated in Fig. 3.8.
2. Processing is lawful (following other applicable legal regulations).
3. Processing represents a real and present interest. This means that data cannot just be collected for speculative reasons (i.e. for some future use). Scenarios like the engagement in *conventional direct marketing and other forms of marketing or advertisement* or the provision of *IT and network security* are two examples where the legitimate interests argument can be used as a valid legal ground for the processing. If legitimate interests are used as a legal ground, however, then a company needs to provide information about its data processing activity in a transparent way. Also, customers must be given an option to object to such processing in case they believe that their freedom is negatively affected by it.

3.5.2   Implementation of Privacy Controls

Technical controls, available in our context, deal mainly with anonymisation to avoid location data being personally identifiable, pseudonymisation for enhancing security and transparency to ensure informed consent. Location anonymity is given, when the location information is dissociated from an individual, while pseudonymisation links location information to a pseudonym that is disconnected from the individual [30]. Though pseudonymisation does not provide adequate promises of privacy and is considered as personally identifiable data, it is still considered as a recommended technique for ensuring security of personal data along with other techniques like encryption. In the following, we discuss possible privacy controls for the outlined scenarios in Sect. 3.4. Note that in all these scenarios, the location sensing is taking part on the provider side and not on the user side. Location-sensing system providers should consider implementing these controls to minimise the associated privacy threats.

3.5.2.1   Privacy Controls for RFID Systems

Blocking Unknown RFID Reads to Ensure Data Minimisation

To ensure data minimisation, companies should collect only relevant data. Thus, RFID tags, which are not associated with the company should not be read. This control can simply be implemented by maintaining *a whitelist* of the inventory of tagged items that the company owns or issues. More specifically, such a whitelist includes Electronic Product Codes (EPCs) that are supposed to be in the store according to the inventory system. This would lead to discarding the reads of unknown or unexpected EPCs that were not deactivated by other organisations and could potentially identify individuals. Also, the EPCs of sold items should also not be read, if the purpose to use RFID in store was mainly to enhance the visibility of the inventory.

Deactivating/Destroying Integrated Tags in Sold Products to Safeguard Customers' Confidentiality

As discussed in Sect. 3.3.2.1, through estimation of constellations, RFID tags may provide unique personal identifiers related to customers. Thus, to prevent such threats, companies should either destroy or deactivate the tags. A lot of different techniques have been discussed in the literature for achieve this [11, 47].

Securing RFID Enabled Loyalty Cards

If a company decides to put RFID tags in their loyalty cards then the deactivation of such tags is not longer an option. Read range of 1–10 m also amplifies the associated privacy threats. In that case, companies should ensure that the chips on the cards are only using Near-Field Communication (NFC) [65] where the read range is no more than 1 m and there exist ways to prevent unauthorised reads. Since, read range for NFC is less than 1 m, it becomes comparatively difficult to read tags in an unauthorised way.

Anonymisation of Tag Reads for RFID Enabled Interaction Points

As RFID data from RFID enabled interaction points (e.g., smart fitting rooms, kiosks etc.) can be correlated to items that customers bought (indirectly identifying them). Hence, RFID data has to be considered as personal data. Thus, companies can either anonymise the data and remove any personally identifiable information or obtain an informed consent from the customers for the processing of such data. For ensuring that the RFID data is anonymous, temporal cloaking (e.g., reducing the time granularity from seconds to days) can be applied on the read RFID enabled items suggested by Gruteser and Grunwald [40]. By making the time information less precise, the data can be turned into k-*anonymous* data. *k-anonymous* means that individuals'information is sufficiently imprecise in order to make them indistinguishable from at least $k - 1$ other individuals. In this way, the system can still count how often a given article type was brought to interaction points (e.g., fitting rooms) in a broader time range (e.g., day, week). If temporal information is cloaked to days, this measure hides the information of the number and times of visits to the interaction points per individual, maintaining anonymity of visitors per day. By doing this, the individual level information is lost but marketing insights about the ratio of fitting room visits and effective sales can still be collected on an item level. Note that in settings with a high variety of article types, it might be possible that the assignment of the fact whether a customer was in the fitting room could still be reconstructed (e.g., if a rare item was sold on 1 day, and there was also one visit of that outlier item in the fitting room). In that case, it is possible to increase time censoring to weekly, monthly, or even coarser granularity [40].

Ensuring Transparency for RFID Enabled Interaction Points

To avoid complications with the identifiability of outliers, companies can also rely on obtaining consent from customers. In that case, informing customers about RFID readers would be essential to ensure transparency. It can be easily achieved by using RFID logos at customer interaction points, such as shop entries. To provide a choice, there should be an option to use a non-RFID based interaction point. Alternatively, the RFID readers could only read the tags after the customer has confirmed that he or she wishes to use the underlying smart services.

3.5.2.2 Privacy Controls for RFID Real-Time Locating Systems

The main purpose of RTLS is to analyse the movement of tagged items. Thus, unless there is a specific purpose to identify and track customers using the technology, substantial efforts should be made by companies to avoid identifiability of customers in a RTLS dataset.

Discarding Historical Data to Ensure Storage Limitation and Data Minimisation

By only storing the latest position of tags, customer behaviour cannot be reconstructed. Furthermore, the RTLS information system should not store information about EPC tags that have been sold, or EPCs of other stores. This can be easily implemented with a whitelist mechanism. In this way, only the whereabouts of items currently available for sale in the shop are recorded.

Anonymising RFID Location Data

To make this data anonymous, companies would need to destroy the linkability between the RTLS and POS data sets. Here, the relation is not only based on temporal information (relating RFID read timestamps with POS timestamps) but also spatial information (relating different baskets of items moving around). For obstructing a correlation of space and time at checkout, temporal and spatial cloaking can be applied [30] without losing valuable information regarding position of items. A discussion of privacy-preserving techniques with provable privacy guarantees is presented in Chap. 5.

3.5.2.3 Privacy Controls for Wi-Fi Locating Systems

Wi-Fi based locating systems can be used for assisting customers for in-store navigation, analysis of most crowded and least crowded parts of store etc. Paths are uniquely identified by the MAC-address of the device which is considered as identifiable data. Simple *pseudonymisation* by replacing the MAC address by an identifier (e.g., a hash value of the address) does not suffice, as the data can still be correlated to individual's sales data through spatio-temporal overlaps of the purchase and the visit of the point of sale. However, even applying temporal and spatial cloaking for the visits on a daily basis is not enough as by looking at data created by multiple visits of a returning customer, the anonymity of that customer can be compromised. For example, consider the case where temporal cloaking to a daily resolution is applied. Looking at only 1 day of records, the customer is hidden in the anonymity of all the customers that visited that day. However, when looking at multiple days when that customer made purchases, the customer's identity is only hidden in the *intersection* of the sets of customers on these multiple days, which in turn allows singling out customers. Thus, if companies are interested in analysing location traces of identifiable customers then an informed consent must be obtained. Otherwise, anonymisation can also be used as it also supports a number of applications.

Anonymising WiFi Location Data

As even through spatio-temporal cloaking customers can be uniquely identified, we suggest using new identifiers for every *visit* instead of a single identifier per MAC address to disable possible linkability. Also, for anonymising location data, different techniques are defined in literature. The approach by Tang et al. [79] ensures privacy as individual's data is not stored but only anonymous visits to areas are recorded. An application of this method to Wi-Fi location-sensing technologies means to only count visits in areas or transitions from one area to another instead of entire uniquely identifiable paths. Further methods include the framework by Duckham and Kulik [30], where the authors propose to obfuscate information, that is, increase the imprecision of location information.

3.6   Conclusion

In this chapter, we first explained how location can be inferred by means of trilateration or fingerprinting. Subsequently, we discussed potential technologies that allow us to perform location-sensing. Then, we used the example of a brick-and-mortar retail organisation to illustrate and discuss the use cases empowered by location-sensing technologies. We found that many previously existing use cases benefit from this additional form of information, while some entirely novel use cases are only possible through location-sensing technologies.

From these use cases, we turned to the threats of privacy that location-sensing technologies entail and exemplified them in the retail domain after an introduction into an existing taxonomy and categorization of threats. Therefore, we discussed popular location-sensing scenarios ranging from only collecting location data at fixed positions to real-time locating systems that can surveil the entire store area. Finally, we discussed controls to mitigate the identified privacy threats from the perspective of the location-sensing system provider and presented controls that the users have in this setting.

We live in times, where new technologies pop up at an increasing rate and outperform previous technologies in terms of accuracy and efficiency, sometimes by magnitudes. It is difficult to cope with the privacy implications of these novel technologies, as even within a single technology the potential use cases become apparent only as time progresses. Furthermore, new combinations of technologies allow for unforeseen use cases. For example, there are first supermarkets, where the system fully automatically detects the picked up items and the customers' identities by facial recognition to entirely automate checkout and avoid queueing [87].

Therefore, adherence to existing and upcoming legislations, and responsible use of collected data of organisations is of utmost importance. When new information gathering systems are implemented in organisations, we need to ensure that privacy is built in by design, because afterwards it might be too late, and privacy breaches can dearly cost an organisation in both reputation, trust and also by legal fines. Thus, we urge the implementers of novel technologies and the users to consider privacy and ethics throughout their systems and processes.

References

1. ISO/IEC 29100:2011 - Information technology – Security techniques – Privacy framework. 2011.
2. G. D. Abowd, A. K. Dey, P. J. Brown, N. Davies, M. Smith, and P. Steggles. Towards a better understanding of context and context-awareness. In *International Symposium on Handheld and Ubiquitous Computing*, pages 304–307. Springer, 1999.
3. S. Adler, S. Schmitt, K. Wolter, and M. Kyas. A survey of experimental evaluation in indoor localization research. In *Indoor Positioning and Indoor Navigation (IPIN), 2015 International Conference on*, pages 1–10. IEEE, 2015.
4. G. Adomavicius and A. Tuzhilin. Toward the next generation of recommender systems: A survey of the state-of-the-art and possible extensions. *IEEE transactions on knowledge and data engineering*, 17(6):734–749, 2005.
5. M. Afanasyev, T. Chen, G. M. Voelker, and A. C. Snoeren. Usage patterns in an urban WiFi network. *IEEE/ACM Trans. Netw.*, 18(5):1359–1372, 2010.
6. K. Albrecht and L. C. McIntyre. Scandal: Wal-mart, p &g involved in secret RFID testing, consumers against supermarket privacy invasion and numbering (caspian). Retrieved on 10-01-2017 from http://www.spychips.com/press-releases/broken-arrow.html.
7. Y. Andreu-Cabedo, P. Castellano, S. Colantonio, G. Coppini, R. Favilla, D. Germanese, G. Giannakakis, D. Giorgi, M. Larsson, P. Marraccini, et al. Mirror mirror on the wall an intelligent multisensory mirror for well-being self-assessment. In *2015 IEEE International Conference on Multimedia and Expo (ICME)*, pages 1–6. IEEE, 2015.
8. Anonymous. Silverpush launches cross-device ad targeting with unique audio beacon technology. Retrieved on 14-01-2017 from http://www.steamfeed.com/silverpush-launches-cross-device-ad-targeting-with-unique-audio-beacon-technology/.
9. Article 29 Data Protection Working Party. Opinion 05/2014 on Anonymisation Techniques, 2014.
10. Article 29 Data Protection Working Party. Opinion 06/2014 on Notion of legitimate interests of the data controller under Article 7 of Directive 95/46/EC, 2014.
11. J. Ayoade. Roadmap to solving security and privacy concerns in rfid systems. *Computer Law & Security Review*, 23(6):555–561, 2007.
12. M. Azizyan, I. Constandache, and R. Roy Choudhury. Surroundsense: mobile phone localization via ambience fingerprinting. In *Proceedings of the 15th annual international conference on Mobile computing and networking*, pages 261–272. ACM, 2009.
13. R. Bajaj, S. L. Ranaweera, and D. P. Agrawal. GPS: location-tracking technology. *Computer*, 35(4):92–94, 2002.
14. M. Baldauf, S. Dustdar, and F. Rosenberg. A survey on context-aware systems. *International Journal of Ad Hoc and Ubiquitous Computing*, 2(4):263–277, 2007.

15. A. R. Beresford and F. Stajano. Location privacy in pervasive computing. *IEEE Pervasive computing*, 2(1):46–55, 2003.
16. A. Bibby. Invasion of the privacy snatchers. *Financial Times, January*, 9, 2006.
17. R. R. Burke. The third wave of marketing intelligence. In *Retailing in the 21st Century*, pages 113–125. Springer, 2006.
18. J. D. Cai. Business intelligence by connecting real-time indoor location to sales records. In *International Conference on Web-Age Information Management*, pages 817–823. Springer, 2014.
19. M. T. Capizzi and R. Ferguson. Loyalty trends for the twenty–first century. *Journal of Consumer Marketing*, 22(2):72–80, 2005.
20. G. Castelli, A. Rosi, M. Mamei, and F. Zambonelli. A simple model and infrastructure for context-aware browsing of the world. In *Fifth Annual IEEE International Conference on Pervasive Computing and Communications (PerCom'07)*, pages 229–238. IEEE, 2007.
21. A. Cavoukian. Privacy guidelines for RFID information systems (RFID privacy guidelines), 2006. Information and Privacy Comissioner/Ontario, Toronto.
22. A. Chattopadhyay and A. R. Harish. Analysis of low range indoor location tracking techniques using passive UHF RFID tags. In *2008 IEEE Radio and Wireless Symposium*, 2008.
23. I. Cil. Consumption universes based supermarket layout through association rule mining and multidimensional scaling. *Expert Systems with Applications*, 39(10):8611–8625, 2012.
24. P. Cohan. How nordstrom uses WiFi to spy on shoppers. Retreived on 13-01-2017 from http://www.forbes.com/sites/petercohan/2013/05/09/how-nordstrom-and-home-depot-use-wifi-to-spy-on-shoppers/#2421d0ec3bf9.
25. M. Cunche. I know your MAC address: Targeted tracking of individual using Wi-Fi. *Journal of Computer Virology and Hacking Techniques*, 10(4):219–227, 2014.
26. Y.-A. De Montjoye, C. A. Hidalgo, M. Verleysen, and V. D. Blondel. Unique in the crowd: The privacy bounds of human mobility. *Scientific reports*, 3, 2013.
27. D. Delen, B. C. Hardgrave, and R. Sharda. Rfid for better supply-chain management through enhanced information visibility. *Production and Operations Management*, 16(5):613–624, 2007.
28. L. Demir. Wi-Fi tracking : what about privacy. Master's thesis, M2 SCCI Security, Cryptolo-gyand Coding of Information - UFR IMAG, Sept. 2013. https://hal.inria.fr/hal-00859013.
29. A. K. Dey. Understanding and using context. *Personal Ubiquitous Comput.*, 5(1):4–7, Jan. 2001.
30. M. Duckham and L. Kulik. A formal model of obfuscation and negotiation for location privacy. In *International Conference on Pervasive Computing*, pages 152–170. Springer, 2005.
31. C. Dwork. Differential privacy: A survey of results. In *International Conference on Theory and Applications of Models of Computation*, pages 1–19. Springer, 2008.
32. C. Dwork, F. McSherry, K. Nissim, and A. Smith. Calibrating noise to sensitivity in private data analysis. In *Theory of Cryptography Conference*, pages 265–284. Springer, 2006.
33. Eileen P. Kelly G. Scott Erickson. RFID tags: commercial applications v. privacy rights. *Industrial Management & Data Systems*, 105(6):703–713, 2005.
34. European Emergency Number Association. Caller location in support of emergency services. *EENA Operations Document*, (2), 2014.
35. Z. Farid, R. Nordin, and M. Ismail. Recent advances in wireless indoor localization techniques and system. *Journal of Computer Networks and Communications*, 2013, 2013.
36. B. Friedman, E. Felten, and L. I. Millett. Informed consent online: A conceptual model and design principles. *University of Washington Computer Science & Engineering Technical Report 00–12-2*, 2000.
37. B. Friedman, P. Lin, and J. K. Miller. Informed consent by design. *Security and Usability*, (2001):503–530, 2005.
38. S. L. Garfinkel, A. Juels, and R. Pappu. RFID privacy: An overview of problems and proposed solutions. *IEEE Security & Privacy*, 3(3):34–43, 2005.
39. A. Ghosh, T. Roughgarden, and M. Sundararajan. Universally utility-maximizing privacy mechanisms. *SIAM Journal on Computing*, 41(6):1673–1693, 2012.

40. M. Gruteser and D. Grunwald. Anonymous usage of location-based services through spatial and temporal cloaking. In *Proceedings of the 1st international conference on Mobile systems, applications and services*, pages 31–42. ACM, 2003.
41. Y. Gu, A. Lo, and I. Niemegeers. A survey of indoor positioning systems for wireless personal networks. *IEEE Communications surveys & tutorials*, 11(1):13–32, 2009.
42. M. Hazas, J. Scott, and J. Krumm. Location-aware computing comes of age. *IEEE Computer*, 37(2):95–97, 2004.
43. M. Hazas, J. Scott, and J. Krumm. Location-aware computing comes of age. *Computer*, 37(2):95–97, 2004.
44. Information technology – Security techniques-Information security risk management. Standard, International Organization for Standardization, Geneva, CH, 2008.
45. A. Juels. RFID security and privacy: A research survey. *IEEE journal on selected areas in communications*, 24(2):381–394, 2006.
46. Kalyan S. Pasupathy, Thomas R. Hellmich. How RFID technology improves hospital care, 31-12-2015. Retrieved on 14-01-2017 from https://hbr.org/2015/12/how-rfid-technology-improves-hospital-care.
47. G. Karjoth and P. A. Moskowitz. Disabling RFID tags with visible confirmation: clipped tags are silenced. In *Proceedings of the 2005 ACM workshop on Privacy in the electronic society*, pages 27–30. ACM, 2005.
48. V. Kopytoff. For retailers, tracking shoppers brings new insights. Retrieved on 13-01-2017 from https://www.technologyreview.com/s/520811/stores-sniff-out-smartphones-to-follow-shoppers/.
49. D. Kravets. Tracking school children with RFID tags? it's all about the benjamins. Retrieved on 14-01-2017 from https://www.wired.com/2012/09/rfid-chip-student-monitoring/.
50. J. Krumm, editor. *Ubiquitous computing fundamentals*. CRC Press, 2016.
51. P. Kumar, M. U. Kalwani, and M. Dada. The impact of waiting time guarantees on customers' waiting experiences. *Marketing science*, 16(4):295–314, 1997.
52. A. LaMarca, Y. Chawathe, S. Consolvo, J. Hightower, I. E. Smith, J. Scott, T. Sohn, J. Howard, J. Hughes, F. Potter, J. Tabert, P. Powledge, G. Borriello, and B. N. Schilit. Place lab: Device positioning using radio beacons in the wild. In *Pervasive Computing, Third International Conference, PERVASIVE 2005, Munich, Germany, May 8-13, 2005, Proceedings*, pages 116–133, 2005.
53. M. Langheinrich. A survey of RFID privacy approaches. *Personal and Ubiquitous Computing*, 13(6):413–421, 2009.
54. J. S. Larson, E. T. Bradlow, and P. S. Fader. An exploratory look at supermarket shopping paths. *International Journal of research in Marketing*, 22(4):395–414, 2005.
55. M. Levy, B. A. Weitz, and D. Grewal. *Retailing management*. Irwin/McGraw-Hill New York, 1998.
56. H. Liu, H. Darabi, P. Banerjee, and J. Liu. Survey of wireless indoor positioning techniques and systems. *IEEE Transactions on Systems, Man, and Cybernetics, Part C (Applications and Reviews)*, 37(6):1067–1080, Nov 2007.
57. H. Liu, H. Darabi, P. P. Banerjee, and J. Liu. Survey of wireless indoor positioning techniques and systems. *IEEE Trans. Systems, Man, and Cybernetics, Part C*, 37(6):1067–1080, 2007.
58. H. Liu, Y. Gan, J. Yang, S. Sidhom, Y. Wang, Y. Chen, and F. Ye. Push the limit of WiFi based localization for smartphones. In *The 18th Annual International Conference on Mobile Computing and Networking, Mobicom'12, Istanbul, Turkey, August 22–26, 2012*, pages 305–316, 2012.
59. A. Machanavajjhala, D. Kifer, J. Gehrke, and M. Venkitasubramaniam. l-diversity: Privacy beyond k-anonymity. *ACM Transactions on Knowledge Discovery from Data (TKDD)*, 1(1):3, 2007.
60. C. Martin. How beacons are changing the shopping experience. *Harvard Bus. Rev*, 2014.
61. A. S. Mattila and J. Wirtz. Congruency of scent and music as a driver of in-store evaluations and behavior. *Journal of retailing*, 77(2):273–289, 2001.
62. A. Moore. Defining privacy. *Journal of Social Philosophy*, 39(3):411–428, 2008.

63. D. Namiot. Context-aware browsing–a practical approach. In *2012 Sixth International Conference on Next Generation Mobile Applications, Services and Technologies*, pages 18–23. IEEE, 2012.

64. A. Narayanan and V. Shmatikov. Robust de-anonymization of large sparse datasets. In *2008 IEEE Symposium on Security and Privacy (sp 2008)*, pages 111–125. IEEE, 2008.

65. P. V. Nikitin, K. Rao, and S. Lazar. An overview of near field UHF RFID. In *IEEE international Conference on RFID*, volume 167. Citeseer, 2007.

66. A. Novotny and S. Spiekermann. Personal information markets and privacy: a new model to solve the controversy. 2012.

67. P. Ohm. Broken promises of privacy: Responding to the surprising failure of anonymization. *UCLA law review*, 57:1701, 2010.

68. R. Prasad and M. Ruggieri. *Applied satellite navigation-using GPS, GALILEO and augmentation systems*. 2005.

69. N. M. Puccinelli, R. C. Goodstein, D. Grewal, R. Price, P. Raghubir, and D. Stewart. Customer experience management in retailing: Understanding the buying process. *Journal of Retailing*, 85(1):15–30, 2009. Enhancing the Retail Customer Experience.

70. V. Rastogi, D. Suciu, and S. Hong. The boundary between privacy and utility in data publishing. In *Proceedings of the 33rd international conference on Very large data bases*, pages 531–542. VLDB Endowment, 2007.

71. F. Ricci, L. Rokach, and B. Shapira. *Introduction to recommender systems handbook*. Springer, 2011.

72. D. Y. Sha and L. Guo-Liang. Improving service quality of retail store by innovative digital content technology. In *2012 IEEE International Conference on Computer Science and Automation Engineering*, pages 655–660, June 2012.

73. D. J. Solove. A taxonomy of privacy. *University of Pennsylvania law review*, pages 477–564, 2006.

74. D. J. Solove. Understanding privacy. *Harvard University Press, GWU Law School Public Law Research Paper*, (420), May 2008.

75. J. Song, C. T. Haas, and C. H. Caldas. A proximity-based method for locating RFID tagged objects. *Advanced Engineering Informatics*, 21(4):367–376, 2007.

76. R. A. Spinello. Privacy rights in the information economy. *Business Ethics Quarterly*, 8(4):723–742, 1998.

77. steinunn. Discover a museum with BLE app for indoor location – Video from Eldheimar, 2014. Retrieved on 14-01-2017 from https://locatify.com/blog/discover-a-museum-with-ble-app-for-indoor-location-video-from-eldheimar/.

78. L. Sweeney. k-anonymity: A model for protecting privacy. *International Journal of Uncertainty, Fuzziness and Knowledge-Based Systems*, 10(05):557–570, 2002.

79. K. P. Tang, P. Keyani, J. Fogarty, and J. I. Hong. Putting people in their place: An anonymous and privacy-sensitive approach to collecting sensed data in location-based applications. In *Proceedings of the SIGCHI Conference on Human Factors in Computing Systems*, CHI '06, pages 93–102, New York, NY, USA, 2006. ACM.

80. Z. Technologies. Analysis of iOS 8 MAC Randomization on Locationing. http://mpact.zebra.com/documents/iOS8-White-Paper.pdf, 2015. Zebra Whitepaper.

81. The European Parliament And The Council Of The EU. Regulation (eu) 2016/679 on the protection of natural persons with regard to the processing of personal data and on the free movement of such data, and repealing directive 95/46/ec (general data protection regulation). *Official Journal of the European Union*, L119/1, 2016.

82. The European Parliament, the Council and the Commission. Eu directive 95/46/ec on the protection of individuals with regard to the processing of personal data and on the free movement of such data: Eu directive 95/46/ec. pages 0031–0050, 1995.

83. F. Thiesse. RFID, privacy and the perception of risk: A strategic framework. *J. Strategic Inf. Sys.*, 16(2):214–232, 2007.

84. F. Thomas and L. Ros. Revisiting trilateration for robot localization. *IEEE Trans. Robotics*, 21(1):93–101, 2005.

85. J. P. Van Den Berg. A literature survey on planning and control of warehousing systems. *IIE Transactions*, 31(8):751–762, 1999.
86. R. Want. An introduction to RFID technology. *IEEE Pervasive Computing*, 5(1):25–33, 2006.
87. A. Wee. A supermarket without a checkout line-Amazon go, 2016. Zing Gadget http://en. zinggadget.com/a-supermarket-without-a-checkout-line-amazon-go.
88. S. A. Weis. Security and privacy in radio-frequency identification devices. Master's thesis, Massachusetts Institute of Technology, 2003.
89. A. Yaeli, P. Bak, G. Feigenblat, S. Nadler, H. Roitman, G. Saadoun, H. J. Ship, D. Cohen, O. Fuchs, S. Ofek-Koifman, et al. Understanding customer behavior using indoor location analysis and visualization. *IBM Journal of Research and Development*, 58(5/6):3–1, 2014.
90. C. Yang and H.-R. Shao. WiFi-based indoor positioning. *IEEE Communications Magazine*, 53(3):150–157, 2015.
91. D. Zanetti, P. Sachs, and S. Capkun. On the practicality of UHF RFID fingerprinting: How real is the RFID tracking problem? In S. Fischer-Hübner and N. Hopper, editors, *Privacy Enhancing Technologies - 11th International Symposium, PETS 2011, Waterloo, ON, Canada, July 27–29, 2011. Proceedings*, volume 6794 of *Lecture Notes in Computer Science*, pages 97–116. Springer, 2011.

Part II
Main Research Directions in Mobility Data Privacy

Chapter 4
Privacy Protection in Location-Based Services: A Survey

Claudio Bettini

Abstract Location awareness has enabled efficient and accurate geo-localised Internet services. Mobile apps exploiting these services have changed our way of navigating and searching for resources in geographical space. This chapter provides a classification of location based services (LBS) and illustrates the privacy aspects involved in releasing our location information as part of a service request. It includes a discussion about legal obligations of the LBS provider and about ways to specify personal location privacy preferences. The chapter also provides a systematic survey of the main approaches that have been proposed for protecting the user's privacy while using these services.

4.1 Introduction

A large majority of smartphone users take advantage of apps that provide Location Based Services (LBS), from weather forecast to map based navigation and search of nearby resources.[1]

LBS can be generally described as Internet based services that offer functionalities enabled by the geo-localization of the device issuing the service request. They belong to the family of context-aware services with the timestamped location and/or the user trajectory acting as the context information. The service provisioning often involves performing some spatio-temporal data processing that may consider the relative distance between users, their velocities, the nearby resources, as well as spatial constraints such as road networks, or real time events like traffic conditions. LBS provisioning may also involve data analysis as, for example, discovering the users' frequent trajectories.

---

[1]Pew Research Center http://www.pewinternet.org/2013/09/12/location-based-services/.

C. Bettini (✉)
Dipartimento di Informatica, University of Milan, Milan, Italy
e-mail: claudio.bettini@unimi.it

© Springer Nature Switzerland AG 2018
A. Gkoulalas-Divanis, C. Bettini (eds.), *Handbook of Mobile Data Privacy*,
https://doi.org/10.1007/978-3-319-98161-1_4

73

LBS are not exclusive to mobile devices. Indeed a LBS, for example a local weather forecast, can also be invoked from a desktop PC obtaining the user position by IP address geo-coding. However, there is no doubt that their huge popularity is due to mobile users. Indeed the integration of global navigation satellite receivers (e.g., GPS) into even the cheapest smartphone, and the availability of very effective outdoor positioning services that combine GPS with methods based on cellular and WiFi signal, have made LBS very effective and available to every mobile user. Indoor positioning is less consolidated than outdoor positioning, but research advances, exploring the use of technologies like WiFi fingerprinting, BT beacons, UWB, ultrasound and more, are promising to reach a very high precision and new LBS are being deployed taking advantage of indoor as well as integrated outdoor and indoor positioning.

In order to understand the possible privacy threats in using LBS it is important to briefly review the different types of LBS and their main properties.

4.1.1 A Classification of LBS

We can divide LBS into two broad categories according to their sharing model: (a) personal services, and (b) social network services. The former provide information to the user that asked for the service, and typically share the user location information only with the service provider. The latter are intended to share the user location information also with a group of users and to receive information based on other users position and/or the relative distance between users.

4.1.1.1 Personal LBS

Personal LBS can be grouped as follows:

- Navigational services. They typically provide instructions to reach a destination based on the user position.
- Resource discovery services. They provide nearby points of interest (ATM, gas station, shop, . . .) in response to a user location.
- Local traffic/news/weather services. They provide local information based on the user position.
- Emergency services. They can send operators to the location from which the request was issued (e.g., road assistance).
- Proximity marketing services. They send ads to a user based on her proximity to shops or even items on sale.
- Location based personalization/advertising services. Different information/ads are sent to users depending on their location.

Note that the last two groups follow a *push* model while the others are usually based on a *pull* model.

4.1.1.2 Social Network LBS

Social network LBS can be grouped as follows:

- Geo-SN posting/check-in services. They allow users to associate a timestamped position to a resource that they share (e.g., a picture, video, text) or simply share their current position (check-in). They may also offer resource discovery services based on the density (or other property) of check-ins.
- Friend finder services. They provide information about the proximity of contacts or other users of a Geo-SN.
- Workforce management services. They allow coordination and optimization of mobile workers and items based on their position (e.g., taxi or packages in logistics).
- Location based games. They engage participants in a game that involves users as well as resources geographical positions.

The above categorization is definitely not exhaustive, but it provides a good coverage of currently available LBS.

4.1.2 Privacy Threats in LBS

In this chapter we formally define a privacy threat as follows.

A privacy threat occurs whenever an unauthorised entity can associate with high probability the identity of an individual with private information about that individual.

In the context of *personal LBS* such a threat occurs when the information contained in one or more requests issued by a given user can be used, possibly associated with external information, to associate the user identity with the private information.

In the context of *social network LBS* the above association can be obtained also from requests or postings performed by individuals different from the one involved in the threat [56].

According to some country regulation (e.g., the EU GDPR) any information that is specifically associated with an individual should be considered private, while other regulations refer to specific types of information. However, most guidelines for privacy risk assessment highlight the risk involved in revealing information on political, sexual, and religious orientation, health, financial assets, or closeness to specific individuals or organizations. LBS services play a role in this context because both identity and private information can be directly or indirectly released through a single or a sequence of LBS requests. LBS requests can reveal, for example, (a) information on the specific location of individuals at specific times, (b) movement patterns (specific routes at specific times and their frequency), (c) requests for sensitive services (closest temple for a specific religious worship),

(d) personal points of interest (home, workplace, frequent visits to specific shops, clubs, or institutions). Moreover the above information can also be used to infer when the individual is not where it is supposed to be (*absence privacy* [56]), where it is likely to be at a given hour of a given day, or when and how frequently the individual met other individuals (*co-location privacy* [56]).

Unauthorised use of this information exposes the user to several types of *privacy violation risks* including unsolicited advertising, discrimination, loss of reputation, family and work related issues (with divorce and getting fired among the outcomes), stalking and even exposure to robberies based on absence privacy violations.

4.1.2.1 Adversaries

In the privacy protection literature, as well as in the following of this chapter, the unauthorised entity that can acquire some data exchanged as part of the LBS and that may pose a privacy threat is often called the *adversary*. The LBS service provider can be considered an adversary if he uses the acquired information in any way different from what the user has agreed upon. An external entity that tries to break into the communication channel or into the service provider IT infrastructure and get LBS requests in transit or stored in a database is another adversary. In social network LBS, a generic user can be considered an adversary, for example when he can access a geo-localised post involving a user without his explicit consent. As it will be clear in Sect. 4.3 many privacy protection techniques require accurate modeling of the adversary in order to provide guarantees about their effectiveness.

4.1.2.2 Online Versus Offline Data Release

The literature on privacy protection for LBS has considered separately the problem of protecting privacy at the time a LBS request is sent to the service, and the problem of privately releasing data from a database where the history of requests from different users has been stored.

The first case is named *online data release* and the service provider, as well as any entity that may get access to the content of the issued requests are considered adversaries. The second case is named *offline data release* and the adversaries are the third parties or any entity that obtain data extracted from the database of requests stored at the service provider. This can be statistical data (e.g., aggregated counts on visited locations) or a set of individual data records. Such data is used, for example, for profiling, mobile advertising, or to refine online marketing strategies.

Online data release is intuitively more challenging, since a mechanism should be applied at the client side or through a trusted proxy server at the time of each request issued to the service. This case is also characterised by a limited knowledge available to this mechanism about the whole set of requests and about the location of other service users; information that may be useful for some protection techniques.

4.1.2.3   Single Versus Multiple Data Release

In both online and offline data release we should distinguish the case of a single data release from the case of multiple data releases. In the case of tabular data (e.g., databases) it is well known that the privacy guarantee that a protection technique can provide on a specific release of the data cannot be considered valid when more data is released. Similarly for sanitised answers to database queries. Despite each query answer may be considered privacy preserving in isolation, this is not the case when considering them together. Intuitively this is due to the correlations between the releases. It should be clear that this problem is present also in LBS offline data release, since the dataset involved in each release is indeed (spatio-temporal) tabular data.

   In the case of online data release we can model a threat as a single release when we assume that the adversary can get access only to a single LBS request, or to LBS requests from the same user that cannot be considered correlated (someone assumes this is the case when only sporadic requests are exposed). There are however LBS that require frequent updates of the user location so that their spatio-temporal correlation is clear and can lead to trajectory identification. These cases can be classified as online multiple (or repeated/continuous) data release.

4.1.3   Analysing Privacy Threats

From the definition of privacy threat it is clear how relevant is to understand how the information exchanged as part of the service can be used to identify the individual, to infer private information, and to connect the two to obtain the association that actually leads to the threat.

   We have also highlighted the relevance of identifying and modeling the possible adversaries depending on the sharing model, the service architecture, and the type of data release. The prior knowledge that these adversaries may have could be joined with information obtained from the LBS in order to perform a privacy violation. When multiple adversaries may be involved, we should also consider the case of their possible collusion.

   As explained later in the chapter, most LBS privacy protection techniques mitigate the risk of privacy threats at an additional cost in terms of computational and communication overhead and/or decrease of service utility (when location information is obfuscated to decrease its sensitivity). Since in LBS service utility is usually dependent on the precision of provided location information, an important parameter to consider when analysing LBS privacy threats is the *precision*, in terms of time and location, required by the service and how the service utility degrades when the precision decreases.

   Table 4.1 reports a simplified classification of LBS in terms of required precision, single or continuous release, need of explicit identification and main adversaries (sharing parties). Here we only refer to online data release.

Table 4.1 A classification of location-based services (SP = Service Provider)

LBS type	Req. precision	Continuous	Explicit identity	Adversaries
POI services	High	No	No	SP
Weather/news	Low	No	No	SP
Navigation	High	Yes	No	SP
GeoSN posts	High	No	Yes	SP, users

The service required precision is particularly important to verify the adherence to the principle of *minimization* recommended by many regulators as we will explain in Sect. 4.2.1.

Regarding re-identification, a number of LBS require authentication and profiling, and in these cases we can assume that each request can be straightforwardly associated with an individual, at least by the service provider. In GeoSN, even in case of using pseudonyms, the information existing in public profiles is often sufficient to re-identify individuals. For anonymous services we should carefully consider the re-identification power of the location information transmitted as part of the request. Continuous or repeated requests may also be exploited to re-identify since certain frequent trajectories may be unique to individuals and joined with external information may lead to their identity.

4.1.4 Chapter Organisation

The rest of this chapter is organised as follows. In Sect. 4.2 we discuss regulation compliance and personal preferences as the main requirements for LBS privacy preservation. In Sect. 4.3 we provide an overview of the privacy protection techniques that have been proposed in the literature. We conclude the chapter in Sect. 4.4.

4.2 Compliance with Data Protection Regulation and Individual Privacy Preferences

When designing a LBS or when assessing the privacy impact of an already deployed LBS we need to consider the requirements imposed by the data protection regulation in the countries where the LBS is deployed. Indeed, location privacy protection, similarly to more general privacy protection, is regulated in many countries. In addition to adhering to legal obligations, the design and implementation of LBS should consider also user preferences since different individuals may have very different opinions about sharing their whereabouts with service providers and other users. From a service provider point of view, offering personalised privacy control and transparency through an effective interface may be a competitive advantage.

In the following we briefly introduce these topics providing some references for the interested reader.

4.2.1   A Legal Perspective

The regulation framework for handling geo-location data is fragmented within and across countries, a property unfortunately shared by other types of privacy [30], but particularly true for this type of data. A preliminary analysis of the type of location data being handled should determine if the data is associated with a specific individual and if it is acquired as part of *network traffic* data. In the first case regulations about personal data protection apply, and in the second regulations about data traffic in telecommunication networks also apply. In the following we provide a brief overview of the regulation concerning personal geo-location data, mostly focusing on US and EU.

In the recent past, some specific recommendations for geo-location services have been issued by the EU WP29 Working party, a group of experts and privacy regulation authorities by the EU member states [48]. However, the recommendations did not evolve into a specific regulation but were rather considered in the general data protection regulation. Indeed, in the EU, the principles that should guide the design of a LBS can be extracted from the General Data Protection Regulation (GDPR), approved in May 2016 [47]. Some general principles of the GDPR straightforwardly apply to geo-location data; for example, *privacy by design*, *data portability*, the need of an *informed user consent* (with few exceptions), and the right of the individuals to obtain, update, and even delete their own data. Some other principles can be relatively easily interpreted considering specifically geo-location data; for example, the location privacy interpretation of the *data minimization* principle says that the timestamped geographical position of a user should be acquired and stored only at the precision required for the service being offered. This is well exemplified by localised weather/news services that do not require high precision. Indeed, in a recent recommendation on data processing at work [49], the EU WP29 Working party states that "The information registered from the ongoing monitoring, as well as the information that is shown to the employer, should be minimized as much as possible. Employees should have the possibility to temporarily shut off location tracking, if justified by the circumstances. Solutions that for example track vehicles can be designed to register the position data without presenting it to the employer." Similarly to minimization, *privacy by default* requires the initial location privacy settings of a LBS to be the most protective among the ones available for the specific service. A useful starting point for browsing information about the EU privacy legislation is the *Protection of personal data* web page on the EU official website.[2]

---

[2]http://ec.europa.eu/justice/data-protection.

In the US, privacy is mostly regulated sector by sector. Regarding location privacy, several U.S. states have enacted laws establishing personal rights. However, current U.S. statute at the federal level does not provide clear protection of geolocation information. The bill known as Geolocation Privacy and Surveillance Act (GPS Act) has been proposed and discussed in congress. As a general principle, the act prohibits companies from collecting or disclosing geolocation information from an electronic communications device without the user's consent. It provides exceptions for parents tracking their children, emergency services, law enforcement, and other cases. Regulations have also been proposed to specifically prohibit development and distribution of "stalking apps," establish an Anti-Stalking Fund at the Department of Justice, and take other steps to prevent geolocation-enabled violence against women. A useful starting point for browsing information about the US is the Geolocation Privacy Legislation page on the GPS.org website.[3]

4.2.2   Privacy Preferences

A successful LBS should not only comply with the applicable regulation, but it should also consider user privacy preferences, and in particular location sharing preferences. Recent experimental studies suggest that the LBS user population has rich location privacy preferences, with a number of critical dimensions, including time of day, day of week, and location [5]. Clearly, another important dimension is who the information is shared with and the precision of the temporal and spatial information being shared. Indeed, the sensitivity of being in a location at a given time is often dependent on the semantics associated with the place, and this can be perceived differently by different individuals.

User preferences can also change over time, not only because users may become more confident in the service and trust more other users or the service provider, but because personal privacy preferences can change based on specific context (e.g., being on a tourist trip with respect to being at home, work or shopping).

The study in [36] also highlights differences between users in different countries in willingness to share location at "home" and at "work" and differences in the granularity of disclosures people feel comfortable with. Several formalisms to represent location privacy policies have been proposed (see e.g., [54]). However, the complexity of control mechanisms offered to LBS users has a clear trade-off with user experience aspects. Complex policy specification interfaces may easily lead to users relying on default settings. Examples of graphical interfaces to set temporal and spatial privacy preferences in mobile location sharing apps can be found in Locaccino [55] and PCube [18]. Some recent efforts have also focused on how to minimize the user intervention in setting and updating privacy preferences in mobile apps using machine learning and other techniques [17, 37].

---

[3]http://www.gps.gov/policy/legislation/gps-act/.

4.3   Methods and Techniques for Privacy Protection in LBS

In this section, we present and discuss basic methods and techniques for protecting user privacy while using LBS. We restrict ourselves to methods that can be applied *online* since the focus of this chapter is on providing privacy-preserving online LBS, such as services to find points of interest (POI), friend finders in geo-social networks, or online navigation services rather than protecting from the offline analysis of collected data stemming, for instance, from mobility traces collected by a mobile network operator.

Previous surveys on location privacy preserving techniques include [7, 8, 20, 31, 33, 35, 58]. In this section we present a systematized updated view of the research literature in this field.

Protecting user privacy while using LBS implies considering the LBS provider as one of the potential adversaries. Privacy can be protected in two fundamentally different ways:

- **Location-based k-anonymity**. A first approach is to hide the identity of the user since user anonymity would guarantee also the user privacy. Even assuming that we adopt effective anonymization techniques for IP addresses and other general information contained in the LBS queries, the spatio-temporal data contained in the queries can sometimes re-identify the user. In Sect. 4.3.1 we review the research work that has focused on avoiding this re-identification.
- **Sensitive location obfuscation**. Since anonymization is difficult to achieve and sometimes LBS require user identification for using the service, a second possibility to prevent private data leakage is to *restrict the amount of private information being released while interacting with the LBS*. The potentially sensitive information that is specifically released through LBS is the information about the whereabouts of the user: their location at given times. The location may be sensitive by itself (e.g., because the user was supposed to be somewhere else) or may indirectly reveal other sensitive information (e.g., religious or political orientation). In Sect. 4.3.2 we discuss techniques aimed at reducing the sensitivity of the spatio-temporal data in LBS queries by *obfuscating* that information in different ways while trying to preserve the quality of service.

Besides anonymization and obfuscation techniques, we discuss two other classes of approaches: The first uses **cryptographic methods** while the second follows the ideas behind **differential privacy** to devise techniques that provide quantifiable probabilistic guarantees independently from the background knowledge that the adversary may have.

4.3.1   Location-Based k-Anonymity

Anonymity is a general concept not restricted to LBS. So it is a valid question to ask why anonymization for LBS should be different from, say, anonymously visiting

Web pages. The answer is in the fact that location information can be used to de-anonymize users by serving as a so-called *location-based quasi-identifier* [9], which joined with some background knowledge that the adversary may have or acquire, can reveal the user identity. For instance, consider a mobile user called Alice searching for a hospital in her vicinity through a POI finder service. To this end, Alice transmits "anonymously", i.e., without explicitly specifying her true identity in the LBS request, her IP address—as part of the TCP connection—and current location to the LBS to enable the LBS to search for nearby hospitals. In this example, an obvious pseudo-identifier is the IP address of the mobile device of Alice. If the LBS provider has access to a database storing the IP-address-to-customer mapping stored at the Internet service provider of Alice, the LBS can reveal the identity of Alice and might as well conclude that Alice has health problems. Such an attack exploiting the network address can be avoided by using an anonymization service like TOR [13] based on onion routing [24], which forwards the request through a chain of anonymization servers, each one changing the sender IP address for the next "hop" in order not to reveal the initial sender's IP address. However, for an LBS this common anonymization of network addresses is insufficient since the user location might as well serve as quasi-identifier. For instance, assume that Alice sends the request from her home location. Using an easily available map and address book or telephone directory, the LBS can map the home location to Alice's identity, and if Alice is the only person living at that place, this mapping will be unique.

A solution to this problem, which has been applied first by Gruteser and Grunwald to LBS in [25], is the adaptation to the LBS context of the principle of *k-anonymity*. In general, the principle of *k*-anonymity requires that the individual (here Alice) must be indistinguishable from $k - 1$ other individuals such that the probability of identifying her is $\frac{1}{k}$. Applying *k*-anonymity in the above example requires the geographic user position to be enlarged to a *cloaking region* including $k - 1$ other users before sending the request to the LBS. This region can be calculated—in the simplest case—by a trusted anonymization service knowing all user positions. Then, even in the worst case in which the untrusted LBS provider can identify all the *k* users in the reported area, he can only tell that one of them searched for a hospital, and there is only a chance of $\frac{1}{k}$ that this user was Alice.

Obviously, one major challenge of spatial *k*-anonymity is the calculation of the cloaking region defining the anonymity set containing *k* users. Intuitively, the size of the area has an impact on the service precision and/or the anonymisation cost. Many different approaches have been proposed to address this problem. Several approaches use a hierarchical spatial partitioning like quadtrees to associate users with cloaking areas [4, 19, 25, 45]. Other approaches use space-filling curves [23] or nearest neighbor (NN) queries with randomization [32] to find groups of *k* users.

LBS queries generally include timestamps and indeed *location-based quasi-identifiers* are formally defined considering spatio-temporal information [9]. This naturally suggest defenses that generalize time as well as space leading to spatio-temporal cloaking algorithms. An example of temporal cloaking applied in addition to spatial cloaking is the CliqueCloak approach, which proposes to temporarily defer LBS queries [19].

Seeing all these different approaches, the question arises what essential properties are required by a *secure* LBS k-anonymity approach? While we know perfect security is not achievable, we can summarize the important properties that should be fulfilled:

- a cloaking algorithm satisfying reciprocity;
- a mechanism to take correlations among multiple requests into account;
- a trusted anonymizing service or a distributed approach;
- a formal proof that the mechanism is safe with respect to specific assumptions on the adversary's background knowledge.

The last property is the most critical simply because it is difficult to make realistic assumptions on the adversary's background knowledge. This is the main motivation behind the investigation of methods based on the notion of differential privacy that aim at solutions with probabilistic guarantees independent from the adversary's background knowledge. They will be discussed later on, while we now focus on the first three properties.

4.3.1.1 Reciprocity

Considering the first property, it is reasonable to assume that the adversary knows the algorithm for calculating the cloaking region. Then, all the spatial cloaking algorithms should prevent the adversary to use the resulting cloaking region and the algorithm to exclude possible locations of the individual within the region (or equivalently exclude any of the k individuals in the anonymity set). This property has been independently identified and named *reciprocity* by Kalnis et al. [32] and *inversion* by Mascetti et al. [43]. Intuitively, a cloaking algorithm C satisfies this property if each point contained in any cloaking region r computed by C is mapped to r itself.

In the case of k-anonymity, if reciprocity is not fulfilled, the adversary could identify the actual query issuer by executing the algorithm for each of the k users' position and comparing the resulting cloaked region with the one actually received.

4.3.1.2 Correlation Among Multiple Requests

Besides only considering a single query, also *correlation attacks* based on comparing cloaking regions and anonymity sets from *multiple* subsequent queries have been considered. Bettini et al. [9] first identified attacks based on correlating subsequent requests from the same anonymous user and intersecting the corresponding anonymity sets; They introduced the notion of *Historical k-anonymity*. Cloaking algorithms satisfying this property have been designed [43], despite experimental evaluations show that they can deal with very limited temporal sequences of requests after which they need mechanisms to safely change pseudonyms or use other methods to break the correlation.

4.3.1.3 The Diversity Problem

A different form of correlation, based on observing queries from *different* users, also shows the limits of the basic approaches to spatial k-anonymity. The reason follows the motivation for the introduction of l-diversity in data privacy [40]. Bettini et al. [6] illustrated the diversity problem in location k-anonymity by the following example and proposed a way for a trusted anonymiser to compute anonymity sets that avoids this problem.

Consider the LBS user Jane, that is in her office inside a building and she does not want her identity to be associated with the LBS she is requesting. Hence she uses a location k-anonymiser to avoid being re-identified through the coordinates of her own office. Suppose it is known that only Jane and Tom are inside that building at that time. Since she is happy with 2-anonymity, her location is cloaked to the area of the whole building. Suppose that by chance, Tom also asks for the same 2-anonymous service. Since the algorithm satisfies reciprocity the same cloaked region is used in the two requests.Then, the LBS provider, or any adversary that can see the anonymised queries, even if he does not know which of the two requests was sent by Jane will know that Jane (and Tom) asked for that (sensitive) service.

A slightly different notion of location l-diversity is introduced by Bamba et al. [4] for the case in which the location information in the LBS queries is not re-identifying but it is *sensitive* information.

Consider the LBS user Bob searching for a nearby taxi through a POI finder. Assume that Bob is currently located at a hospital. Even if there are $k - 1$ other patients at the hospital, a LBS query from Bob's current location (the hospital) would reveal that Bob might have health problems. The problem here is that all k users are located at the same sensitive location (hospital).

To avoid such inference, their PrivacyGrid approach applies spatial cloaking so that l different symbolic addresses (e.g., hospitals and *other* types of locations) are within the same cloaking region.

4.3.1.4 Trusted Anonymiser or Distributed Anonymization

Using a centralized service to calculate a cloaking region containing k potential users requires this service to be trusted, and introduces a potential bottleneck and single-point of failure with respect to availability and—more importantly—privacy. In particular, a compromised anonymization service reveals all true user positions. Therefore, decentralized approaches for spatial k-anonymity based on Hilbert space-filling curves have been proposed by Ghinita et al. [22, 23], to calculate anonymity sets distributedly between a set of "peers". Further distributed peer-to-peer approaches have been proposed, e.g., based on measuring the distance to other peers using WiFi signal strength and a scheme for distributively calculating the cloaking region by peers without revealing precise location information to other peers [29].

Among the proposals that do not require a trusted anonymiser, Kido et al. [34] proposed to locally generate, for each LBS query, a set of position dummies (*fake locations*) as the locations of other users and sending them to the LBS together with the true location. This is equivalent to issue multiple LBS queries for each original query. The results form the service are then locally filtered to retain only the ones related to the correct user location. The intention of the authors is to achieve a form of k-anonymity using $k - 1$ dummies, but the limited local knowledge does not guarantee that the generated fake positions are actual position of other potential users. Moreover, the adversary may use public information about geographical and street network constraints to exclude some of the dummy positions. The SybilQuery technique by Shankar et al. [51] follows the same approach, but it improves the quality of the dummies by using locations with similar traffic conditions, exploiting a database containing historic traffic, and traffic restrictions like one-way streets. However, the technique has similar limits.

4.3.2 Protecting Location Information by Obfuscation and Perturbation

In contrast to the anonymization techniques discussed in the previous section, obfuscation techniques do not try to hide the identity of the user. Indeed, there are several LBS requiring authentication and others for which the identity may be easily derived from other information in the request. Instead, these techniques aim at blurring or perturbing the location information contained in LBS requests because of its potential sensitivity.

As an example, the precision of location information can be decreased by translating precise point coordinates to geographic regions; Analogously, the precision of the temporal information usually associated with location can be decreased by converting precise timestamps into time intervals. Note that, as opposed to anonymity approaches, in this case the location is not enlarged in order to include other potential users, but to decrease the sensitivity, for example by including different types of semantic locations. This is a fundamental difference between the two protection approaches.

Although most approaches reduce precision by using areas that contain the user location, some approaches also reduce accuracy by sending a fake location, which might be specified very precisely, e.g., as a point coordinate, but deviates from the true user location.

Sending to a LBS inaccurate location and time information may impact the quality of service (QoS) of the LBS. For instance, searching for the nearest restaurant through an LBS might yield different (imprecise) results when given a larger obfuscation area rather than the precise position of the user. However, the QoS of the final result can be improved by post-processing imprecise or inaccurate results returned by the LBS by the same agent that performed the obfuscation, as for example, the app on the user's smartphone.

As a positive property compared to anonymization, obfuscation typically does not require an additional trusted infrastructure like, for instance, a trusted anonymization service as used by the centralized k-anonymity approaches discussed above. Instead, obfuscation (and answer filtering) can be performed on the user device alone, possibly assisted by some locally available information like maps.

Similarly to anonymity, an obfuscation defense should satisfy reciprocity, i.e., we must assume that the adversary knows the obfuscation algorithm, and it should be proved that this knowledge would not lead to any privacy leak. For instance, assume a naive deterministic obfuscation algorithm that simply creates a circular area centered at the true user position. Obviously, in this example an adversary can simply revert the obfuscation since the distribution of the user position is very skewed with a 100% chance of the user being at the center of the circle. Therefore, one of the challenges in designing a defense is to devise an algorithm withstanding attacks calculating the probability density function of the user location.

Jensen et al. [31] provide a good survey of obfuscation based techniques. In the following we summarise these techniques by grouping them as follows:

- Query enlargement techniques
- Dummy based techniques
- Coordinate transformation techniques

4.3.2.1 Query Enlargement Techniques

We group in this class all techniques that instead of including a specific location (and time) as part of the LBS query, they include a geographical area (and a time instant/interval) often called *obfuscation area*.

Ideally, the distribution of the user position within the spatio-temporal obfuscation area should be uniform not to give the adversary any hint about where the user might be located. However, this is not trivial to achieve since spatial constraints like streets, buildings, lakes, or forests might increase or decrease the probability of users being located in some parts of the obfuscation area. Correlations with other queries could also rule out some spatio-temporal areas.

A representative obfuscation approach based on transforming user positions in circular areas by applying a set of operators is the one proposed by Ardagna et al. [3]. The obfuscation operators can enlarge the circle, shifting the center of the circle, and shrinking the circle; the effects of applying these operators on the probability distribution of the user position is analyzed. Randomness can be introduced, for instance, by shifting the circle into a random direction. That work has been also extended by the authors considering background knowledge such as maps that might assist an adversary to find locations within the obfuscation area where the user might be located with higher probability. The obfuscation area is adapted, for instance, by increasing the radius of the circle, to compensate for the constraints given by a map leading to a non-uniform distribution over the obfuscation area.

Damiani et al. [12] also considered background knowledge in their obfuscation approach modeling the fact that positions are not uniformly distributed. Beyond this, they also considered the semantics of locations for calculating obfuscation areas, leading to different sensitivity of locations from a user's perspective. For instance, a user might not want to disclose that he is currently in a hospital, thus, the obfuscation area should include other non-sensitive locations leading to a low probability of the user being located in a hospital.

Spatial k-Group-Nearest-Neighbor (kGNN) queries over obfuscated location information are in the focus of [27]. Such a kGNN returns the "meeting point" minimizing the aggregated distance to all group members. For instance, an LBS could propose a restaurant minimizing the travel distance of a geographically distributed group of people. The privacy objective here is not to reveal the precise location to the LBS nor to other group members. To this end, each group member obfuscates his location by a rectangular area, which is sent to the LBS. Given imprecise locations of all group members, the LBS can only calculate a candidate set for the kGNN. This candidate set is post-processed by each group member sequentially to calculate the final kGNN.

A computational method that can be used both for anonymization and obfuscation through query enlargement has been presented by Mascetti et al. [42]. The method is agnostic about the semantics of the generalization function (for anonymity the semantics concerns the number of candidate individuals in the region, for obfuscation it may be the size of the area, the type of the area, the number of different pubs, etc...). Moreover, as opposed to most reciprocity-safe methods for finding generalized regions, it does not partition the space but uses an efficient bottom-up approach to find for each LBS query its generalised spatial region, called *Safebox*.

The trade-off between QoS and privacy as achieved by obfuscation is studied by Cheng et al. in [1]. Obviously, the definition of QoS essentially depends on the service offered by the LBS. In this work, the authors consider spatial range queries as a primitive frequently used by LBS. The authors assume that both the location of the query issuer as well as the locations of the queried objects, are obfuscated. In that case, answers to the range query are probabilistic, since some obfuscated locations might overlap with the queried range, and a precise answer by the LBS about the object being inside or outside the queried range is not possible.

4.3.2.2 Dummy Based Techniques

The idea of generating fake user positions proposed by Kido et al. [34] for k-anonymity can also be used to hide a possibly sensitive user location. The generation algorithm in this case has a different goal: the fake position should not resemble the position where another potential user is located, but it should be a non-sensitive location where the actual user could be. For example, instead of reporting a medical facility, the address of a grocery store across the street (or a set of such locations) is reported as the current location in the LBS query. Several dummy generation algorithms in this category are proposed by Lu et al. [39].

Related to dummy locations is the approach proposed by Duckham and Kulnik [14]; They show how to apply obfuscation to graph models. Graph vertices model locations, including the current user location, while edges model connections between locations such as roads, which can also have a weight to model some notion of distance. In their approach, the current location is obfuscated by sending as part of the LBS query a set of vertices representing dummy locations plus the actual location. Clearly, the more elements are contained in the set, the more imprecise is the obfuscated location. The LBS answers to proximity queries asking for the closest resource by performing computations on the graph model. The authors propose a negotiation protocol by which the LBS can ask for a smaller set of candidate locations in order to improve the quality of service.

Finally, the SpaceTwist approach by Yiu et al. [61] addresses the location privacy problem in answering k-Nearest-Neighbor (kNN) queries with a dummy-based *progressive retrieval technique*. Indeed, it generates a single dummy location called *anchor* and communicates only that location to the LBS. The distance of the dummy location from the real one is a parameter and it determines the achieved level of privacy. The SpaceTwist algorithm incrementally queries the LBS about the nearest objects for the same given anchor. These results are then filtered on client-side to find the actual k nearest neighbors for the true user position.

4.3.2.3   Coordinate Transformation Techniques

Another obfuscation method that has been explored is *coordinate transformations* [26]. Instead of creating obfuscation areas or dummy locations, coordinate transformations change the complete coordinate reference system using geometric transformations such that transformed coordinates cannot be interpreted by the adversary with respect to a "real-world" location on earth. However, the transformation should still allow for the LBS to answer the queries. For example, a friend finder service should still be able to evaluate proximity, i.e., the transformation should, at least approximately, preserve the distance. For instance, in [26] the authors outline how to use coordinate transformations for implementing basic spatial queries such as position queries, spatial range queries, and to detect spatial events such as "on entering area" or "on meeting" events. The idea is that the LBS managing mobile user positions performs query processing on transformed coordinates, while the transformation rules serve as shared secret between a user and other users or services with whom the user wants to share his location.

The essential challenge for coordinate transformation approaches is that an adversary can exploit background knowledge like maps and spatial distributions of locations to revert the transformation, i.e., to find the original location on earth given the transformed coordinates. In [38], the authors analyze distance preserving transformations as proposed in privacy preserving data mining. They conclude that approximate locations of users can be inferred based only on partial relative distance information and publicly available background knowledge about mobile object distributions. A specific attack to LBS protected by distance preserving transformations has been shown to be practical by Mascetti et al. [41].

4.3.3 PIR and Cryptographic Approaches

Private information retrieval (PIR) and cryptographic methods, namely, encryption, cryptographic hashing, secret sharing, and secure multi-party computation have also been considered to implement privacy-aware LBS. The basic objectives of these approaches are the same as for the approaches discussed above, namely, anonymity and sensitive location protection. However, by applying proven cryptographic methods, these approaches strive for stronger, provable privacy guarantees. The essential challenge is to allow for efficient processing of spatial queries at the LBS provider, although location information is not available in plain text to the provider (encryption and hashing methods), or despite the computational complexity of the cryptographic method (private information retrieval, secure multi-party computation).

Ghinita et al. in [21]. apply the concept of *Private Information Retrieval* (PIR) to an LBS implementing spatial nearest neighbor (NN) queries. The general idea of PIR is to privately retrieve data from a database without revealing which information has actually been requested. Applied to spatial NN queries, the goal is to retrieve the objects (POIs) nearest to the query issuer without revealing to the LBS which spatial region has actually been queried by the user. A naive solution would be to query the whole database, i.e., all POIs, however, obviously the overhead would be very high. Informally, PIR reduces this overhead by sending an encrypted query to the LBS not revealing what entry has been queried, but allowing the LBS to return a result significantly smaller than the whole database to the client, which then can be used by the client to calculate the value of the actually queried database entry. For mapping POIs to database entries, the authors use space-filling curves to preserve the spatial proximity required by NN queries. Khoshgozaran and Shahabi [33] provide a comprehensive survey of PIR approaches to LBS privacy preservation. Despite the solid theory, the PIR techniques have not yet been proven practical and scalable mainly for efficiency reasons.

In the application area of social network LBS, and in particular friend-finders, *secure multi-party computation* (SMC) has been used to implement protocols for computing proximity [62]. More generally, the basic objective of SMC is to jointly calculate a known function (e.g., proximity) by n participants, each participant providing a secret input to the function (e.g., position), without revealing the secret input to the other participants.

A cryptographic approach targeting location privacy in friend-finder services has been proposed by Mascetti et al. in [44]. The objective of this approach is to allow participants to issue queries to a central service for finding all friends within a given distance, while hiding to the service provider any information about their position and proximity of other users. Their method also allow the user to control the precision of the location information released to friends. To this end, proximity is computed by using a combination of cryptographic hash functions and SMC exploiting the commutative property of an encryption function. Location information is encrypted at different levels of granularity so that, according to privacy preferences, friends will only be able to infer the user's position with a given approximation. The system has been implemented in a prototype app, called *PCube*, that has been available both for iOS and Android devices.

Another cryptographic approach is based on the concept of *secret sharing*. The basic idea of secret sharing is to split a secret into a number of shares, say n. The secret can be revealed if a certain number of shares, say t, are known (so-called (t, n)-threshold scheme [50]). This concept is applied in [57] to implement a distributed location service managing locations of a user population and providing a set of LBSs with location information. To this end, locations—which can be geographic or symbolic locations—are defined as secrets. n shares are generated per location and distributed among n different servers. Consequently, in order to reveal the location, an adversary has to break into t servers, thus, avoiding a single point of failure. Moreover, by using a *multi-secret sharing scheme*, this approach supports providing location information of different precision levels, corresponding to a multi-secret, to different LBSs querying the location service.

4.3.4 Differential Privacy Approaches

Considering the difficulty of providing formal privacy guarantees independent from background knowledge for anonymity and obfuscation based approaches as well as the costs and applicability limitations of cryptographic approaches, a new type of methods has been proposed inspired by the success of the *differential privacy* notion in statistical databases.

Differential privacy has been introduced by Dwork [15] in statistical databases as a general method for the privacy preserving analysis of tabular personal data. The intuitive idea behind differential privacy is the following: Given two databases that differ only for the second including an additional record about an individual that is not present in the first, the information separately extracted from the two databases with a differential privacy method will not be significantly different. In other words the result of the analysis will be independent from the presence of information about the specific individual, hence it cannot be used in any way to violate her privacy. The way this result is achieved is by probabilistically inserting noise in the data. We refer to the original paper and to the rich literature on this topic, including other contributions in this book, for a formal definition and technical properties.

Differential privacy had a significant impact also on location privacy with natural applications to the offline analysis of location data, as in answering counting queries on a large dataset of user positions [11]. He et al. have also shown how to use differential privacy methods to synthesize mobility data based on raw GPS trajectories of individuals while ensuring strong privacy protection [28].

4.3.4.1   Differentially Private Methods for LBS

A more challenging task has been adapting the principles of differential privacy to online data release in LBS. (D, ϵ)-location privacy, illustrated by Elsalamouny and Gambs [16] results from adapting the adjacency relation in the standard differential privacy to the domain of locations. Two locations are considered "adjacent" if the distance between them is less than a predefined value D. In this context, a mechanism satisfies (D, ϵ)-location privacy if the (log of) the ratio between the probabilities of obtaining a certain output, from any two adjacent locations is at most ϵ. This property guarantees that the distinguishability between the location of the user and all the points that are adjacent is always restricted to a certain level quantified by ϵ.

A similar extension of differential privacy introduced by Andres et al. [2] is ϵ-geo-indistinguishability in which the bound on the distinguishability between two arbitrary positions increases linearly with the distance d between them. This means that the (logarithm of) the ratio between the probabilities of obtaining a certain output from two locations is at most d, which provides a low level of distinguishability (i.e., high privacy) between neighboring positions. In contrast, a higher level of distinguishability (i.e., low privacy) occurs for points that are further apart.

Analogously to the original differential privacy proposal, the way to achieve these properties is by inserting noise. In the LBS case this is done by *probabilistically determining a fake location that replaces the real location* when performing the LBS query. Different randomization functions can be used as long as they allow to prove the desired differential properties [16]. In the original proposal of ϵ-geo-indistinguishability *planar Laplacian* noise is used. The investigation of alternative randomization functions with more favorable trade-offs between privacy and utility is an active research area.

Finally, analogously to what we have seen for anonymity and obfuscation, differential privacy methods also have to deal with multiple (sequential) release of data and, more generally, with correlations that an adversary may exploit. A composition theorem for differential privacy says that we should consider the sum of the ϵ values associated with each release in the sequence. When considering LBS that require frequent or continuous queries this seems to imply that we would quickly reach unacceptable values of ϵ. A result consistent with what has been experimentally observed with the spatial cloaking for anonymity. An Interesting work on protecting locations from temporal correlations under differential privacy has been done by Xiao et al. [59, 60].

4.3.4.2 Analysing Trade-Offs Between Protection and Utility

The major critic to differential privacy is on the practical utility of the resulting mechanisms since keeping the ϵ parameter low happens at the expenses of the utility of the resulting query answers, which in the domain of LBS is the quality of service. A number of research efforts are directed to investigate this problem [10]. A natural question that still has no clear answer is what value of ϵ provides a good level of privacy. More theoretical work considers as good values close to 1, while applications seem to use quite higher values.

As part of research on finding optimal trade-offs between privacy and utility, Shokri [52] proposes a game theoretic approach to find an optimal location protecting mechanism while respecting each individual user's service quality requirements. Protection is achieved by a combination of differential privacy and distortion functions.

The application of game theory cited above for finding optimal trade-offs between privacy and quality of service has also been extended to deal with multiple releases, i.e., sets of queries that may reveal location traces [53].

4.4 Conclusions

In this chapter we provided a classification of the many location based services that are being offered today, and we illustrated the privacy threats that their users may face when using these services.

An important message that we would like to convey is that in order to understand if a given privacy protection method is adequate for a given service, it is necessary to carefully analyse the service in terms of the information being exchanged, the service architecture and the different parties to which the information is exposed, as well as to evaluate the requirements in terms of location data accuracy in order not to degrade the service quality. We also highlighted the importance of modeling the adversaries in terms of their access to LBS queries (single or multiple queries, sporadic or continuous) and in terms of the prior knowledge that they may have or acquire, including the knowledge of the privacy preserving algorithm and parameters.

In this chapter we also briefly reviewed the legal framework and personal privacy preferences as hard and soft requirements to be considered in the design of a defense technique. Finally, we provided a survey of the technical solutions proposed for on-line protection of LBS queries, hence focusing on techniques that aim to protect personal data before they reach the service provider, as opposed to offline techniques that aim at protecting the release of datasets from LBS providers to third parties, typically for statistical analysis. Other chapters in this book provide a deeper coverage of some of the approaches illustrated in Sect. 4.3 when applied to specific categories of LBS.

Overall, we can conclude observing that the online location privacy protection problem is a very challenging one, especially if considering the protection of trajectories as revealed by sequences of correlated LBS queries. The difficulty is mostly due to the uniqueness property of human trajectories [46] and to modeling realistic assumptions about the prior knowledge of the adversary.

Despite the protection proposals based on the notion of differential privacy have the advantage of providing provable probabilistic guarantees independent from the adversary's knowledge, their utility in terms of quality of service for many LBS is still to be demonstrated. An interesting research direction would be considering new probabilistic methods to insert location noise based on specific LBS deployment contexts, user preferences, and adversary model. Some of the research results obtained by the anonymization and obfuscation approaches may turn out to be applicable.

Acknowledgements We would like to thank Frank Dürr for providing a preliminary draft of the description of some of the established defense techniques as analysed in his survey [58].

References

1. Preserving user location privacy in mobile data management infrastructures. *Privacy Enhancing Technologies*, LNCS 4258:393–412, 2006.
2. M. E. Andrés, N. E. Bordenabe, K. Chatzikokolakis, and C. Palamidessi. Geo-Indistinguishability: Differential Privacy for Location-Based Systems. In *Proceedings of ACM SIGSAC Conference on Computer and Communications Security (CCS)*, 2013.
3. C. A. Ardagna, M. Cremonini, S. D. Capitani, and P. Samarati. An Obfuscation-based Approach for Protecting Location Privacy. *IEEE Transactions on Dependable and Secure Computing (TDSC)*, 8(1):13–27, 2011.
4. B. Bamba, L. Liu, P. Pesti, and T. Wang. Supporting anonymous location queries in mobile environments with PrivacyGrid. In *Proc. Int. WWW Conf.*, pages 237–246, 2008.
5. M. Benisch, P. G. Kelley, N. Sadeh, and L. F. Cranor. Capturing location-privacy preferences: quantifying accuracy and user-burden tradeoffs. *Personal and Ubiquitous Computing*, 15(7):679–694, 2011.
6. C. Bettini, S. Jajodia, and L. Pareschi. Anonymity and diversity in LBS: A preliminary investigation. In *IEEE International Conference on Pervasive Computing and Communications Workshops*, 2007.
7. C. Bettini, S. Mascetti, D. Freni, X. S. Wang, and S. Jajodia. Privacy and anonymity in Location Data Management. In F. Bonchi and E. Ferrari, editors, *Privacy-Aware Knowledge Discovery: Novel Applications and New Techniques*. Chapman & Hall, 2010.
8. C. Bettini, S. Mascetti, X. Wang, D. Freni, and S. Jajodia. Anonymity and historical-anonymity in location-based services. In *Privacy in location-based applications*, volume LNCS 5599. Springer Berlin Heidelberg, 2009.
9. C. Bettini, X. Wang, and S. Jajodia. Protecting privacy against location-based personal identification. In *Proceedings of Secure Data Management*, volume 3674 LNCS. Springer, 2005.
10. K. Chatzikokolakis, E. Elsalamouny, and C. Palamidessi. Efficient Utility Improvement for Location Privacy. *Proceedings on Privacy Enhancing Technologies (PoPET)*, 2017(4):210–231, 2017.

11. G. Cormode, C. Procopiuc, D. Srivastava, E. Shen, and T. Yu. Differentially private spatial decompositions. In *Proceedings - International Conference on Data Engineering*, pages 20–31, 2012.
12. M. L. Damiani, E. Bertino, and C. Silvestri. The PROBE framework for the personalized cloaking of private locations. *Transactions on Data Privacy*, 3(2):123–148, 2010.
13. R. Dingledine, N. Mathewson, and P. Syverson. Tor: The second-generation onion router. Technical report, Naval Research Lab Washington DC, 2004.
14. M. Duckham and L. Kulik. A Formal Model of Obfuscation and Negotiation for Location Privacy. In *International Conference on Pervasive Computing*, pages 152–170. Springer Berlin Heidelberg, 2005.
15. C. Dwork. Differential privacy: A survey of results. In *International Conference on Theory and Applications of Models of Computation*, pages 1–19. Springer, 2008.
16. E. Elsalamouny and S. Gambs. Differential Privacy Models for Location- Based Services. *Transactions on Data Privacy*, 9:15–48, 2016.
17. K. Fawaz and K. G. Shin. Location privacy protection for smartphone users. In *ACM Conference on Computer and Communications Security*, 2014.
18. D. Freni, S. Mascetti, C. Bettini, and M. Cozzi. Pcube: A system to evaluate and test privacy-preserving proximity services. In *2010 Eleventh International Conference on Mobile Data Management*, pages 273–275, May 2010.
19. B. Gedik and Ling Liu. Location Privacy in Mobile Systems: A Personalized Anonymization Model. In *25th IEEE International Conference on Distributed Computing Systems (ICDCS'05)*, pages 620–629, 2005.
20. G. Ghinita. Privacy for location-based services. *Synthesis Lectures on Information Security, Privacy & Trust*, 4(1):1–85, 2013.
21. G. Ghinita, P. Kalnis, A. Khoshgozaran, C. Shahabi, and K.-L. Tan. Private queries in location based services: anonymizers are not necessary. In *Proceedings of the 2008 ACM SIGMOD international conference on Management of data*, pages 121–132. ACM, 2008.
22. G. Ghinita, P. Kalnis, and S. Skiadopoulos. MOBIHIDE: a mobilea peer-to-peer system for anonymous location-based queries. *Advances in spatial and temporal databases*, pages 221–238, 2007.
23. G. Ghinita, P. Kalnis, and S. Skiadopoulos. PRIVE: anonymous location-based queries in distributed mobile systems. In *Proceedings of the 16th international conference on World Wide Web*, pages 371–380. ACM, 2007.
24. D. Goldschlag, M. Reed, and P. Syverson. Onion routing. *Communications of the ACM*, 42(2):39–41, 1999.
25. M. Gruteser and D. Grunwald. Anonymous Usage of Location-Based Services Through Spatial and Temporal Cloaking. In *Proceedings of the 1st international conference on Mobile systems, applications and services - MobiSys '03*, pages 31–42, 2003.
26. A. Gutscher. Coordinate transformation-a solution for the privacy problem of location based services? In *Parallel and Distributed Processing Symposium, 2006. IPDPS 2006. 20th International*, pages 7—pp. IEEE, 2006.
27. T. Hashem, L. Kulik, and R. Zhang. Privacy preserving group nearest neighbor queries. In *Proceedings of the 13th International Conference on Extending Database Technology*, pages 489–500. ACM, 2010.
28. X. He, G. Cormode, A. Machanavajjhala, C. M. Procopiuc, and D. Srivastava. DPT: Differentially Private Trajectory Synthesis Using Hierarchical Reference Systems. *Proceedings of the VLDB Endowment*, 8(11):1154–1165, 2015.
29. H. Hu and J. Xu. Non-exposure location anonymity. In *Proceedings - International Conference on Data Engineering*, pages 1120–1131, 2009.
30. R. P. Jay. Data protection & privacy in 31 jurisdictions worldwide. Gideon Roberton, Law Business Research Ltd, 2015.
31. C. S. Jensen, H. Lu, and M. L. Yiu. Location privacy techniques in client-server architectures. In *Privacy in location-based applications*, pages 31–58. Springer, 2009.

32. P. Kalnis, G. Ghinita, K. Mouratidis, and D. Papadias. Preventing location-based identity infer-
    ence in anonymous spatial queries. *IEEE Transactions on Knowledge and Data Engineering*,
    19(12):1719–1733, 2007.
33. A. Khoshgozaran and C. Shahabi. Private information retrieval techniques for enabling
    location privacy in location-based services. In *Privacy in Location-Based Applications*, volume
    0831505, pages 59–83. Springer Berlin Heidelberg, 2009.
34. H. Kido, Y. Yanagisawa, and T. Satoh. Protection of location privacy using dummies for
    location-based services. In *21st International Conference on Data Engineering Workshops.*,
    page 1248. IEEE, 2005.
35. J. Krumm. A survey of computational location privacy. *Personal and Ubiquitous Computing*,
    13(6):391–399, 2009.
36. J. Lin, M. Benisch, N. Sadeh, J. Niu, J. Hong, B. Lu, and S. Guo. A comparative study of
    location-sharing privacy preferences in the United States and China. *Personal and Ubiquitous
    Computing*, 17(4):697–711, Apr 2013.
37. B. Liu, M. S. Andersen, F. Schaub, H. Almuhimedi, S. Zhang, N. Sadeh, A. Acquisti, and
    Y. Agarwal. Follow my recommendations: A personalized privacy assistant for mobile app
    permissions. In *Symposium on Usable Privacy and Security*.
38. K. Liu, C. Giannella, and H. Kargupta. An attacker's view of distance preserving maps for
    privacy preserving data mining. *Knowledge Discovery in Databases: PKDD 2006*, pages 297–
    308, 2006.
39. H. Lu, C. S. Jensen, and M. L. Yiu. PAD: Privacy-Area Aware, Dummy-Based Location
    Privacy in Mobile Services. In *Proceedings of the CM International Workshop on Data
    Engineering for Wireless and Mobile Access (MobiDE)*, pages 16–23, 2008.
40. A. Machanavajjhala, J. Gehrke, D. Kifer, and M. Venkitasubramaniam. l-diversity : Privacy
    beyond k-anonymity. In *Proceedings - International Conference on Data Engineering*, 2006.
41. S. Mascetti, L. Bertolaja, and C. Bettini. A practical location privacy attack in proximity
    services. In *Proceedings - IEEE International Conference on Mobile Data Management*, 2013.
42. S. Mascetti, L. Bertolaja, and C. Bettini. SafeBox : adaptable spatio-temporal generalization
    for location privacy protection. *Transactions on Data Privacy*, 7:131–163, 2014.
43. S. Mascetti, C. Bettini, D. Freni, and X. S. Wang. Spatial generalisation algorithms for LBS
    privacy preservation. *Journal of Location Based Services*, 1(3):179–207, 2007.
44. S. Mascetti, D. Freni, C. Bettini, X. Wang, and S. Jajodia. Privacy in geo-social networks:
    Proximity notification with untrusted service providers and curious buddies. *VLDB Journal*,
    20(4), 2011.
45. M. F. Mokbel, C.-Y. Chow, and W. G. Aref. The new casper: Query processing for location
    services without compromising privacy. In *Proceedings of the 32nd international conference
    on Very large data bases*, pages 763–774. VLDB Endowment, 2006.
46. Y.-a. D. Montjoye, M. Verleysen, and V. D. Blondel. Unique in the Crowd: The privacy bounds
    of human mobility. *Scientific Reports*, 3(1376):1–5, 2013.
47. E. Parliament and Council. General data protection regulation - 2016/679. Technical report,
    European Commission, 2016.
48. A. D. P. W. Party. Opinion 13/2011 on geolocation services on smart mobile devices. Technical
    report, European Commission, 2011.
49. E. A. D. P. W. Party. Opinion 2/2017 on data processing at work - wp249. Technical report,
    European Commission, 2017.
50. A. Shamir. How to share a secret. *Communications of the ACM*, 22(11):612–613, 1979.
51. P. Shankar, V. Ganapathy, and L. Iftode. Privately querying location-based services with
    SybilQuery. In *Proceedings of the 11th international conference on Ubiquitous computing*,
    pages 31–40. ACM, 2009.
52. R. Shokri. Privacy Games : Optimal User-Centric Data Obfuscation. In *Proceedings on Privacy
    Enhancing Technologies*, pages 299–315, 2015.
53. R. Shokri, G. Theodorakopoulos, and C. Troncoso. *ACM Transactions on Privacy and Security
    (TOPS)*.

54. E. Snekkenes. Concepts for personal location privacy policies. In *Proceedings of the 3rd ACM conference on Electronic Commerce*, pages 48–57. ACM, 2001.
55. E. Toch, J. Cranshaw, P. Hankes-Drielsma, J. Springfield, P. G. Kelley, L. Cranor, J. Hong, and N. Sadeh. Locaccino: A privacy-centric location sharing application. In *Proceedings of the 12th ACM International Conference Adjunct Papers on Ubiquitous Computing - Adjunct*, UbiComp '10 Adjunct, pages 381–382, New York, NY, USA, 2010. ACM.
56. C. R. Vicente, D. Freni, C. Bettini, and C. S. Jensen. Location-related privacy in geo-social networks. *IEEE Internet Computing*, 15(3):20–27, 2011.
57. M. Wernke, F. Durr, and K. Rothermel. PShare: Position sharing for location privacy based on multi-secret sharing. In *2012 IEEE International Conference on Pervasive Computing and Communications*, pages 153–161, 2012.
58. M. Wernke, P. Skvortsov, F. Durr, and K. Rothermel. A classification of location privacy attacks and approaches. *Personal and Ubiquitous Computing*, 18(1):163–175, 2014.
59. Y. Xiao and L. Xiong. Protecting Locations with Differential Privacy under Temporal Correlations. In *Proceedings of the 22nd ACM SIGSAC Conference on Computer and Communications Security (CCS)*, pages 1298–1309, 2015.
60. Y. Xiao, L. Xiong, S. Zhang, and Y. Cao. LocLok: Location Cloaking with Differential Privacy via Hidden Markov Model. *Proceedings of the VLDB Endowment (VLDB)*, 10(12):1901–1904, 2017.
61. M. L. Yiu, C. S. Jensen, X. Huang, and H. Lu. SpaceTwist: Managing the Trade-Offs Among Location Privacy, Query Performance, and Query Accuracy in Mobile Services. In *IEEE International Conference on Data Engineering (ICDE)*, pages 366–375, 2008.
62. G. Zhong, I. Goldberg, and U. Hengartner. Louis, lester and pierre: Three protocols for location privacy. In *Privacy Enhancing Technologies*, pages 62–76. Springer, 2007.

Chapter 5
Analyzing Your Location Data with Provable Privacy Guarantees

Ashwin Machanavajjhala and Xi He

Abstract The ubiquity of smartphones and wearable devices coupled with the ability to sense locations through these devices has brought location privacy into the forefront of public debate. Location information is actively collected to help improve ad targeting, provide useful services to users (e.g., traffic prediction), or study human mobility/activity patterns and correlate them to the health of individuals. In this chapter, we highlight the privacy concerns in large-scale collections of location data from user-centric mobile devices and explain how simple cloaking based techniques might be ineffective. This motivates the need for algorithms that collect and analyze location data with formal provable privacy guarantees. We discuss the state of the art in specifying formal privacy guarantees for location data, as well as algorithms that achieve these formal privacy guarantees. We conclude with open research directions in this area.

5.1 Introduction

The advancement of location-sensing technology such as GPS together with mobile devices has brought forth numerous location-based applications to track, record and share individuals' locations. Long sequences of detailed location records about individuals are passively collected by organizations or actively shared by individuals. Analysis of this giant collection of location data for the benefits of individuals, business and society has been the focus of many research studies and applications. For instance, the human location patterns learned from taxi trips can help the discovery of important crossroads in a road network [54] and encourage vehicle pooling [45]. Location data has also been found predictive of human purchasing behavior [3], emergency behavior following large-scale disaster [47], and epidemiological patterns [50], and hence improves existing prediction models in

A. Machanavajjhala · X. He (✉)
Duke University, Durham, NC, USA
e-mail: ashwin@cs.duke.edu; hexi88@cs.duke.edu

© Springer Nature Switzerland AG 2018
A. Gkoulalas-Divanis, C. Bettini (eds.), *Handbook of Mobile Data Privacy*,
https://doi.org/10.1007/978-3-319-98161-1_5

many fields. These location-based studies can potentially enhance our understanding of human behavior and foster the development of tools to facilitate our life. To realize these potentials, location data has to be made available to the interested researchers or analysts. However, this data releasing process may reveal sensitive properties of individuals.

What are the special properties about location data of individuals compared to general tabular databases that make privacy protection challenging? The first property is that individuals' location data are highly identifiable. Montjoye et al. [14] showed using mobility data of 1.5 million individuals over a period of 15 months that approximately 95% of the individuals in this dataset can be uniquely identified by 4 spatio-temporal points, and the uniqueness of the location data decays insignificantly as their spatial and temporal resolution coarsens. The second property of location data is that individual's location patterns exhibit high predictability. Song et al. [46] found a more than 90% potential predictability in the future whereabouts of each individual despite heterogeneous travel patterns among the population.

Based on these properties, what kind of privacy guarantees can we hope to achieve when releasing location trajectories? A good notion of privacy for location data should consider adversaries with background knowledge. Even if the adversary knows a small set of location points visited by an individual, these points can help the adversary uniquely identify this individual from the location data, and infer the other sensitive locations visited by this individual. Moreover, since location trajectories are highly predictable, adversaries can leverage correlations between points in a user's trajectory to infer sensitive information even if the locations are coarsened [1, 28, 40, 55], or perturbed [4]. For instance, sensitive locations such as home and work addresses can be discerned easily based on the frequency with which locations are visited [5, 21], and perturbed locations can be reconstructed based on temporal correlations within the sequence of locations [53]. Next, linking these location records of individuals to public information can further reveal more about these individuals such as their health status based on their visits to hospitals. Ma et al. [35] showed that an adversary can infer an extended view of a user including the true identity in an anonymous trace with a small amount of side information with high probability. Thus, it is important to consider adversaries with background knowledge (about points in the trajectory as well as other side information) when quantifying the privacy loss of a method for sharing location trajectories. Finally, many applications (especially in upcoming IoT applications) require users to share their locations multiple times, or even periodically. For every release to be useful, more information about individuals must be disclosed each time. Hence, any method for sharing locations must be able to provide privacy guarantees across multiple releases and not just one release. A graceful degradation of privacy protection is highly desired over multiple releases.

Traditional location privacy preserving practices are mainly based on anonymization. For instance, k-anonymity removes identifiers and coarsens data values such that each individual is indistinguishable from $k - 1$ others. However, these practices fail to achieve the privacy desiderata discussed above. It is well known that k-anonymous releases do not protect against adversaries with background knowledge [38]. Besides, k-anonymous releases do not guarantee privacy under composition; i.e., two k-anonymous release can be combined to learn the sensitive locations of an individual exactly. Moreover, many anonymization algorithms are susceptible to attacks like the minimality attack [51], where the decisions made by the anonymization algorithm reveal information to the attacker. Therefore, in this work, we will present a formal framework that allows the releases or analysis of location data of individuals with provable privacy guarantees that can achieve these desiderata.

The rest of this chapter is organized as follows. Section 5.2 describes an important provable privacy notion that satisfies all these desiderata, known as *differential privacy*, and presents several variants of differential privacy for location data and corresponding algorithms for these variants. Section 5.3 introduces a more general privacy framework, *Pufferfish privacy*, which can capture all the variants of differential privacy for location data discussed in Sect. 5.2, explain the privacy semantics underlying these notions, and allow new and rigorous privacy definitions to be created based on the needs of different applications. A general algorithm to ensure Pufferfish privacy is also presented. However, not all the privacy definitions instantiated under Pufferfish privacy can guarantee privacy under composition. Hence, we present a special class of privacy notions instantiated under this framework, called *Blowfish privacy*, in Sect. 5.4. This privacy class guarantees privacy under composition, and allows users to tune privacy-utility tradeoffs by specifying privacy policies. We conclude in Sect. 5.5 with a discussion of challenges and some open research directions.

5.2 Differential Privacy

Differential privacy was first introduced in 2006 [15] as a promise to ensure the private information of an individual while allowing the learning of useful information about a population. This promise has quickly arisen as the state of the art privacy definition with a rich class of mechanisms satisfying it. Unlike anonymization, this privacy guarantee specifies a provable property of the privacy-preserving mechanisms and satisfies many of the privacy desiderata discussed in the previous section. In this chapter, we define differential privacy (Sect. 5.2.1), discuss variants of this definition in the context of location data (Sect. 5.2.2), and survey algorithms for differentially private release of location data (Sect. 5.2.3).

5.2.1 Definition and Properties of Differential Privacy

Let $\mathcal{I}$ be the set of all possible database instances, and let each database instance be a collection of record values/tuples. The variable r is used to represent a record

Table 5.1 Table of notation

$\mathcal{I}$	The set of possible database instances
D	A database instance belonging to $\mathcal{I}$
$\mathcal{T}$	The domain of tuples/record values in the database
t	A tuple/record value, a value in $\mathcal{T}$
$\mathcal{H}$	The set of all individuals. $\mathcal{H} = \{h_1, h_2, \ldots\}$
r_i	The record associated with individual h_i
$\mathfrak{Data}$	A random variable representing the true dataset (which is unknown to the adversary)
tuples($\mathfrak{Data}$)	The tuples in the database (record values without explicit reference to the identities of individuals)
records($\mathfrak{Data}$)	The identities and record values of individuals in the data
$\mathfrak{M}$	A privacy mechanism: a deterministic or randomized algorithm (ofter used in the context of a privacy definition)
N_{np}	The set of neighboring databases for unbounded differential privacy
N_{np}^{n}	The set of neighboring databases for bounded differential privacy
σ_i	$r_i \in records(\mathfrak{Data})$: The statement that the record r_i belonging to individual h_i is in the data
$\mathbb{S}$	Set of potential secrets. Revealing s or $\neg s$ may be harmful if $s \in \mathbb{S}$
$\mathbb{S}_{pairs}$	Discriminative pairs, $\mathbb{S}_{pairs} \subseteq \mathbb{S} \times \mathbb{S}$
$\mathbb{D}$	The set of evolution scenarios: a conservative collection of plausible data generating distributions
θ	A probability distribution. The probability, under θ, that the data equals D_i is $\Pr(\mathfrak{Data} = D_i \vert \theta)$
Σ	The spatial domain with distance metric $d(\cdot)$

and is associated with an individual h_i in the population $\mathcal{H}$. Let $\mathcal{T}$ be the domain for the record variable r, and a tuple $t \in \mathcal{T}$ be a value taken by a record. The data curator will choose a *privacy definition* and a *privacy mechanism* (algorithm) $\mathfrak{M}$ that satisfies that privacy definition. Then the data curator will apply $\mathfrak{M}$ to the data to obtain a sanitized output $\omega \equiv \mathfrak{M}(\mathfrak{Data})$, where $\mathfrak{Data}$ is the random variable representing the true database instance owned by the data curator which is unknown to the adversary. We use *records*($\mathfrak{Data}$) to denote the set of records in $\mathfrak{Data}$ and $t(\mathfrak{Data})$ to denote the record values (tuples). These notations and the key notations from the rest of this chapter is summarized in Table 5.1.

An algorithm satisfies differential privacy if adding, removing or changing a record in terms of the input does not significantly alter the output of the algorithm. More formally:

Definition 5.1 (Differential Privacy [15, 16]) Given a privacy parameter $\epsilon > 0$, a randomized algorithm $\mathfrak{M}$ satisfies ϵ-differential privacy if for any outputs $\omega \in range(\mathfrak{M})$ and all pairs of datasets D and D' in $\mathcal{I}$ that differ in one record (i.e. D can be derived from D' by either adding or deleting one record), the following holds:

$$\Pr[\mathfrak{M}(D) = \omega] \leq \exp(\epsilon) \Pr[\mathfrak{M}(D') = \omega], \qquad (5.1)$$

where the probability only depends on the randomness in $\mathfrak{M}$.

In this definition, the number of individuals in the database is unknown, and hence this definition is also known as *unbounded* differential privacy. When the number of the individuals is known in the database, the neighboring databases are defined as a pair of databases that differ in the value of only one individual's record, and the remaining individuals all have the same record values. This is also known as *bounded* differential privacy or *indistinguishability*. We represent the set of neighbors for unbounded differential privacy by N_{dp}, and the set of neighbors for unbounded differential privacy by N_{dp}^n, where n is the number of records in the database.

Intuitively, changing an individual's record value to the database for bounded differential privacy (or adding or removing an individual's record for unbounded differential privacy) has little impact on the distribution of the output of a randomized algorithm. The parameter ϵ is usually known as the privacy budget. When ϵ is small, the output distributions of $\mathfrak{M}$ are similar regardless of whether an individual's record value was used in the computation. The definition only applies to randomized algorithms, since it is easy to see that deterministic algorithms cannot satisfy this definition.

Laplace mechanism is an important building block for designing differentially private algorithms.

Definition 5.2 (The Laplace Mechanism) Given any function $f : \mathcal{I} \rightarrow \mathbb{R}^k$, the Laplace mechanism is defined as:

$$\mathfrak{M}_L(D, f(\cdot), \epsilon) = f(D) + (\eta_1, \ldots, \eta_k), \qquad (5.2)$$

where η_i are i.i.d random variables drawn from $Lap(\Delta f/\epsilon)$, and Δf is the l_1-sensitivity of the query f.

The l_1-sensitivity of the query f is a key concept for the Laplace mechanism, defined as the maximum difference in the query output between any two neighboring databases. Formally,

Definition 5.3 (l_1-Sensitivity) The l_1-sensitivity of a function $f : \mathcal{I} \rightarrow \mathbb{R}^k$ is

$$\Delta f = \max_{D, D' \in \mathcal{I}, (D, D') \in N_{dp}(\text{ or } N_{dp}^n)} ||f(D) - f(D')||_1, \qquad (5.3)$$

where $||x - y||_1$ denotes the l_1 norm of the difference between vectors x and y, and is defined as $\sum_i |x[i] - y[i]|$.

Intuitively, the Laplace mechanism adds noise that is large enough to hide the maximum difference in the query output between any two neighboring databases such that adversaries cannot distinguish the neighboring databases from the noisy output. Besides Laplace mechanism, there are many other algorithmic building blocks for differential privacy. Readers may refer to [18] for more details.

An important property of differentially private algorithms is that their composition also satisfies differential privacy.

Theorem 5.1 (Sequential Composition [15, 16]) *Let $D \in \mathcal{I}$ be an input database. Let $\mathfrak{M}_1(\cdot)$ and $\mathfrak{M}_2(\cdot, \cdot)$ be algorithms with independent sources of randomness that satisfy ϵ_1- and ϵ_2-differential privacy, resp. Then an algorithm that outputs both $\mathfrak{M}_1(D) = \omega_1$ and $\mathfrak{M}_2(\omega_1, D) = \omega_2$ satisfies $(\epsilon_1 + \epsilon_2)$-differential privacy.*

If the second algorithm $\mathfrak{M}_2$ does not access the raw data ($\epsilon_2 = 0$), but only applies on the output of the first algorithm, the provable privacy guarantee of the first algorithm after applying $\mathfrak{M}_2$ is unchanging. Formally,

Theorem 5.2 (Post-processing [15, 16]) *Let $D \in \mathcal{I}$ be an input database. Let $\mathfrak{M}_1(\cdot)$ be an algorithm that satisfies ϵ-differential privacy. Then if an algorithm $\mathfrak{M}_2$ is applied to the output of $\mathfrak{M}_1(\cdot)$, then the overall mechanism $\mathfrak{M}_2 \circ \mathfrak{M}_1(\cdot)$ also satisfies ϵ-differential privacy.*

All steps in the post-processing algorithm do not access the raw data, and hence they do not affect the privacy analysis. While it seems intuitive that postprocessing the output of a privacy algorithm should not result in additional privacy loss, there are some privacy metrics, like k-anonymity, that do not satisfy the postprocessing theorem.

Theorem 5.3 (Parallel Composition [15, 16]) *Let $D \in \mathcal{I}$ be an input database. Let $\mathcal{H}_1, \ldots, \mathcal{H}_p$ be disjoint subsets of individuals $\mathcal{H}$; $D \cap \mathcal{H}_i$ denotes the dataset restricted to the individuals in $\mathcal{H}_i$. Let $\mathfrak{M}_i$ be mechanisms that each ensure ϵ_i-differential privacy. Then the sequence of $\mathfrak{M}_i(D \cap \mathcal{H}_i)$ ensures $(\max_i \epsilon_i)$-differential privacy.*

These composition properties are very useful in proving the privacy guarantees of complex algorithms. Sequential composition theorem allows us to decompose an algorithm into a few sequential components, and then analyze each component separately. The parallel composition theorem enables us to analyze an algorithm that works on disjoint partitions of the data. The postprocessing theorem ensures that we only need to analyze the steps in the algorithm that actually touch the private database. Then the overall privacy guarantee of an algorithm over the entire database can be established with the two theorems above. These composition theorems are also very important, as they can address the impossibility result by Dinur and Nissim [51] that a database of size n can be reconstructed with high accuracy from the answers to $n \log(n)^2$ statistical queries even if each answer is perturbed with up to $o(\sqrt{n})$ error. Differential privacy conforms to this negative result as the privacy guarantee degrades as the number of sequential accesses to the data increases (according to the sequential composition result). Nevertheless, unlike k-anonymity, the privacy degradation is gradual and can be theoretically quantified. We would like to note that the sequential composition theorem holds (a) in the worst case and (b) even when the next query or differential private mechanism in the sequence is chosen adversarially and adaptively based on the answers to the previous queries. There are more sophisticated but advanced composition theorems [18]. When the queries are not adaptively chosen, tighter bounds on the privacy loss are known [18, 23, 32].

There are other privacy axioms which differential privacy and other good provable privacy notions can satisfy [36]. These properties make differential privacy an appealing choice for many privacy-aware applications and research.

5.2.2 Variants of Differential Privacy for Location Data

Differential privacy has arisen as a popular choice for privacy sensitive applications that use location data. We map the differential privacy definition to location data as follows. Consider Σ as the spatial domain for the location data with a distance metric, denoted by $d(\cdot)$. The spatial domain is usually a set of latitude-longitude coordinates, or a discretized 2-dimensional space, e.g. a uniform grid over a map. A location database $D \in \mathcal{I}$ consists of individuals with their location data. Each individual $h_i \in \mathcal{H}$ has a variable r_i to represent his or her location trajectories. If the events are recorded at regular time intervals, known as regular trajectories, $r_i[j]$ for $j = 1, 2, \ldots$, represents the jth event of individual h_i which takes a location value from the spatial domain Σ at time point j. Otherwise, each event has a temporal dimension in addition to the space domain, where privacy notions and techniques for regular trajectories can be adapted accordingly.

We will focus on regular trajectories and bounded differential privacy in this section. Neighboring databases for bounded differential privacy N_{np}^n differ in the record value/tuple for the record r_i of a single individual $h_i \in \mathcal{H}$ in the database. We can define neighboring databases in multiple ways for location data of individuals, and they result in distinct privacy notions with different levels of privacy protection. We describe these in detail below: (1) r_i can differ in one event with two different location values; or (2) differ completely in all the events of a single individual; or (3) differ in a short window of consecutive events. Therefore, these choices result in three key variants of differential privacy with details shown below.

- *Event-differential privacy* (**Event-DP**): In event-differential privacy, neighboring databases differ in only one single location (at a single time) of a single individual. Intuitively this definition ensures that the output of an algorithm is insensitive to changing one location at one time point. More formally,

  Definition 5.4 (Neighboring Databases for Event-DP) Databases D and D' are neighbors for event-DP if they differ in a single record r_i which takes values t in D and t' in D' such that

  $$|t| = |t'|, \text{ and if } t[j^*] \neq t'[j^*], \text{ then } \forall j \neq j^*, \ t[j] = t'[j] \tag{5.4}$$

Algorithms designed under this privacy notion are commonly applied in the scenarios where each individual has one or few sensitive events in the database.

- *User-differential privacy* (**User-DP**): In user-differential privacy, neighboring databases differ in the record of a single individual. Intuitively, this definition ensures that the output of an algorithm is insensitive to changing locations of an individual at any time point. More formally,

Definition 5.5 (Neighboring Databases for User-DP) Databases D and D' are neighbors for user-DP if they differ in a single record r_i which takes values t in D and t' in D' such that

$$t \neq t' \tag{5.5}$$

This protection is applicable to scenarios where the entire location sequences are released [10, 26, 37, 39, 56].

- *Window-differential privacy* (w-**event privacy/Window-DP**): This window-level protection takes in a privacy parameter w to specify how the neighboring databases differ. Intuitively, this definition ensures that the output of an algorithm is insensitive to changing of a window of w consecutive events of a single individual. More formally,

Definition 5.6 (Neighboring Databases for w-Event Privacy) Databases D and D' are neighbors for window-DP (or w-event privacy) if they differ in a single record r_i which takes values t in D and t' in D' such that

$$\forall j_1 < j_2, \text{ if } t[j_1] \neq t'[j_1] \ \& \ t[j_2] \neq t'[j_2], \text{ then } j_2 - j_1 + 1 \leq w$$

Hence, any pairs of neighbors that differ in an event window of length at most w are considered window-level neighbors. This variant is typically used in the streaming setting [29]. When the window size w is 1, this definition is equivalent to the event-level differential privacy.

In summary, these variants of differential privacy ensure the output of an algorithm is insensitive to different levels of changes in location data. Based on the levels of changes, user-DP offers the strongest privacy protection, following by w-event privacy and event-DP. User-DP is preferred over other variants when we would like to protect the properties of the entire location trajectory, for instance, to protect the home locations of an individual since it can reappear many times, or to protect the routines of an individual. w-event privacy suits the scenarios where short activities of an individual such as in a day or an hour require protection. Event-DP is applicable for one-time release of a single event of an individual. Moreover, by the sequential composition theorem, an event-DP algorithm simultaneously ensures both w-event DP and user-DP, albeit with a much larger value of ϵ. If an algorithm ensures ϵ-event DP, then it also satisfies $(w \cdot \epsilon, w)$-event DP, and $(T \cdot \epsilon)$-user-DP where T is the maximum possible number of events per individual in the database.

Finally, user-DP implies that only a finite number of queries can be answered, while a potentially infinite number of queries could be answered under event/window-DP.

5.2.3 Differentially Private Algorithms for Location Data

We have seen three variants of differential privacy for location data. Given the same query, algorithms that satisfy these different privacy guarantees can result in different utilities. Thus, in addition to tuning ϵ, one can navigate the privacy-utility tradeoff space by using these variants of differential privacy. We illustrate with the example of point queries. A point counting query across time asks for the number of events in D that occur at location $l \in \Sigma$, denoted by $f(D, l)$, (across all time), where Σ is the spatial domain.[1] Under event-DP, the l_1-sensitivity of $f(D, l)$ is just 1 as neighboring databases differ in at most one location of an individual, and hence the count for location l is affected by at most 1. Adding noise drawn from $Lap(1/\epsilon)$ satisfies ϵ-event-DP and the error in terms of the l_2 norm of the difference between the noisy answer and the true count is $\sqrt{2}/\epsilon$ in expectation. On the other hand, under user-DP, the l_1-sensitivity of $f(D, l)$ is T, where T is the maximum possible number of events per individual in the database. To ensure ϵ-user-DP, the noise added to the query is drawn from $Lap(T/\epsilon)$, and hence the answer has an expected error of $\sqrt{2}T/\epsilon$. Similarly, to ensure w-event privacy, adding noise from $Lap(w/\epsilon)$ to $f(D, l)$ is sufficient. This noise results in an expected error $\sqrt{2}w/\epsilon$, which is smaller than the error under user-DP, but larger than event-DP.

There are many interesting queries for location data, but we will focus on three important settings: (a) answering counting queries on a single snapshot in time; (b) answering counting queries in a streaming fashion; (c) synthesizing location trajectory databases. We will present corresponding algorithms for each setting.

5.2.3.1 Answering Counting Queries on a Single Snapshot in Time

In a snapshot location database, each individual has a single location. Hence, all the three variants of differential privacy described in Sect. 5.2.2 provide equivalent privacy protections. In this setting, besides point counting queries, range counting queries are commonly asked. A range counting query asks for the number of individuals in D within rectangle $R \subseteq \Sigma$, denoted by $f(D, R)$. We represent a set of counting queries by $\{f(D, l_i)\}_i$, where $l_i \in \Sigma$, and represent a set of range counting queries by $\{f(D, R_i)\}_i$, where $R_i \subseteq \Sigma$. A naive way to answer all possible point and range counting queries is to first obtain a differentially private answer to all point counting queries, i.e, $\{f(D, l)\}_{l \in \Sigma}$, using the Laplace mechanism. Then each range counting query can be answered by adding up all noisy counts of points

---

[1]The domain is assumed to be discrete, otherwise it can be discretized.

falling into the rectangle R. The summation step is a post-processing step which does not require the original data, and hence does not change the privacy guarantee. However, this approach injects too much noise to query answers. The expected error for this algorithm is $\sqrt{\frac{8|R|}{\epsilon^2}}$, and it can easily dominate the true count, especially when the range queries span large sparse rectangles.

To improve the query accuracy, many prior work [13, 25, 42, 43, 48, 56] proposed quad-tree based solutions. A quad-tree denoted by T is built from the partitioned spatial domain, where each node of the tree, v, is associated to a sub-region, denoted by $dom(v)$ and a noisy count for the number of events in D falling into that region, denoted by $\tilde{c}(v)$. The set of children of a node v is a partition of the region associated to v. The counts stored in T can be used to answer all possible range counting queries. Given range rectangle R, we first traverse T from the root and initiate the query answer 0. As traversing downwards, each node v is examined whether its associated region intersects with the query rectangle R. The count of v is considered only when $dom(v)$ intersects with R. If $dom(v)$ is fully contained in R, the answer is incremented by $\tilde{c}(v)$. If $dom(v)$ partially intersects R and v is not a leaf node, then every child of v that is not disjoint with R will be visited. If $dom(v)$ partially intersects q and v is a leaf node, then we inspect the data points in $dom(v)$, and the answer is incremented by the number of points contained in q. In this way, a range counting query associated with a large rectangle can be answered with few nodes and hence this approach gives a smaller amount of noise.

Given a fixed tree height h, the l_1 sensitivity for answering all counts in T is $2(h + 1)$. By the Laplace mechanism, adding noise drawn from $Lap(2(h + 1)/\epsilon)$ to the count of each node in T satisfies ϵ-DP. As both sensitivity and hence the amount of noise depends on the tree height h, existing work has made tremendous effort in improving the accuracy by exploring privacy budgeting strategy [13], correlations between noisy counts [13, 25] and pruning tree nodes [42, 43, 48] where the maximum tree height h is given, or by designing algorithms independent of h [56].

- **Optimizations with a given maximum tree height h.** Most of the prior work proposed algorithms [13, 25, 42, 43, 48] with the maximum tree height given. The first key optimization with a given maximum tree height is to distribute different privacy budget to each level of the tree [13] by applying the sequential and parallel composition of DP mechanisms. The intuition behind is that the nodes at a higher level have larger counts and hence are more resistant to perturbation while nodes at a lower level with smaller counts are less prone to noise. The baseline method that adds noise drawn from $Lap(2(h + 1)/\epsilon)$ to all nodes is equivalent to uniform budgeting by split ϵ uniformly across each level. Cormode et al. [13] aim to improve the total error injected to the tree T with given height h, by considering the total error as the sum of the node variances. The variance of the Laplace mechanism with parameter ϵ_i is $Var(Lap(\epsilon_i)) = 2/\epsilon_i^2$. Since the noise is independently generated in each node, the total variance is $Err(q) = \sum_{i=0}^{h} 2n_i/\epsilon_i^2$, where n_i is the number of nodes at height i contributed to the query q and ϵ_i is the privacy budget assigned to the nodes at height i. This error is minimized with the constraint that $\sum_{i=0}^{h} \epsilon_i = \epsilon$ when $\epsilon_i = 2^{(h-i)/3} \epsilon \frac{2^{1/3}-1}{2^{h+1/3}-1}$. This strategy corresponds to a geometric budgeting strategy where nodes at higher level receive smaller budgets, and the budget increases geometrically downwards the tree. Another popular optimization technique considers the consistency correlation between the noisy counts, thus to further reduce the total variance of the noise injected to the tree counts [13, 25]. The last optimization [42, 43, 48] is to prune nodes that have small counts and hence their descendants in the tree. This approach can introduce biased noise, but can reduce the total amount of noise with respect to the true counts.

- **Private partition without the maximum tree height.** Another research direction explores spatial partition without the specification of the maximum tree height. A recent work [56] adds a constant amount of noise (regardless of the maximum tree height) to a bias count of each node. If this noisy count is bigger than the threshold, this node will be further partitioned. After obtaining this partition, only the leaf nodes are published with their noisy counts. The intermediate nodes obtain their counts by summing up the counts of all the leaf nodes under them. This approach is the first algorithm that does not require the maximum tree height as an input. The constant amount of noise instead of height-dependent noise largely improves the accuracy of the query answer. The details of this algorithm can be referred to [56].

There are other data-dependent methods, such as kd-tree to partition the spatial domain based on other mechanisms. These algorithms can be referred to [13].

5.2.3.2 Answering Counting Queries in a Streaming Fashion

For infinite sequences of locations, stream counting queries have been well studied and are defined as a sequence of counting queries $(f(D[j], l))_{j=1,2,\ldots}$ for location $l \in \Sigma$ in database D at time stamp $j = 1, 2, \ldots$. Event-DP is a special case of w-event privacy where $w = 1$. User-DP does not apply here, as there is no bound on the maximum possible length of the sequence. However, an $(\epsilon/2)$-user-DP mechanism can be applied to disjoint subsequences of the stream prefix, where each subsequence has a length w. This ensures (ϵ, w)-event privacy, but this approach is not optimal.

Kellaris et al. [29] proposed a sliding window methodology. The overall mechanism denoted by $\mathfrak{M}$ which takes an input stream with prefix $D[1 : j]$ can be decomposed into j sub-mechanisms $\mathfrak{M}_1, \ldots, \mathfrak{M}_j$. Let each $\mathfrak{M}_k$ for $k \in [1 : j]$ generate independent randomness to achieve ϵ_k-event-DP. If for all $0 \leq j_2 - j_1 \leq w$, $\sum_{k=j_1}^{j_2} \epsilon_k \leq \epsilon$, then the overall mechanism $\mathfrak{M}$ satisfies w-event privacy. This means that if the sum of privacy budgets per sliding w-window is no more than ϵ, the overall mechanism ensures (ϵ, w)-event privacy.

With this sliding window methodology, two baselines were given by [29]: (1) uniformly allocating ϵ/w budget to each event so that the sum of the budget per sliding window is always ϵ; (2) publishing a single event with privacy budget ϵ for every w timestamps. The first baseline does not work well if w is large as each event is given a small budget. The second baseline approximates the other unpublished counts from the released one. If the released count is very different from the others, the overall estimation is very poor. In order to address these shortcomings, the same work [29] proposed to skip publications of counts that are similar to previously released ones.

In the new solutions, each sub-mechanism $\mathfrak{M}_j$ has two parts $\mathfrak{M}_{j,1}$ and $\mathfrak{M}_{j,2}$. The first part $\mathfrak{M}_{j,1}$ differentially privately computes the distance between the current count and the preceding released counts. If the distance is small, the publication of this count is skipped; otherwise then the second part $\mathfrak{M}_{j,2}$ releases the noisy count with part of the remaining budget available for the current sliding window. There are two ways to deal with the privacy budget for events within a window: (1) budget distribution, and (2) budget absorption. Both schemes assign some budget ϵ_1/w to $\mathfrak{M}_{j,1}$ for computing distance differentially privately at time stamp j, and use the remaining budget ϵ_2 for publishing counts, where $\epsilon_2 = \epsilon - \epsilon_1$. We will see how the publication budget ϵ_2 is spent within each sliding window.

- **Budget Distribution**: The publication budget ϵ_2 is distributed in an exponentially decreasing fashion to the timestamps where a publication is decided to occur. Formally, at each timestamp, remaining budget is computed as $\epsilon_{rm} = \epsilon_2 - \sum_{k=j-w+1}^{j-1} \epsilon_{k,2}$, where $\epsilon_{k,2}$ is the privacy budget assigned to each of the last $w - 1$ aggregated statistics. Then a Laplace noise with a budget of $\epsilon_{rm}/2$ is added to the query output. If a publication is skipped, its budget is saved and spent in timestamps falling outside the active window. If there are m publications per window, the sequence of budget can be $\epsilon_2/2, \epsilon_2/4, \ldots, \epsilon_2/2^m$.
- **Budget Absorption**: This scheme uniformly distributes the publication budget to all timestamps. If it decides not to publish at a timestamp based on the noisy distance, the corresponding budget becomes available for future publication. If it decides to publish at a timestamp, it absorbs all the budget that became available from the previous skipped publications. This allows higher accuracy for the current statistics. To ensure the total budget within a window not exceeding the maximum ϵ_2, after the absorption of budgets from previous timestamps, the same amount of budget must be nullified from the immediate succeeding timestamps.

Both mechanisms satisfy w-event privacy. There are no theoretic guarantees that they can do better than the baseline methods, but the experiments in [29] show their superiority over the baselines in most of their settings. In general, this sliding window methodology highly depends on the choice of w for both privacy and utility. It remains an interesting question that what w should be set for each application.

The algorithms discussed so far can achieve w-event privacy (and hence event-DP) in a streaming setting. There are more event-DP algorithms [6–8, 11] developed for streaming setting. Among them, only PeGaSus [11] can simultaneously support a variety of stream processing tasks—counts, sliding windows, event monitoring—over multiple resolutions of the stream, and outperform the other solutions specialized to individual queries. These event-DP algorithms can also be extended to user-DP algorithms when each user has a limited number of contributions to the streaming data. If each user contributes a count of 1 at most l times to the entire streaming setting, then an ϵ-event-DP algorithm can automatically guarantee $l\epsilon$-user-DP. This assumption is valid for certain scenarios. For instance, in a hotel, most customers stay there for a few days. There are also algorithms that summarize or sample user's information so that their contributions to the streaming data is bounded [20], but may result in poorer data quality.

5.2.3.3 Synthesizing Location Trajectory Databases

Synthetic location databases are important for applications and research in city/-traffic planning, epidemiology, and location-driven advertising, especially when the analysis cannot be limited to a set of counting queries. The synthetic data also keeps the same format of the true data such that data analysts do not have to adapt to a new tool for exploring the private data. Synthesizing location databases corresponds to a non-interactive setting. Under this setting, we learn a model first from the original ground truth and then generate a synthetic database from the model. Depending on the privacy definition, the sensitivities of the queries used to compute the sufficient statistics of the model will change. We will focus on user-DP here for databases of location sequences, but the techniques presented can be extended for event/window-DP. Counting queries for sub-sequence of locations are common queries used for building the model. If the entire sequence per user is short and fixed, e.g. home location and work location, the l_1-sensitivity for the sub-sequence counting queries is small, and hence the privacy budget can be split over different sets of queries. Related work can be found in [37, 39, 56]. On the other hand, if sequences are long, a Markov process is commonly considered [10, 26, 56] to model the correlation between the events. Formally,

Definition 5.7 (Markov Process) A sequence of locations $(l_1 l_2 \cdots l_n) \in \Sigma^n$ is said to follow an order ℓ Markov process if for every $\ell \leq j < n, l \in \Sigma$

$$\Pr[l_{i+1} = l | l_1 \cdots l_i] = \Pr[l_{i+1} = l | l_{i-\ell+1} \cdots l_i]. \tag{5.6}$$

We refer to the probability $\Pr[l_{i+1} = l | l_{i-\ell+1} \cdots l_i]$ as a *transition probability* of the Markov process. The collection of transition probabilities for all $x = l_{i-\ell+1} \cdots l_i \in \Sigma^\ell$ can be estimated using the set of all ℓ- and $\ell+1$-gram counts, i.e.

$$\Pr[l_{i+1} = l | l_{i-\ell+1} \cdots l_i] = \frac{f(D, xl)}{f(D, x)}, \tag{5.7}$$

where $f(D, x)$ denotes the number of occurrences of x in the database D. Starting symbols ($\top$) and stopping symbols ($\bot$) are prepended and appended (respectively) to the original trajectories to capture the starting and stopping probabilities in the Markov process. The synthesis of a trajectory begins with a starting symbol ($\top$). Based on the transition probabilities from the Markov process, a next location is sampled continuously till reaching the stopping symbol ($\bot$). This model requires to maintain all ℓ-gram counts for $1 \leq \ell \leq h$, where $h - 1$ is the maximum order of Markov process considered. A *prefix tree T* of heights h is used to store these counts, where nodes in T are $\Sigma^1 \cup \ldots \cup \Sigma^h$, and edges connected each ℓ-gram x to $\ell + 1$-gram xl for all $l \in \Sigma$.

To ensure user-DP, prior approaches add noises drawn from a Laplace distribution to parts of the prefix tree T [10, 56]. These prior work performed well for small domain, and can be applied to continuous spatial domains by discretizing locations (e.g. via a uniform coarse grid). However, they failed to scale to realistic location sequences that span large geographical regions. Though a sufficiently fine discretization of the spatial domain can capture all the mobility patterns in the data, this discretization results in very large domain sizes (of several tens of thousands), and hence making the model fitting procedure very slow and overfitting the data. Moreover, the amount of noise added to ensure differential privacy also grows with the number of nodes in the tree. On the other hand, if a coarse discretization of the space is used for a small prefix tree, then much of the spatial correlation information in the original trajectories is lost. Hence, He et al. [26] proposed an end-to-end system, named Differentially Private Trajectories (*DPT*) to address these challenges. The schematic overview of this system is shown in Fig. 5.1.

DPT discretizes the spatial domain at multiple resolutions to capture different step sizes (see Step 1 in Fig. 5.1). Every resolution has a prefix tree (Step 2). Within each resolution, only movements from each grid cell to neighboring cells in one step are allowed. Though there is a *larger* number of prefix trees, each prefix tree has a much *smaller* branching factor, thus resulting in a big reduction in the number of counts maintained by the model. DPT uses a novel model selection algorithm (Step 3) to set the tree heights and to prune unrealistic resolutions in a differentially private manner. The following steps add noises drawn from the Laplace distribution to the chosen prefix trees (Step 4), and prune adaptively these noisy trees (Step 5) to further improve utility. In the last sampling step (Step 6), a novel postprocessing strategy is applied by DPT to restore the directionality of synthetic trajectories which could be lost due to the noise added to the private model. Based on these optimizations, this end-to-end system can synthesize trajectories spanning large geographical areas with significantly more utility than the prior work [10] and is orders of magnitude faster. These synthetic trajectories have been shown mirroring the original trajectories on three utility metrics—distribution of diameter (i.e., distance traveled), conditional distributions of destinations given starting regions, and frequent patterns. However, synthetic trajectories cannot join with other datasets due to the absence of join keys. Prior work [16] has shown that non-interactive setting can have more error than an interactive setting, main due to the difficulty of supplying utility that has not yet been specified at the time the data synthesis is carried out. Moreover, additional efforts have to be applied to synthetic trajectories such that they are realistic and satisfying real-world constraints.

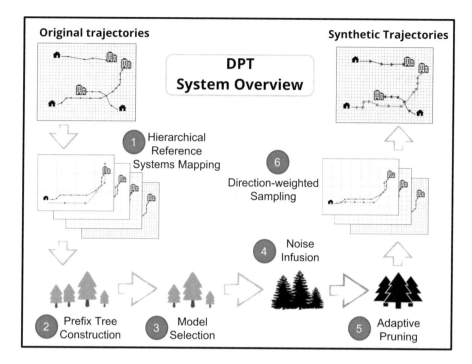

Fig. 5.1 DPT framework overview

In summary, from these prior work [4, 10, 13, 19, 26, 29, 37, 39, 49, 53, 56], we see that differential privacy has been well explored for location data. Applications considered various forms of neighboring databases and hence different algorithms. However, it is not clear that (1) what information is protected via the different specification of neighboring databases; and (2) which algorithms can be used to protect against adversaries with prior background knowledge. Moreover, the algorithms presented so far are designed mainly for counting queries. For location-based services that require location information at a particular time stamp, even event-DP, the weakest privacy notion seen so far, is too strong to provide good accuracy. Hence, we will first see how to quantify the privacy guarantees of algorithms through the lens of secrets and adversaries using a more general framework called *Pufferfish* in Sect. 5.3. Then, we will show how to design algorithms for location-based services that can achieve both accuracy and provable privacy guarantees named as *Blowfish Privacy* under this framework in Sect. 5.4.2.

5.3   Pufferfish Privacy

In this section, we summarize the Pufferfish privacy framework [31] that generalizes differential privacy, helps understand the privacy semantics underlying privacy definitions, and create new privacy definitions customized to the requirements of an application. In Sect. 5.3.1, we introduce how the Pufferfish framework defines privacy in terms of secrets and (the prior knowledge available to) adversaries, rather than neighboring databases. In Sect. 5.3.2, we show that variants of differential privacy are instantiations of the Pufferfish framework, and explain under what assumptions about secrets and adversaries each of these variants ensure *semantic* privacy guarantees. Finally, in Sect. 5.3.3, we describe algorithms for general Pufferfish privacy definitions.

5.3.1   Definition of Pufferfish Framework

Pufferfish framework requires domain expert to specify three components: (1) a set of *potential secrets* $\mathbb{S}$, (2) a set of *discriminative pairs* $\mathbb{S}_{pairs} \subseteq \mathbb{S} \times \mathbb{S}$, and (3) a collection of data *evolution scenarios* $\mathbb{D}$. The specification of these three components in this framework gives a rich class of privacy definitions.

- **The set of potential secrets** $\mathbb{S}$ represents the information that data curator would like to protect. A secret can be specified as a statement such as "*Bob is at location $l \in \mathcal{T}$*", "*Bob is not at location $l \in \mathcal{T}$*". In general, a domain expert should add a statement s to the potential secrets $\mathbb{S}$ if either the claim that s is true or the claim that s is false can be harmful. The resulted $\mathbb{S}$ forms a domain for the *discriminative pairs*, a subset of $\mathbb{S} \times \mathbb{S}$.
- **The set of discriminative pairs** $\mathbb{S}_{pairs}$, is a subset of $\mathbb{S} \times \mathbb{S}$. The role of $\mathbb{S}_{pairs}$ is to tell how to protect the potential secrets $\mathbb{S}$. For any discriminative pair $(s_i, s_j) \in \mathbb{S}_{pairs}$, we would like to guarantee that adversaries are unable to distinguish between the case where s_i is true of the actual data and the case where s_j is true of the actual data. For this reason, s_i and s_j must be mutually exclusive, but not necessarily exhaustive (it could be the case that neither is true). One example of a discriminative pair is ("Bob is at location l_1", "Bob is at location l_2"), where $l_1 \neq l_2$, or ("Bob is at location l_1", "Bob is not at location l_1"), where $l_1, l_2 \in \mathcal{T}$. The set of changes for neighboring databases shown in Sect. 5.2.2 are examples for the set of discriminative pairs.

    This specification allows highly customizable privacy guarantees. For instance, many location-based applications such as OpenPaths [41] and Airbnb [2] state in their policies that user's location information will only be shared or collected at coarse granularity. This property can be specified by pairs of secrets, such as ("Bob is at location $l_1 \in \mathcal{T}$", "Bob is at location $l_2 \in \mathcal{T}$), where l_1 is 21 miles away from l_2. Or if users are fine with releasing their location at city-level, but not at any street level within a city [34], this privacy preference can be expressed via a set of discriminative pairs $\mathbb{S}_{pairs}$ that exclude pairs of secrets like ("Bob is at Durham", "Bob is at New York"), but includes pairs of secrets with nearby places, such as ("Bob is at a cafe in Durham", "Bob is at home in Durham").

• **The evolution scenarios** $\mathbb{D}$ can be viewed as a set of conservative assumptions how the data evolved (or were generated) and about knowledge of potential adversaries. Note that assumptions are absolutely necessary—privacy definitions that can provide privacy guarantees without making any assumptions provide little utility beyond the default approach of releasing nothing at all [17, 30]. In order to release useful information about the database, the domain expert should be able to identify a reasonable set of assumptions. In many cases, they already do this informally [44]. Formally, $\mathbb{D}$ is represented as a set of probability distributions over $\mathcal{I}$ (the possible database instances). Each probability distribution $\theta \in \mathbb{D}$ corresponds to an adversary that we want to protect against and represents that adversary's belief in how the data were generated (incorporating any background knowledge and side information). For $D \in \mathcal{I}$, we use the notation $\Pr(\mathfrak{Data} = D|\theta)$ to represent the probability, under θ, that the true database is D. Below we give some examples of possible choices of $\mathbb{D}$ and their interpretations.

Example 5.1 (No Assumptions) $\mathbb{D}$ can consist of all possible probability distributions over database instances (i.e. including those with arbitrary correlations between records). This corresponds to making no assumptions.

Example 5.2 (Independent Individuals but Markov Model-Based Events) Several work [49, 53] consider that individuals are independent, but the events per individual are correlated by Markov model. The individuals in the database are independent of each other, that is, $\mathbb{D}$ consists of all θ for which

$$\Pr[\mathfrak{Data} = \{r_1, \ldots, r_n|\theta\}] = f_1(r_1) \times f_2(r_2) \times \ldots \times f_n(r_n) \qquad (5.8)$$

for arbitrary $f_1, f_2, \ldots, f_n$. The correlation within the events of an individual is modeled by a transition matrix or a class of transition matrices P_θ for $\mathbb{D}$, where each (l_1, l_2)th entry of this matrix specifies the probability an individual h_i being at location l_2 at time stamp j given the previous $(j-1)$th event, i.e. $\Pr(r_i[j] = l_2|r_i[j-1] = l_1)$.

Readers may refer to [31] for more examples of data evolution scenarios.

To use the Pufferfish framework, the domain expert simply does what he or she does best, and is no longer required to be a privacy expert. After specifying the assumptions explicitly, the corresponding Pufferfish privacy instance is formally stated as follows.

Definition 5.8 (Pufferfish Privacy [31]) Given a set of potential secrets $\mathbb{S}$, a set of discriminative pairs $\mathbb{S}_{pairs}$, a set of data evolution scenarios $\mathbb{D}$, and a privacy parameter $\epsilon > 0$, a potentially randomized algorithm $\mathfrak{M}$ satisfies ϵ-Pufferfish

$(\mathbb{S}, \mathbb{S}_{pairs}, \mathbb{D})$ privacy if (i) for all possible outputs $\omega \in range(\mathfrak{M})$, (ii) for all pairs $(s_i, s_j) \in \mathbb{S}_{pairs}$ of potential secrets, (iii) for all distributions $\theta \in \mathbb{D}$ for which $\Pr(s_i|\theta) \neq 0$ and $\Pr(s_j|\theta) \neq 0$, the following holds:

$$\Pr[\mathfrak{M}(\mathfrak{Data}) = \omega|s_i, \theta] \leq e^{\epsilon} \Pr[\mathfrak{M}(\mathfrak{Data}) = \omega|s_j, \theta] \tag{5.9}$$

$$\Pr[\mathfrak{M}(\mathfrak{Data}) = \omega|s_j, \theta] \leq e^{\epsilon} \Pr[\mathfrak{M}(\mathfrak{Data}) = \omega|s_i, \theta], \tag{5.10}$$

where $\mathfrak{Data}$ is a random variable representing the true dataset (which is unknown to the adversaries).

5.3.2 Relation to Differential Privacy

Recall that the definition of differential privacy is based on neighboring databases by changing an individual's record. This definition is a condition of a randomized algorithm—the output of the randomized algorithm is insensitive to the change of an individual's record to the database. In this definition, there is no mention or assumption of data evolution scenarios known by the adversaries. In this section, we would like to show how to understand and analyze differential privacy in the framework of Pufferfish.

Consider the following specifications. Let $\mathcal{H} = \{h_1, h_2, \ldots, h_N\}$ be the set of all individuals in a population of size N. Define σ_i be the statement $r_i \in records(\mathfrak{Data})$ (i.e. "records r_i belonging to individual h_i is in the data", and let $\sigma_{(i,t)}$ be the statement $r_i \in records(\mathfrak{Data}) \wedge r_i = t$ (i.e. "record r_i belonging to individual h_i has value t and is in the data"). Let the set of secrets and the set of discriminative secret pairs be specified respectively as

$$\mathbb{S} = \{\sigma_{i,t} : h_i \in \mathcal{H}, t \in \mathcal{T}\} \cup \{\neg\sigma_i : h_i \in \mathcal{H}\} \tag{5.11}$$

$$\mathbb{S}_{pairs} = \{(\sigma_{i,t}, \neg\sigma_i) : h_i \in \mathcal{H}, t \in \mathcal{T}\} \tag{5.12}$$

This specification of secret pairs aim to prevent an adversary from distinguishing whether the record r_i associating with h_i is in the data and has the value t v.s. the record about individual h_i is not in the data, for any individual h_i in the population $\mathcal{H}$, and any possible tuple value $t \in \mathcal{T}$. Consider the data evolution scenario $\mathbb{D}$ where all individuals are independent (including their presence/absence in the data and their tuple values if present in the data). This distribution can be specified as

$$\Pr[\mathfrak{Data}|\theta\}] = \prod_{r_i \in records(\mathfrak{Data})} f_i(r_i) \Pr[\sigma_i] \prod_{r_i \notin records(\mathfrak{Data})} (1 - Pr[\sigma_i]), \tag{5.13}$$

where $f_i(r_i)$ is the distribution for the value taken by record r_i of an individual h_i, and $\Pr[\sigma_i]$ is the probability of the record of an individual being in the data.

Theorem 5.4 ([31]) *With the choices of $\mathbb{S}$ and $\mathbb{S}_{pairs}$ defined in Eqs. (5.11) and (5.12), and the set of data evolution scenarios $\mathbb{D}$ as specified in Eq. (5.13), unbounded ϵ-differential privacy is equivalent to ϵ-Pufferfish($\mathbb{S}$, $\mathbb{S}_{pairs}$, $\mathbb{D}$).*

The variants of differential privacy for location data described in Sect. 5.2.2 can all be described under this framework. For instance, event-DP can be shown to be equivalent to a Pufferfish instantiation where: (1) the set of secrets are properties of an individual's location at a single time point; and (2) adversaries may know arbitrary prior knowledge about an individual's location at each time point, but do not know correlations across time points (or across trajectories). On the other hand, user-DP can be shown to be equivalent to an instantiation where: (1) the set of secrets are properties of the entire trajectory; and (2) adversaries may have arbitrary prior knowledge about a user's trajectory (including correlations across time points), but assume that there are no correlations across trajectories.

This means that event-DP algorithms are susceptible to attacks when adversaries know constraints or correlations between consecutive locations in a trajectory. Consider a single user's location sequence, and consider adversaries who know that the individual stayed at the same location for a long period of time, e.g. at home in the evening. Event-DP that adds noise with standard deviation about $1/\epsilon$ to the histogram counts of locations over time cannot hide the evidence of that location. While user-DP does protect against such attacks, it may be an overkill. We can use Pufferfish to design new privacy definitions that match such adversaries. For instance, if one wants to hide properties of individual time points, but handle correlations, one could use the same secrets as Event-DP, but handle more complex adversaries as defined in Example 5.2. Let's name this privacy *Event-MarkovAdversary-Privacy*. There are algorithms like the Markov Quilt Mechanism described next, that can ensure more privacy than event-DP, and more accuracy than user-DP, helping us better tradeoff privacy and utility.

5.3.3 Algorithms for Pufferfish Privacy

We first present a special algorithm for location data which consider adversaries with assumptions shown in Example 5.2, that individuals are independent but sequences of events are correlated and are modeled by Markov model. Then, we will present a general algorithm for Pufferfish privacy.

5.3.3.1 Markov Quilt Mechanism

Markov Quilt Mechanism was proposed by Wang et al. [49]. This mechanism applies Event-MarkovAdversary-Privacy and considers counting queries (with l_1-sensitivity of 1) over a location database over a period of time. Based on the definition of the adversary in Event-MarkovAdversary-Privacy, the sequence of locations in the location trajectory can be modeled as a Bayes net, a chain $X_1 \rightarrow X_2 \rightarrow \ldots X_T$, where each event X_i only depends on its previous event X_{i-1}. Based on this correlation, the impact of X_i on X_{i+1} is more significant than the impact of X_i on X_{i+k} when k is large. Thus it is sufficient to add noise proportional to the number of events that are highly correlated with the event at each time point. The notion of Markov Quilt is based on the set of highly correlated events for a given event. The size of the Markov Quilt depends on the strength of the correlation, and not on the total size of the trajectory. Hence, unlike user-DP which would add noise proportional to the length of the trajectory, Markov Quilt mechanism will add noise proportional to the size of the Markov Quilt which could be much smaller, thus protecting against adversaries who know correlations as well as ensuring low error. The details of this mechanism can be referred to [49].

5.3.3.2 General Algorithm for Pufferfish Privacy

In the Laplace mechanism for differential privacy, the noise to the query output is proportional to the l_1-sensitivity defined in Eq. (5.3), which is the worst case distance between $f(D_1)$ and $f(D_2)$ where D_1 and D_2 are neighboring databases that differ in the value of a single individual. The corresponding concept for a pair of neighboring databases in Pufferfish framework are all possible pairs of databases that differ in a given pair of discriminative secrets $(s_i, s_j) \in \mathbb{S}_{pairs}$. Hence, Wang et al. [49] consider the two distributions given a secret pair $(s_i, s_j) \in \mathbb{S}_{pairs}$, i.e.

$$\mu_{i,\theta} = \Pr(f(\mathfrak{Data}) = \cdot | s_i, \theta)$$

$$\mu_{j,\theta} = \Pr(f(\mathfrak{Data}) = \cdot | s_j, \theta)$$

and apply Wasserstein distance to measure the relevant distance between distributions $\mu_{i,\theta}$ and $\mu_{j,\theta}$. Wasserstein distance is formally defined as below.

Definition 5.9 (∞-Wasserstein Distance [49]) Let $(\mathcal{X}, d)$ be a Radon space, and μ, ν be two probability distribution on $\mathcal{X}$ with finite p-th moment. The ∞-Wasserstein distance between μ, ν with $d(x, y) = |x - y|$:

$$W_\infty(\mu, \nu) = \inf_{\gamma \in \Gamma(\mu, \nu)} \max_{(x,y) \in A} |x - y|, \tag{5.14}$$

where $A = \{(x, y) | \gamma(x, y) \neq 0\}$ is the support of γ, and $\Gamma(\mu, \nu)$ is the set of all couplings γ over μ and ν.

Intuitively, $\gamma \in \Gamma(\mu, \nu)$ is a way to shift probability mass between μ and ν, and $W_\infty(\mu, \nu)$ can be interpreted as the maximum "distance" that any probability mass moves while transforming μ to ν in the most optimal way. Wang et al. [49] proposed a general mechanism for Pufferfish framework Pufferfish framework $(\mathbb{S}, \mathbb{S}_{pairs}, \mathbb{D})$ with privacy budget ϵ and query f. This mechanism first computes the generalized sensitivity, defined as

$$\Delta W_\infty(f, \mathbb{S}, \mathbb{S}_{pairs}, \mathbb{D}) = \max_{(s_i, s_j) \in \mathbb{S}_{pairs}} \quad \max_{\theta \in \mathbb{D} \mid \Pr(s_i|\theta) \neq 0, \Pr(s_j|\theta) \neq 0} W_\infty(\mu_{i,\theta}, \mu_{j,\theta})$$

$$(5.15)$$

This generalized sensitivity iterates over all possible secret pairs in $\mathbb{S}_{pairs}$ and data evolution scenarios $\theta \in \mathbb{D}$ and computes the inf-Wasserstein distance between the distributions given each secret and θ. Similar to Laplace mechanism, Wasserstein mechanism adds noise that is proportional to the general sensitivity of the given function can guarantee ϵ-Pufferfish privacy. Here is the formal statement.

Definition 5.10 (Wasserstein Mechanism) Given any function $f : \mathcal{I} \to \mathbb{R}^k$, the Wasserstein mechanism is defined as :

$$\mathfrak{M}_W(D, f(\cdot), \mathbb{S}, \mathbb{S}_{pairs}, \mathbb{D}, \epsilon) = f(D) + (\eta_1, \ldots, \eta_k), \qquad (5.16)$$

where η_i are i.i.d random variables drawn from $Lap(\Delta W_\infty(f, \mathbb{S}, \mathbb{S}_{pairs}, \mathbb{D})/\epsilon)$.

Wang et al. [49] showed that Wasserstein mechanism provides ϵ-Pufferfish privacy in the framework $(\mathbb{S}, \mathbb{S}_{pairs}, \mathbb{D})$. This mechanism is also shown with a smaller sensitivity parameter than the l_1-sensitivity of query f under group differential privacy if f is L-Lipschitz query [49], and hence can result in higher accuracy for query f.

5.4 Blowfish Privacy

Though Pufferfish framework provides a wide variety of privacy definitions, the domain experts are required to specify adversarial knowledge as sets of complex probability distributions, and this framework does not always result in composable privacy definitions [31]. Hence, we introduce a simple but useful class of privacy definitions named Blowfish privacy [22, 27] that addresses limitations of the general framework of Pufferfish. The definition and properties of Blowfish privacy are presented in Sect. 5.4.1. We also show in Sect. 5.4.2 algorithms for special instances of Blowfish privacy and also a general algorithm for Blowfish privacy.

5.4.1 Definition and Properties of Blowfish Privacy

The key building block of an instantiation of Blowfish privacy is named *policy graph*. A policy graph is a graph representation of $\mathbb{S}_{pairs}$, pairs of domain values in $\mathcal{T}$ that an adversary must not be able to distinguish.

Definition 5.11 (Policy Graph [22, 27]) A policy graph is a graph $G = (V, E)$ with $V \subseteq \mathcal{T} \cup \{\perp\}$, where $\perp$ is the name of a special vertex, and $E \subseteq (\mathcal{T} \cup \{\perp\}) \times (\mathcal{T} \cup \{\perp\})$.

An edge $(u, v) \in E$ corresponds to a pair of domain values that an adversary should not be able to distinguish between. $\perp$ is a dummy value not in $\mathcal{T}$, and an edge $(u, \perp) \in E$ means that an adversary should not be able to distinguish between the presence of a tuple with value u, or the absence of the tuple from the database. If a policy graph does not include $\perp$, we can focus on databases with fixed known size. Based on this policy graph G, we re-define the concept of neighboring databases of differential privacy in the following way.

Definition 5.12 (Neighbors [22, 27]) Consider a policy graph $G = (V, E)$. Let D_i and D_j be two databases in $\mathcal{I}$. D_i and D_j are neighbors, denoted by $(D_i, D_j) \in N(P)$, iff exactly one of the following is true:

- D_i and D_j differ in the value of exactly one entry such that $(u, v) \in E$, where u is the value of the entry in D_i and v is the value of the entry in D_j;
- D_i differs from D_j in the presence or absence of exactly one entry, with value u, such that $(u, \perp) \in E$.

Example 5.3 (Event-DP) Recall event-DP in Sect. 5.2.2 considers neighboring databases differ in a single event. The set of discriminative pairs for event-level neighbors can be specified as

$$\mathbb{S}_{pairs}^{event} = \{(r_i = t, r_i = t') | h_i \in \mathcal{H}; \ t, t' \in \Sigma^*, \ |t| = |t'|, \tag{5.17}$$

$$\forall j^*, \text{if } t[j^*] \neq t'[j^*], \text{ then } \forall j \neq j^*, \ t[j] = t'[j]\}.$$

Hence, the policy graph of blowfish privacy considers all possible sequence of events as vertices V, and adds an edge to any pair of event sequences with the same length differing in one event. This policy graph results in a set of neighboring databases for event-DP.

Example 5.4 (Geo-indistinguishability) This is a special case of event-DP, where discriminative pairs differ in only one event. Additionally, secrets form pairs if the location they differ in are close to each other. More formally,

$$\mathbb{S}_{pairs}^{event,\theta} = \{(r_i = t, r_i = t') | h_i \in \mathcal{H}, \ t, t' \in \Sigma^*, \ |t| = |t'|, \tag{5.18}$$

$$\forall j^*, \text{if } d(t[j^*], t'[j^*]) \leq \theta, \text{ then } \forall j \neq j^*, \ t[j] = t'[j]\}$$

This captures variants of event DP proposed in prior work like Geo-indistinguishability [4] and (θ, ϵ)-location privacy [19], each discriminative secret pair differ not only in a single event, but the difference in location value of the event is bounded by a given distance θ. Compared to event-level discriminative pairs $\mathbb{S}_{pairs}^{event}$, $\mathbb{S}_{pairs}^{event,\theta}$ protects a smaller set of discriminative secret pairs with the same privacy guarantee. This results in a sparser policy graph, as fewer pairs of secrets are connected by an edge. It is easy to see that the policy graph for $\mathbb{S}_{pairs}^{event,\theta}$ is a subgraph of the policy graph for $\mathbb{S}_{pairs}^{event}$. Correspondingly, the set of neighbors protected by geo-indistinguishability is a subset of neighbors protected by event-DP. Hence, geo-indistinguishability provides a weaker guarantee than event-DP.

Besides G, Blowfish privacy policy in [27] includes $\mathcal{I}_Q$ which denotes the set of databases that are possible under the public constraints Q that are known about the database. The constraints in Q makes a subset of the possible database instances impossible, and the rest of possible database instances are denoted by $\mathcal{I}_Q$. The presence of the constraints will make some neighboring databases no longer possible. For instance, due to temporal constraints, certain sequences of locations are impossible. Below is an example that considers such temporal constraints.

Example 5.5 (δ-Location Set Based Differential Privacy) The privacy definition proposed by Xiao et al [53] considers temporal constraints in the database, and these constraints are also known as the data evolution scenarios in the Pufferfish framework. The data generation model $\mathbb{D}$ is represented by a hidden Markov model (HMM) which consists of a single transition matrix P_θ and an emission probability P_θ^e. The prior distribution for an individual h_i being at location l at timestamp j given the previous $(j-1)$ events $\Pr(r_i[j] = l|l_{j-1} \ldots l_1)$ can be derived from P_θ and P_θ^e, and can eliminate unlikely secrets from $\mathbb{S}$. The remaining possible locations are specified by a new term called δ-location set. Formally, for any $j \in [1, 2, \ldots]$, the δ-location set at time point j, is defined as a set containing minimum number of locations that have prior distribution sum no less than $1 - \delta$, i.e.

$$\Delta X_j = \min\{l| \sum_l \Pr(r_i[j] = l|l_{j-1} \ldots l_1) \geq 1 - \delta\}. \tag{5.19}$$

At any time point j, a randomized mechanism $\mathfrak{M}$ satisfies ϵ-differential privacy on δ-location set ΔX_j, if for any output ω_j and any two locations l_1 and l_2 in ΔX_j, the following holds: $\Pr(\mathfrak{M}(l_1) = \omega) \leq e^\epsilon \Pr(\mathfrak{M}(l_2) = \omega)$.

This privacy definition is also known as δ-*location set based differential privacy*, and can be perceived as a special case of Blowfish privacy at each timestamp, where the neighboring databases at each timestamp differ. The δ-location set removes all impossible database instances based on the data evolution scenarios, and hence guarantees a stronger privacy against an adversary who knows this evolution scenario than an event-DP algorithm which assumes no correlation between events.

Readers may refer to [27] for the general version of Blowfish neighbors with constraints. For more general policy graph with constraints, we can define Blowfish privacy as follows.

Definition 5.13 (Blowfish Privacy [22, 27]) Let $\epsilon > 0$ be a real number and policy $P = (\mathcal{T}, G, \mathcal{I}_Q)$ be a policy. A randomized mechanism M satisfies (ϵ, P)-Blowfish privacy if for every pair of neighboring databases $(D_i, D_j) \in N(P)$, and every set of outputs $S \subseteq range(M)$, we have

$$Pr[\mathfrak{M}(D_i) \in S] \leq e^\epsilon Pr[\mathfrak{M}(D_j) \in S] \tag{5.20}$$

If we consider policy without constraints, we simplify our notation for Blowfish privacy as (ϵ, G)-Blowfish privacy.

Now the privacy guarantee is not only controlled by the privacy parameter ϵ, but also by the policy graph G. Consider two databases $D_1 = D \cup \{u\}$ and $D_2 = D \cup \{v\}$ that differ in one tuple. Given a mechanism $\mathfrak{M}$ that satisfies (ϵ, G)-Blowfish privacy, we have

$$Pr[\mathfrak{M}(D_i) \in S] \leq e^{\epsilon \cdot d_G(u,v)} Pr[\mathfrak{M}(D_j) \in S] \qquad (5.21)$$

where $d_G(u, v)$ is the shortest distance between u, v in G. This implies that an adversary may better distinguish pairs of nodes farther apart in the graph than those that are closer. Similarly, an adversary can distinguish between u, v with probability 1, when u and v appear in disjoint components of G, where $d_G(u, v) \rightarrow \infty$. Note that when G is a complete graph K, then (ϵ, K)-Blowfish privacy is equivalent to ϵ-differential privacy.

Theorem 5.5 (Sequential Composition [27]) *Let $P = (\mathcal{T}, G, \mathcal{I}_Q)$ be a policy and $D \in \mathcal{I}_Q$ be an input database. Let $\mathfrak{M}_1(\cdot)$ and $\mathfrak{M}_2(\cdot, \cdot)$ be algorithms with independent sources of randomness that satisfy (ϵ_1, P) and (ϵ_2, P)-Blowfish privacy, resp. Then an algorithm that outputs both $\mathfrak{M}_1(D) = \omega_1$ and $\mathfrak{M}_2(\omega_1, D) = \omega_2$ satisfies $(\epsilon_1 + \epsilon_2, P)$-Blowfish privacy.*

Algorithms that satisfy Blowfish also satisfy the postprocessing theorem (like Theorem 5.2) and a restricted form of parallel composition. We refer the reader to [27] for details.

5.4.2 Mechanisms for Blowfish Privacy

Blowfish privacy generalized (bounded and unbounded) differential privacy. In fact, we can show that any algorithm that satisfies ϵ-bounded DP (or $\epsilon/2$-unbounded DP) also satisfies (ϵ, G)-Blowfish privacy for any policy graph G (when the definition has no constraints Q). Thus, in the absence of constraints, differentially private algorithms can be used to satisfy Blowfish privacy definitions. However, leveraging the policy can lead to algorithms that provide more accuracy than DP algorithms as we will see in the rest of this section.

5.4.2.1 Releasing Perturbed Locations

In many Location-Based Services (LBSs), an actual location needs to be shared. While one can design event-DP algorithms using variants of randomized response to release perturbed locations, they would have poor utility. Due to the large domain size of locations, the probability that one would report a point close to the true location would be vanishingly small. On the other hand, we can develop methods to release perturbed locations with high accuracy under Blowfish policies corresponding to Geo-indistinguishability [4] and δ-location set differential privacy [53]. The details of the algorithms are described below.

- **Geo-indistinguishable mechanism** was designed by Andres et al. [4] to ensure ϵ-geo-indistinguishability. The location domain Σ is modeled as the Euclidean plane equipped with the standard notion of Euclidean distance. For the ideal case of the continuous plane, where $\Sigma = \mathbb{R}^2$. Given a true location $r_i[j] = l$, where $l \in \mathbb{R}^2$ and privacy parameter ϵ, the mechanism would like to draw a location $l \in \mathbb{R}^2$ with probability function

$$D_\epsilon(l_0)(l) = \frac{\epsilon^2}{2\pi} e^{-\epsilon d(l_0, l)}. \tag{5.22}$$

  It is easy to show that this mechanism satisfies ϵ-geo-indistinguishability. The actual sampling process takes place in a system of polar coordinates that centered at l. Equation 5.22 can be transformed into PDF of the polar laplacian centered in the origin l_0, where $D_{\epsilon, R}(r) = e^2 r e^{-\epsilon r}$, $D_{\epsilon, \Theta}(\theta) = \frac{1}{2\pi}$. Based on the PDF, the angle θ can be drawn uniformly $[0, 2\pi]$. For radius r, we first draw z uniformly in $[0, 1)$ and set $r = C_\epsilon^{-1}(z)$, where $C_\epsilon(r) = 1 - (1 + \epsilon z)e^{\epsilon z}$ is the cumulative function for $D_{\epsilon, R}(r)$. Readers may refer to [4] for the adjusted mechanism for discrete coordinates.

  However, this mechanism is susceptible to attacks when the adversary knows correlations across time points in a trajectory[9, 53]. Hence, this temporal correlation-based attacks motivates the following mechanism.
- **Planar Isotropic Mechanism** was proposed by Xiao et al. [53] to ensure δ-location set based differential privacy (Example 5.5), where the sequence of locations are correlated by a Hidden Markov model. As l_1-sensitivity fails to capture the geometric sensitivity in multidimensional space, Xiao et al. [53] proposed a new notion, *sensitivity hull* to bound the change in the query output caused by the modification of an event. The sensitivity hull of a query f is defined as the convex hull of Δf where Δf is the set of $f(l_1) - f(l_2)$ for any pair l_1 and l_2 in δ-location set ΔX. This sensitivity hull was further transformed into isotropic position to ensure optimal solution of K-norm Mechanism [24] for 2-dimensional space. K-norm Mechanism is defined as below.

Definition 5.14 (K-norm Mechanism) Given a linear function $F : R^N \to R^d$ and its sensitivity hull K, a mechanism is K-norm mechanism if for any output z, the following holds:

$$Pr(z) = \frac{1}{\Gamma(d+1)VOL(K/\epsilon)} exp(-\epsilon \|z - Fx^*\|_K), \tag{5.23}$$

where Fx^* is the true answer, $\|\cdot\|_K$ is the (Minkowski) norm of K, $\Gamma()$ is Gamma function and $VOL()$ indicates volume.

The releasing of location at timestamp j can be summarized as four steps: (1) the sensitivity hull K is computed from the given δ-location set ΔX at timestamp j, (2) the sensitivity hull K is then transformed to its isotropic position K_I to ensure the optimal solution for K-norm mechanism, (3) a sample is picked from K_i and perturbed to z' by K-norm mechanism; (4) finally this perturbed sample z' is transformed to z in the original space and published. This mechanism has been shown in [53] with error $O\left(\frac{1}{\epsilon}\sqrt{AREA(K)}\right)$ at most, and this is the lower bound of any mechanism that satisfies δ-location set based differential privacy.

5.4.2.2   Aggregate Perturbation for Count Queries (General Algorithms for Blowfish Privacy)

In [22, 27], Blowfish private mechanisms were designed to answer aggregate queries under different policy graphs. Each policy graph can instantiate a new notion of neighboring databases. Rather than re-designing a new algorithm for each notion of neighboring databases, [22] showed a transformational equivalence between a large class of Blowfish private algorithms and standard differentially private algorithms and for many policy graphs. This equivalence can be stated as follows: for policy graph G, there exists a transformation of the workload and database $(W, x) \rightarrow (W_G, x_G)$ such that $Wx = W_G x_G$, and a mechanism $\mathfrak{M}$ is an (ϵ, G)-Blowfish private mechanism for answering workload W on input x if and only if $\mathfrak{M}$ is also an ϵ-differentially private mechanism for answering W_G on x_G. This result does not hold in general, but [22] showed that under a class of mechanism called *matrix mechanism*, transformational equivalence holds for any policy graph.

Equivalence for Matrix Mechanism. Matrix mechanism framework was designed for optimally answering a workload of linear queries [33]. Some workloads W have a high sensitivity, but they can be answered with low error by answering a different *strategy* query workload A such that (a) A has a low sensitivity Δ_A, and (b) rows in W can be reconstructed using a small number of rows in A.

In particular, let A be a $p \times k$ matrix, and A^+ denote its Moore-Penrose pseudoinverse, such that $WAA^+ = W$. The matrix mechanism is given by the following:

$$\mathfrak{M}_A(W, x) = Wx + WA^+ Lap(\Delta_A/\epsilon)^p \qquad (5.24)$$

where, $Lap(\lambda)^p$ denotes p independent random variables drawn from the Laplace distribution with scale λ. The corresponding Blowfish specific sensitivity of a workload, $\Delta_w(G)$ is defined as follows:

Definition 5.15 (Policy Specific l_1 Sensitivity) The l_1 policy specific sensitivity of a query matrix W with respect to policy graph G is

$$\Delta_w(G) = \max_{x,x' \in N(G)} ||Wx - Wx'||_1 \qquad (5.25)$$

Let P_G be a matrix that satisfies the following properties.

- P_G has $|V| - 1$ rows and $|E|$ columns.
- Let $W_G = W P_G$. Then $\Delta_w(G) = \Delta_{W_G}$. i.e. the sensitivity of workload W under Blowfish policy G is the same as the sensitivity of W_G under differential privacy.
- P_G has full row rank (and therefore a right inverse P_G^{-1}). For vector x, we let x_G denote $P_G^{-1} x$.

Given such a P_G, the following theorem is true.

Theorem 5.6 ([22]) *Let G be a Blowfish policy graph and W be a workload. Suppose P_G exists with the properties given above. Then the matrix mechanism given by Eq. (5.24) is both a (ϵ, G)-Blowfish private mechanism for answering W on x and an ϵ-differentially private algorithm for answering W_G on x_G. Since $Wx = W_G x_G$, the mechanism has the same error in both instances.*

We illustrate the strategies proposed in [22] with the example of answering range counting queries over two dimensional location for *distance-threshold* policy graphs. These graphs are based on similar secret specification as Geo-indistinguishability, by considering the set of discriminative secrets

$$\mathbb{S}_{pairs}^{\theta} = \{(r_i[j] = l, r_i[j] = l') \mid d(l, l') \leq \theta,\ l, l' \in \Sigma,\ j = 1, 2, \ldots\}. \qquad (5.26)$$

Particularly for a grid-based location domain of size $k \times k$, this class of policy graph $G_{k^2}^{\theta}$ is defined based on the l_1-distance in the domain $[k]^2$, where $[k]$ denotes the set of integers between 1 and k (inclusive). The vertices in G are the grid cells. There is an edge u, v in E if and only if $|u - v|_1 \leq \theta$.

Consider rectangle range counting query $q([x, y], [x', y'])$. When $\theta = 1$, the transformed query $q_{G_{k^2}^1}$ is the sum of four disjoint range counting queries in the transformed domain. Hence, the strategy for answering the transformed query workload would be to answer $2(k - 1)$ one-dimensional range count queries under differential privacy. The error per query under $(\epsilon, G_{k^2}^1)$-Blowfish privacy for all rectangle range counting queries is $O(\log^3 k/\epsilon^2)$ while the best known data independent strategy answering the same workload with ϵ-differential privacy is the Privelet strategy [52] with a much larger asymptotic error of $O(\log^6 k/\epsilon^2)$ per query.

When $\theta > 1$, the policy graph is more complex. The algorithm proposed leverages subgraph approximation and can achieve an error of $O(\log^3 k \log^3 \theta/\epsilon^2)$ per query under $(\epsilon, G_{k^2}^{\theta})$, which is still better than using Privelet ($O(\log^6 k/\epsilon^2)$ per query) when $\log \theta$ is small compared to $\log k$.

5.5 Conclusions and Open Challenges

In this chapter, we identified desiderata that algorithms for privacy-preserving release and analysis of location data must ensure. These include ensuring provable privacy guarantees of an individual's properties even when adversaries have strong prior knowledge and satisfying composition properties that allow for a graceful degradation of privacy even under multiple releases of the data. We described variants of differential privacy and algorithms that satisfy these variants for the tasks of answering queries over a single-time snapshot of location data, continuous queries over location streams and releasing synthetic location trajectory databases. We also presented Pufferfish, a framework for defining privacy that allows us to reason about the privacy semantics underlying the variants of differential privacy. We concluded by describing instantiations of Pufferfish that allow for ensuring privacy when adversaries may know correlations within the location stream, and described a subclass of Pufferfish, called Blowfish, that satisfies composition theorems.

While a wide range of provable privacy definitions are available today, it is still unclear which of these definitions are applicable to a given scenario, and whether the algorithms satisfying these definitions allow realistic analysis of location data with acceptable errors. There are a number of algorithms known for geo-indistinguishability and event-DP, but these approaches constitute the weakest privacy guarantees. More work is needed to identify practical solutions for location-based applications under stronger privacy notions.

Location trajectories have inherent correlations, both within a single trajectory and across trajectories. We have seen examples of the former, and some solutions to handle these correlations when they take a specific form. However, there is little work that acknowledges and handles correlations across individuals. For instance, it is known that when the location trajectories of two individuals are similar, then they are highly likely to have strong social connections [12]. Whether techniques like the Markov Quilt mechanism for handling correlations will be applied to such cases is an interesting open question.

All of the work presented assumes that (a) users require the same level of privacy, and (b) users are able to specify privacy levels in terms of the privacy parameter ϵ. The former is clearly not true in the real world. There is work that suggests that users have different privacy seeking behaviors depending on their demographic attributes as well as based on their context. Moreover, it is not clear whether or not users will be able to express their privacy preferences in terms of an ϵ privacy parameter. These challenges will motivate new interesting privacy research to further advance the state-of-the-art techniques, and ensure their adoption in real systems.

References

1. O. Abul, F. Bonchi, and M. Nanni. Never walk alone: Uncertainty for anonymity in moving objects databases. In *Proceedings of the 24th International Conference on Data Engineering, ICDE 2008, April 7–12, 2008, Cancún, México*, pages 376–385, 2008.
2. Airbnb. Airbnb privacy policy, 2017. https://www.airbnb.com/terms/privacy_policy.
3. S. Altman, N. Sivo, E. Tana, and B. Knapp. Location-based advertising message serving for mobile communication devices, 2008. US Patent App. 11/931,113.
4. M. E. Andrés, N. E. Bordenabe, K. Chatzikokolakis, and C. Palamidessi. Geo-indistinguishability: Differential privacy for location-based systems. In *Proceedings of the 2013 ACM SIGSAC Conference on Computer & Communications Security*, CCS '13, pages 901–914, New York, NY, USA, 2013. ACM.
5. Atockar. Riding with the stars: Passenger privacy in the nyc taxicab dataset, 2014. https://research.neustar.biz/author/atockar/.
6. J. Bolot, N. Fawaz, S. Muthukrishnan, A. Nikolov, and N. Taft. Private decayed predicate sums on streams. In *Proceedings of the 16th International Conference on Database Theory*, ICDT '13, pages 284–295, New York, NY, USA, 2013. ACM.
7. J. Cao, Q. Xiao, G. Ghinita, N. Li, E. Bertino, and K.-L. Tan. Efficient and accurate strategies for differentially-private sliding window queries. In *Proceedings of the 16th International Conference on Extending Database Technology*, EDBT '13, pages 191–202, New York, NY, USA, 2013. ACM.
8. T.-H. H. Chan, E. Shi, and D. Song. Private and continual release of statistics. *ACM Trans. Inf. Syst. Secur.*, 14(3):26:1–26:24, Nov. 2011.
9. K. Chatzikokolakis, C. Palamidessi, and M. Stronati. *A Predictive Differentially-Private Mechanism for Mobility Traces*, pages 21–41. Springer International Publishing, Cham, 2014.
10. R. Chen, G. Acs, and C. Castelluccia. Differentially private sequential data publication via variable-length n-grams. In *Proceedings of the 2012 ACM Conference on Computer and Communications Security*, CCS '12, pages 638–649, New York, NY, USA, 2012. ACM.
11. Y. Chen, A. Machanavajjhala, M. Hay, and G. Miklau. Pegasus: Data-adaptive differentially private stream processing. In *Proceedings of the 2017 ACM SIGSAC Conference on Computer and Communications Security, CCS 2017, Dallas, TX, USA, October 30 - November 03, 2017*, pages 1375–1388, 2017.
12. E. Cho, S. A. Myers, and J. Leskovec. Friendship and mobility: user movement in location-based social networks. In *ACM SIGKDD Conference*, KDD '11, 2011.
13. G. Cormode, C. M. Procopiuc, D. Srivastava, E. Shen, and T. Yu. Differentially private spatial decompositions. In *IEEE 28th International Conference on Data Engineering (ICDE 2012), Washington, DC, USA (Arlington, Virginia), 1–5 April, 2012*, pages 20–31, 2012.
14. Y.-A. de Montjoye, C. A. Hidalgo, M. Verleysen, and V. D. Blondel. Unique in the crowd: The privacy bounds of human mobility. *Scientific Reports*, 3, 2013.
15. C. Dwork. Differential privacy. In *IN ICALP*, pages 1–12. Springer, 2006.
16. C. Dwork, F. Mcsherry, K. Nissim, and A. Smith. Calibrating noise to sensitivity in private data analysis. In *In Proceedings of the 3rd Theory of Cryptography Conference*, pages 265–284. Springer, 2006.
17. C. Dwork and M. Naor. On the difficulties of discloure prevention in statistical databases or the case for differential privacy. *Journal of Privacy and Confidentiality*, 2010.
18. C. Dwork and A. Roth. The algorithmic foundations of differential privacy. Technical report, Theoretical Computer Science, 2013.
19. E. ElSalamouny and S. Gambs. Differential privacy models for location-based services. *Transactions on Data Privacy*, 9(1):15–48, 2016.
20. L. Fan and L. Xiong. Real-time aggregate monitoring with differential privacy. In *Proceedings of the 21st ACM International Conference on Information and Knowledge Management*, CIKM '12, pages 2169–2173. ACM, 2012.

21. S. Gambs, M.-O. Killijian, and M. N. n. del Prado Cortez. Show me how you move and i will tell you who you are. In *Proceedings of the 3rd ACM SIGSPATIAL International Workshop on Security and Privacy in GIS and LBS*, SPRINGL '10, pages 34–41, New York, NY, USA, 2010. ACM.
22. S. Haney, A. Machanavajjhala, and B. Ding. Design of policy-aware differentially private algorithms. *Proc. VLDB Endow.*, 9(4):264–275, Dec. 2015.
23. M. Hardt and K. Talwar. On the geometry of differential privacy. In *Proceedings of the Forty-second ACM Symposium on Theory of Computing*, STOC '10, 2010.
24. M. Hardt and K. Talwar. On the geometry of differential privacy. In *Proceedings of the 42nd ACM Symposium on Theory of Computing, STOC 2010, Cambridge, Massachusetts, USA, 5–8 June 2010*, pages 705–714, 2010.
25. M. Hay, V. Rastogi, G. Miklau, and D. Suciu. Boosting the accuracy of differentially private histograms through consistency. *PVLDB*, 3(1):1021–1032, 2010.
26. X. He, G. Cormode, A. Machanavajjhala, C. M. Procopiuc, and D. Srivastava. Dpt: Differentially private trajectory synthesis using hierarchical reference systems. *Proc. VLDB Endow.*, 8(11):1154–1165, July 2015.
27. X. He, A. Machanavajjhala, and B. Ding. Blowfish privacy: Tuning privacy-utility trade-offs using policies. In *Proceedings of the 2014 ACM SIGMOD International Conference on Management of Data*, SIGMOD '14, pages 1447–1458, New York, NY, USA, 2014. ACM.
28. H. Hu, J. Xu, S. T. On, J. Du, and J. K.-Y. Ng. Privacy-aware location data publishing. *ACM Trans. Database Syst.*, 35(3):18:1–18:42, July 2010.
29. G. Kellaris, S. Papadopoulos, X. Xiao, and D. Papadias. Differentially private event sequences over infinite streams. *Proc. VLDB Endow.*, 7(12):1155–1166, Aug. 2014.
30. D. Kifer and A. Machanavajjhala. No free lunch in data privacy. In *Proceedings of the 2011 ACM SIGMOD International Conference on Management of Data*, SIGMOD '11, pages 193–204, New York, NY, USA, 2011. ACM.
31. D. Kifer and A. Machanavajjhala. Pufferfish: A framework for mathematical privacy definitions. *ACM Trans. Database Syst.*, 39(1):3:1–3:36, Jan. 2014.
32. C. Li, M. Hay, V. Rastogi, G. Miklau, and A. McGregor. Optimizing linear counting queries under differential privacy. In *Proceedings of the Twenty-ninth ACM SIGMOD-SIGACT-SIGART Symposium on Principles of Database Systems*, PODS '10, 2010.
33. C. Li, M. Hay, V. Rastogi, G. Miklau, and A. McGregor. Optimizing linear counting queries under differential privacy. In *Proceedings of the Twenty-Ninth ACM SIGMOD-SIGACT-SIGART Symposium on Principles of Database Systems, PODS 2010, June 6–11, 2010, Indianapolis, Indiana, USA*, pages 123–134, 2010.
34. J. Lin, M. Benisch, N. Sadeh, J. Niu, J. Hong, B. Lu, and S. Guo. A comparative study of location-sharing privacy preferences in the united states and china. *Personal Ubiquitous Comput.*, 17(4):697–711, Apr. 2013.
35. C. Y. T. Ma, D. K. Y. Yau, N. K. Yip, and N. S. V. Rao. Privacy vulnerability of published anonymous mobility traces. *IEEE/ACM Trans. Netw.*, 21(3):720–733, June 2013.
36. A. Machanavajjhala and D. Kifer. Designing statistical privacy for your data. *Commun. ACM*, 58(3):58–67, Feb. 2015.
37. A. Machanavajjhala, D. Kifer, J. M. Abowd, J. Gehrke, and L. Vilhuber. Privacy: Theory meets practice on the map. In *Proceedings of the 24th International Conference on Data Engineering, ICDE 2008, April 7–12, 2008, Cancún, México*, pages 277–286, 2008.
38. A. Machanavajjhala, D. Kifer, J. Gehrke, and M. Venkitasubramaniam. L-diversity: Privacy beyond k-anonymity. *ACM Trans. Knowl. Discov. Data*, 1(1), Mar. 2007.
39. D. J. Mir, S. Isaacman, R. Cáceres, M. Martonosi, and R. N. Wright. DP-WHERE: differentially private modeling of human mobility. In *Proceedings of the 2013 IEEE International Conference on Big Data, 6–9 October 2013, Santa Clara, CA, USA*, pages 580–588, 2013.
40. A. Monreale, G. L. Andrienko, N. V. Andrienko, F. Giannotti, D. Pedreschi, S. Rinzivillo, and S. Wrobel. Movement data anonymity through generalization. *Trans. Data Privacy*, 3(2):91–121, 2010.
41. OpenPaths. Openpaths privacy policy, 2012. https://openpaths.cc/privacy.

42. W. H. Qardaji, W. Yang, and N. Li. Differentially private grids for geospatial data. In *29th IEEE International Conference on Data Engineering, ICDE 2013, Brisbane, Australia, April 8–12, 2013*, pages 757–768, 2013.
43. W. H. Qardaji, W. Yang, and N. Li. Understanding hierarchical methods for differentially private histograms. *PVLDB*, 6(14):1954–1965, 2013.
44. S. Sankararaman, G. Obozinski, M. I. Jordan, and E. Halperin. Genomic privacy and limits of individual detection in a pool. *Nature Genetics*, 2009.
45. P. Santi, G. Resta, M. Szell, S. Sobolevsky, S. H. Strogatz, and C. Ratti. Quantifying the benefits of vehicle pooling with shareability networks. *Proceedings of the National Academy of Sciences*, 111(37):13290–13294, 2014.
46. C. Song, Z. Qu, N. Blumm, and A.-L. Barabási. Limits of predictability in human mobility. *Science*, 2010.
47. X. Song, Q. Zhang, Y. Sekimoto, and R. Shibasaki. Prediction of human emergency behavior and their mobility following large-scale disaster. In *The 20th ACM SIGKDD International Conference on Knowledge Discovery and Data Mining, KDD '14, New York, NY, USA - August 24 - 27, 2014*, pages 5–14, 2014.
48. D. Su, J. Cao, N. Li, E. Bertino, and H. Jin. Differentially private k-means clustering. *CoRR*, abs/1504.05998, 2015.
49. Y. Wang, S. Song, and K. Chaudhuri. Privacy-preserving analysis of correlated data. *CoRR*, abs/1603.03977, 2016.
50. A. Wesolowski, C. J. E. Metcalf, N. Eagle, J. Kombich, B. T. Grenfell, O. N. Bjørnstad, J. Lessler, A. J. Tatem, and C. O. Buckee. Quantifying seasonal population fluxes driving rubella transmission dynamics using mobile phone data. *Proceedings of the National Academy of Sciences*, 112(35):11114–11119, 2015.
51. R. C.-W. Wong, A. W.-C. Fu, K. Wang, and J. Pei. Minimality attack in privacy preserving data publishing. In *Proceedings of the 33rd International Conference on Very Large Data Bases*, VLDB '07, pages 543–554. VLDB Endowment, 2007.
52. X. Xiao, G. Wang, and J. Gehrke. Differential privacy via wavelet transforms. In *Proceedings of the 26th International Conference on Data Engineering, ICDE 2010, March 1–6, 2010, Long Beach, California, USA*, pages 225–236, 2010.
53. Y. Xiao and L. Xiong. Protecting locations with differential privacy under temporal correlations. In *Proceedings of the 22nd ACM SIGSAC Conference on Computer and Communications Security, Denver, CO, USA, October 12–6, 2015*, pages 1298–1309, 2015.
54. M. Xu, J. Wu, Y. Du, H. Wang, G. Qi, K. Hu, and Y. Xiao. Discovery of important crossroads in road network using massive taxi trajectories. *CoRR*, abs/1407.2506, 2014.
55. R. Yarovoy, F. Bonchi, L. V. S. Lakshmanan, and W. H. Wang. Anonymizing moving objects: how to hide a MOB in a crowd? In *EDBT 2009, 12th International Conference on Extending Database Technology, Saint Petersburg, Russia, March 24–26, 2009, Proceedings*, pages 72–83, 2009.
56. J. Zhang, X. Xiao, and X. Xie. Privtree: A differentially private algorithm for hierarchical decompositions. In *Proceedings of the 2016 International Conference on Management of Data*, SIGMOD '16, pages 155–170, New York, NY, USA, 2016. ACM.

Chapter 6
Opportunities and Risks of Delegating Sensing Tasks to the Crowd

Delphine Reinhardt and Frank Dürr

Abstract Mobile phones and tablets have long become ubiquitous with billions of devices sold worldwide. Equipped with a myriad of embedded sensors, these devices have enabled the rise of a new sensing paradigm: participatory sensing. While different terminologies, such as crowdsensing or mobile sensing, are used to define and refine different facets of this new paradigm, they share a common denominator—volunteers collect sensors readings using their personal devices as sensor platforms. The delegation of sensing tasks to a wide public offers multiple opportunities from the perspectives of applications, end users, and participants. However, the introduction of volunteers in the sensing loop also introduces some risks for these stakeholders. In this chapter, we hence provide an overview of existing applications and detail both the opportunities and risks raised by the contributions of volunteers to the sensing process.

6.1 Introduction

Since the introduction of the paradigm of *participatory sensing*, different terminologies have been proposed depending on, e.g., the applied sensing modalities or the monitored subjects. For example, *opportunistic sensing* [21] automatically triggers the sensing process, and hence does not require the participants' involvement as compared to participatory sensing. *Spatial crowdsourcing* [100] especially emphasizes location-based sensing tasks. For environment-centric applications, *urban sensing* [21] and *participatory urbanism* [134] have been proposed, while *citizen sensing* [19], *people-centric sensing* [4, 21], and *community sensing* [102] focus on the participants and the related communities. Despite the existence of

D. Reinhardt (✉)
University of Göttingen, Göttingen, Germany
e-mail: reinhardt@cs.uni-goettingen.de

F. Dürr
University of Stuttgart, Stuttgart, Germany
e-mail: frank.duerr@ipvs.uni-stuttgart.de

© Springer Nature Switzerland AG 2018
A. Gkoulalas-Divanis, C. Bettini (eds.), *Handbook of Mobile Data Privacy*,
https://doi.org/10.1007/978-3-319-98161-1_6

multiple terminologies, the underlying principle remains the same: Applications using participants' mobile devices as sensors (or as data sink for interfaced sensors) collect sensor data. Without loss of generality, we use the generic term *crowdsensing applications* [72] to refer to such applications within the scope of this chapter.

In parallel to the introduction of different terminologies, the enthusiasm of researchers for this field has continuously grown. Leveraging personal mobile devices as sensing platforms has opened the doors to the development of a myriad of novel applications, ranging from people-centric to environment-centric applications. We hence give different examples of these applications and highlight their common architecture in Sect. 6.2. As compared to previous sensing paradigms, such as wireless sensor networks, crowdsensing applications benefit from both the deployment of the devices at large scale and the on-board available resources. As a result, multiple opportunities are offered to both application developers and end users as detailed in Sect. 6.3. They include the possible collection of an unprecedented number of sensor readings following the mobility patterns of contributing participants. The participants' mobility expands the sensing coverage and may provide additional information about their interactions with their environment and other participants. By contributing with their own devices, participants reduce the costs of the sensing platforms to virtually zero for the application developers. Participants building the crowd can also draw benefits from their contribution to these applications. Depending on the nature of the applications and their modalities, participants may obtain incentives for their contributions, such as additional revenues, being part of a community, or discovering new landscapes.

Unfortunately, crowdsensing applications not only offer benefits, but may also involve risks for all stakeholders, i.e., the participants, the application developers, and the end users. In Sect. 6.4, we adopt the participants' perspective and address potential risks that they may encounter when contributing to crowdsensing applications. We especially consider the threats to the participants' privacy and how these threats can be mitigated by applying selected privacy-preserving mechanisms in Sect. 6.4.1. We further consider risks in terms of additional resource consumption and highlight different methods in Sect. 6.4.2 that can be applied to save both battery lifetime and data volume. Next, we consider potential risks for application developers. These risks are mainly caused by the open nature of the crowdsensing applications. By distributing the sensing tasks to the crowd, application developers primarily need to ensure that a sufficient number of sensor readings of satisfying quality will be delivered in a timely manner to match the application's requirements. This means that potential participants must first engage in crowdsensing applications and be encouraged to contribute data on the long term. We therefore give an overview of existing incentives targeting these goals in Sect. 6.5.1. In Sect. 6.5.2, we further highlight different techniques applied to optimize both the quantity and quality of delivered sensor readings. We finally adopt the perspective of end users and discuss the associated risks in Sect. 6.6.

6.2 Crowdsensing Background

In this section, we first introduce a common architecture for crowdsensing systems in Sect. 6.2.1 defining the different building blocks and components that can be found in most crowdsensing systems as well as the stakeholders involved in realizing such a crowdsensing system. Based on this common crowdsensing architecture, we describe a set of typical crowdsensing applications in Sect. 6.2.2 to highlight the broad scope of application areas that benefit from the crowdsensing paradigm.

6.2.1 Crowdsensing System Architecture

Figure 6.1 shows a common system architecture typical for many crowdsensing systems. This architecture consists of a user-facing *frontend* and a *backend* server infrastructure detailed in the following.

6.2.1.1 Frontend: Mobile Sensors

The *crowdsensing frontend* is implemented by *mobile devices of volunteering users* participating in crowdsensing. The major functionality of these mobile devices is to act as *mobile sensors* capturing sensor data. The type of sensor data that can be captured by mobile devices depends on the sensors integrated with the mobile device. Considering a typical smartphone, we can observe that these commodity devices are equipped with various sensors such as positioning sensors, cameras, microphones, accelerometers, gyroscopes, etc. In particular their ability to determine their position outdoors and indoors through GPS, Bluetooth (Apple iBeacon, Google Eddystone), and cellular or Wi-Fi networks (cell id, fingerprinting) allows for capturing location data (e.g., movement trajectories) and geo-located sensor information tagged with positions.

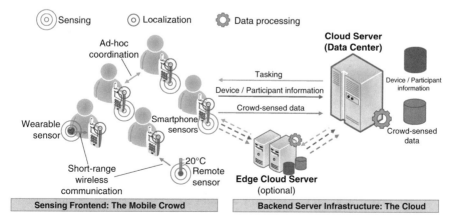

Fig. 6.1 Crowdsensing architecture

Besides smartphones, also other mobile devices can be utilized for crowdsensing. In particular, wearable devices such as fitness trackers, smart watches, or smart glasses are highly relevant for crowdsensing since they integrate sensors for measuring personal information like the users' heart rate. These devices can either be stand-alone mobile devices with direct Internet connection through mobile communication technologies, or they can be connected via short-range wireless communication technologies like Bluetooth to another mobile device. For example, a wearable fitness tracker device can be connected to a smartphone.

Alternatively, a crowd of mobile devices like smartphones can serve as gateways to connect a large set of inexpensive, battery-operated fixed sensors installed in the environment to the Internet in an opportunistic fashion [171]. To this end, the fixed sensors only need to be equipped with energy-efficient short-range wireless communication technologies like Bluetooth Low Energy (BLE) to send sensor data to mobile devices passing by the sensors. The mobile devices then forward these sensor readings to the crowdsensing backend service.

Note that in contrast to a dedicated sensor network owned by a single entity, the sensors and mobile devices of a crowdsensing system are owned by many individual and mostly private participants, i.e., the crowd. Thus, participants keep control over their devices and use them concurrently for other tasks besides sensing like making phone calls or executing other apps. This model have technical implications as shown in Sect. 6.3.

Sensing can either be performed as a background task without requiring user interaction, such as recording a movement trace to map a road network, or by actively involving the user to perform certain tasks like taking a picture of a certain building to texture a crowd-sensed 3D city model. The latter model also allows for leveraging participants themselves as sensors. For example, participants can indicate how safe they feel across cities in uSafe [42].

6.2.1.2   Backend Server Infrastructure

The *backend server infrastructure* manages the crowdsensing system and is responsible for collecting, storing, and processing the gathered sensor data as required by the specific crowdsensing application.

Processing includes the filtering and fusion of data, for example, to remove outliers, and to increase the precision and accuracy of information. Moreover, raw sensor data can be processed into higher-level information. For example, movement traces collected by mobile users using GPS can be processed into a street map. In certain scenarios, not only the sensing tasks but also processing tasks are out-sourced to the crowd. For example, the OpenStreetMap (OSM) community project [3] collects GPS traces from users, and users also generate maps from the collected data, which are then uploaded again to the OSM servers.

Besides collecting and processing sensor data, the backend crowdsensing service can also coordinate the process of crowdsensing by tasking devices. To this end, the mobile devices might transmit device or participant information such as device/user positions, available sensors, remaining energy, or any other context information to the backend server. Tasking can also be supported by the user reputation managed in a reputation system. With this information, the backend server can task suitable devices/participants with relevant sensing tasks like sensing the temperature in a certain area where the mobile device is currently located [11, 137, 172, 173]. Besides this central coordination through the backend service, distributed coordination protocols can also be implemented to let devices coordinate sensing in an ad-hoc fashion [172].

The backend server is typically implemented by a powerful infrastructure, often in the "cloud" to scale with the number of participants. Local servers, often referred to as the "edge cloud" or "fog", can also be deployed to decrease the load onto both the central cloud infrastructure and network as well as reduce latency [151].

The backend infrastructure can be operated by different parties like companies offering a commercial application or service, research institutes collecting data for their studies, or communities like the OSM project mentioned above acting in an altruistic fashion.

6.2.2   Crowdsensing Application Examples

After the introduction of our crowdsensing architecture in Sect. 6.2.1, we next describe examples of crowdsensing applications in order to highlight their diversity and show the versatility of the underlying paradigm.

Due to the ability of mobile devices to sense their position, the first large class of applications deals with collecting *geographic and geo-located information* about the environment. Collected mobility traces of participants, possibly augmented with further information like images from cameras, can be used to create outdoor maps [3, 10] as well as indoor floor plans [6, 73, 90, 136] of a priori unknown environments. Besides solely using participant positions to derive geographic map information, other sensor information can be combined with positions to create various geo-located information. One prominent example is the creation of noise maps from geo-located noise samples captured by microphones of mobile devices to document noise pollution in cities and along roads [13, 14, 58, 94, 114, 117, 126, 144]. Similarly, weather conditions [132] and air pollution [56, 82, 84, 94, 108, 120, 134, 156] can be captured using additional sensors attached to mobile device. A Wi-Fi map showing the coverage and quality of wireless networks can be inferred from crowd-sensed data [34, 70, 175]. Moreover, smart traffic systems can be supported by monitoring the state of traffic signals [101].

Crowdsensing applications can also monitor participants' mobility patterns. By doing so, these applications can infer and predict the number of people present at a certain locations in the case of large scale events like concerts or festivals [16, 154]. Similar techniques can be deployed in emergency scenarios to detect the presence of physical and digital activities as proposed in [115]. At a smaller scale, *QTime* [169] aims at optimizing the users' waiting time at supermarket checkouts by determining when and where the queues are the shortest. Moreover, real-time information about public transport can be inferred based on such applications [69, 184].

Combining location information with other crowd-sensed information can also be used to share information within communities. For example, users can share richer information by integrating sensor-based content [76, 125, 130]. The communities built around the applications can be further exploited to analyze and optimize the distribution of public information using FlierMeet [80] or document prices in local groceries [53] and petrol stations [61].

Another large class of crowdsensing applications is focusing on sensing information about people rather than the environment. Within this class, many applications focus on monitoring and documenting the participants' health. Prominent examples are fitness tracker applications monitoring physical activities including sports activities [64–67, 83, 152, 155]. In this application area, many commercial applications are already available such as RunKeeper [71], Endomondo [68], and Nike+ Running [133]. Other applications assist in documenting eating habits and diet behaviors [8, 54, 131, 147], or exposure to air pollution [12, 94, 127, 140, 179]. Similarly, monitoring stress conditions is a target of further applications [113]. Moreover, applications can assist patients suffering from depression and tinnitus as proposed in [155] and [141, 142], respectively. In general, all of these approaches relieve participants from the tedious task of manual documentation, and various factors impacting the health of participants can be captured. This not only allows gaining detailed insights about individuals, but also provides the data basis for statistical analysis of whole populations of participants.

In summary, our overview of crowdsensing applications shows that the crowdsensing paradigm is highly versatile and can be applied in a wide range of application domains.

6.3 Crowdsensing Opportunities

The selected set of applications described in the previous section draw different benefits from the crowdsensing paradigm. In this section, we hence analyze and highlight the opportunities and benefits stemming from crowdsensing from both the global application perspective as well as the individual end user or crowd perspective.

6.3.1 Application and System Perspective

We start by adopting the global perspective of the crowdsensing applications. The most essential and common goal of all crowdsensing applications is to collect sensor data. With respect to the collected sensor data, crowdsensing improves two essential properties of the collected data, namely, their *quantity* and *quality*. Moreover, crowdsensing not only produces data of higher quantity and quality, it also does so at lower *cost and time*. In what follows, we hence discuss these different aspects in detail.

6.3.1.1 Data Quantity

Sensor data quantity refers to the volume of gathered data. On the one hand, the positive effect of crowdsensing onto the volume of gathered sensor data is due to the *tremendous number of personal devices* owned by the crowd of users participating in sensing, and thus acting as data sources. The scale might become clear by considering the following numbers for the year 2016 from [45]:

- 8 billion mobile devices and connections, where smart devices represented 46% of the total number of mobile devices and connections. In the year 2016 alone, 429 million mobile devices and connections were added, where smartphones accounted for most of that growth,
- 325 million wearable devices, of these, 11 million with embedded cellular connection (other devices might connect to other mobile devices like smartphones through short-range wireless communication technologies like Bluetooth).

Although these numbers are already very impressive, the growth predicted in [45] until 2021 shows the steep gradient of the smart mobile device population: 11.6 billion mobile-connected devices, of which 74.7% will be smart devices. It is not hard to imagine that even if only a fraction of these devices participate in crowdsensing, they can gather an unprecedented volume of sensor data.

Moreover, the volume of sensor data captured by mobile devices also constantly increases due to new developments in sensor technologies. For example, geographic location belongs to one of the most important sensing modalities and illustrates this evolution. For several years now, commodity smartphones integrate GPS functionality measuring longitude, latitude, and altitude values, i.e., only three values, at a maximum rate of few Hertz. In contrast, today's mobile devices also integrate depth sensors to capture detailed 3D point clouds. For example, Google's Tango tablet integrates a depth sensor with 320×180 points resolution at a rate of 5 Hz. Thus, instead of few coordinates per second, a single device can capture tens of thousands of coordinates per second, increasing the volume of sensor data captured by each device by orders of magnitude. One might argue that professional sensors like laser scanners produce even more data due to their higher resolution and sampling rate. However, we need to keep in mind that with crowdsensing the total data volume needs to be multiplied by the size of the device population participating in sensing, which increases the amount of captured data easily by several orders of magnitude.

6.3.1.2 Data Quality

Besides increasing the volume of sensor data, crowdsensing can also increase the quality of sensor data. There are different aspects of data quality to be considered as discussed next.

Precision and Accuracy

If we talk about quality, we first need to discuss the precision and accuracy of gathered sensor data. Precision refers to the repeatability or reproducibility of data, whereas accuracy refers to the deviation of the sampled value from the true value. First of all, both quality metrics depend on the sensors. Although there is constant improvement, typically, sensors integrated with devices like smartphones are less precise than sensors integrated into professional measuring equipment. Taking again the example of the depth sensor of a commodity Tango tablet sensing 3D point clouds and a professional laser scanner, they differ by orders of magnitude in both precision and accuracy. Moreover, there is another essential property of crowdsensing to be considered, namely, the trustworthiness of personal devices. These devices may deliver incorrect sensor readings as detailed in Sect. 6.5.2, thus directly impacting the accuracy of reported sensor data.

So both aspects, low-quality sensors as well as trustworthiness, seem to *negatively* impact the quality of sensed data. So how can we claim that crowdsensing still delivers precise and accurate values? The answer is *redundancy*. In many situations with a dense distribution of crowdsensing devices, many sensor values will be reported rather than only a single value. Although individual values might be inaccurate or imprecise (or both), by post-processing redundant samples using, for example, statistical methods, we can improve the quality by filtering data to remove noise or outliers. A good example are mapping approaches generating indoor floor plans from crowd-sensed data [6, 73, 90, 136] like movement trajectories. Although each single trajectory is imperfect since, for example, gyroscopes of cheap inertial measurement units tend to drift, step detection and/or step length are inaccurate, etc., an accurate floor plan can be generated by considering *all* trajectories covering the same location (and possibly further knowledge like building outlines). This example also directly leads to another quality aspect, namely geographic coverage of data.

Coverage

Having highly precise and accurate data are often not sufficient for large-scale applications, if these data are only available at scattered locations. *Spatial-temporal coverage* of sensor data is hence a crucial quality metric. Although it might be relatively easy to achieve high precision and accuracy with professional measuring equipment, given the required overhead in cost for professional sensing (see Sect. 6.3.1.3), it can only be used at few locations. In contrast, crowdsensing utilizes many personal user devices roaming around a wide geographic area. Especially in urban areas, this leads to high spatial-temporal coverage. It is fair to point out that this coverage might greatly vary between, e.g., urban and rural areas or at day and night. However, many crowdsensing applications benefit from the correlation between user density and relevance of a location. In other words, if an area is interesting for users, there might be numerous users located at this area. Again, building maps based on crowdsensing is a good example. Locations visited by many users generate a large volume of data and can thus be mapped accurately. Similarly, a crowdsensing system for detecting delays in public transport includes many interested users as well as mobile sensors, since both coincide. Consequently, crowdsensing is able in many cases to deliver more data exactly where needed. This is also observed in peer-to-peer systems, where more resource demand from peers coincides with more resources offered by peers.

Besides being able to provide data from large geographic areas, crowdsensing also naturally captures data about relationships between people, and between people and their environment. For example, social relationships can be inferred from the pattern of people meeting. This pattern can be detected from location or proximity data captured by positioning systems or short-range communication technologies. As a result, the notion of coverage can be extended from the geographic domain to other domains, such as the social domain.

Velocity

Another data quality criterion is the timely availability of data. Note that this requirement can be orthogonal to the temporal coverage requirement. Temporal coverage requires data to be available for any time, while timely availability means that data should be available as fast as possible, ideally immediately. To meet the later requirement, *online* sensing methods are necessary. They hence are opposed to capturing data offline, post-processing it, and then after some time making it available.

Due to the availability of fast mobile communication technologies and cheap data plans (flat rates), modern mobile devices are often "always on(line)". Thus, sensed data can be made available online within short time, and "flows" at a high "velocity" as streams of sensor data or streams of events. In particular, this allows for the fast automatic reaction to situations in the physical world to control so-called *cyber-physical systems* online. A popular example is a traffic management system controlled by "floating car data" capturing the speed of cars (sensors), equipped with mobile communication technologies to stream their position and speed on the road network.

Variety

Another interesting feature of crowdsensing is the possibility to capture a *variety* of data with the same devices. Considering again the example of a laser scanner vs. a smartphone, the laser scanner can just capture 3D point clouds, whereas a Tango tablet can capture as well sound (microphone), 2D images and videos (standard camera), and possibly anything else captured by further sensors, e.g., connected through Bluetooth like health-related data of users from a fitness tracker. In other words, today's mobile devices are already multi-purpose sensor platforms, and the trend towards wearable computing will further increase the variety of sensor modalities that can be captured by users.

6.3.1.3   Cost and Time

Another aspect that is essentially related to the coverage quality metric is the *cost of the sensing infrastructure*. Covering a large geographic area with a dense "classic" sensor network of dedicated fixed sensors obviously implies both, large capital investment cost for buying and installing the sensors as well as operational cost for maintaining the infrastructure like monitoring and replacing faulty sensors, replacing batteries of battery-powered sensors, etc. These high cost might make the implementation of applications unattractive, or prevent them from being implemented in the first place, e.g., by small start-up companies or communities with small budget. Moreover, deploying a fixed sensor network for a new application also takes significant time. Consequently, it slows down the speed of bringing new sensing applications into the market. Crowdsensing systems do not suffer from these problems since they employ a readily available base of mobile devices from the crowd. Implementing a new crowdsensing application might be as easy as implementing an app, putting it in an app store, and setting up a backend system, e.g., in a cloud, to collect and process the collected sensor data. The sensing hardware is operated and constantly replaced by newer models bought by participants. Obviously, avoiding the time for deploying a sensing infrastructure significantly reduces the time to market. Besides reducing the installation time, existing applications also benefit from the fact that sensors are already deployed as soon as unexpected events like accidents, traffic jams, or problems in public transportation need to be monitored.

In addition to reducing both deployment cost and time, mobile applications can be developed very quickly and easily. Powerful platforms like Android or iOS and development environments are available. The popularity of these platforms also has led to a large community of developers who are familiar with these tools and platforms. Moreover, these platforms are highly popular from an end-user's perspective and come with app stores and marketplaces reaching billions of users. Thus, the time to gather a critical mass of participants is minimized.

Before heading on to the user perspective, we have shown so far that crowdsensing may not only deliver sensor data of potentially high-quality data at low cost, but also at a high volume and velocity, as well as great variety. These three "V" are known as the fundamental defining parameters of *big data*. Sometimes, the three "V" are extended by two more "V", namely, veracity (or alternatively validity) and value. Veracity or validity refer to the uncertainty of data, which is obviously the case for data sensed by a crowd of participants of heterogeneous trustworthiness using commodity hardware and sensors as discussed above. Moreover, considering the various applications introduced in Sect. 6.2.2 that are benefiting from crowd-sensed data, it is also easy to see that the data collected by the crowd is of high value for many applications, companies, and communities. We can therefore conclude that crowdsensing is a good candidate to capture *high-quality big data at low cost* to provide data, for instance, to machine learning algorithms relying on large quantities of data or the online control of cyber-physical systems relying on timely data.

Being able to deliver big data is challenging and requires processing and storing these large volumes of data coming in at high speed. Thus, *scalability* becomes a major requirement. Typically, a crowdsensing system might start small and then dynamically grow with the popularity of the application. Investing in a dedicated powerful backend infrastructure designed for the maximum size is therefore not cost efficient. Cloud computing with the capability to scale resources on demand and "pay-as-you-go" pricing models are hence good candidates to host the crowdsensing backend. Scalability and managing big data are however not specific to crowdsensing and generic solutions need therefore to be found.

6.3.2 Participants and Crowd Perspective

After having shown the benefits from the perspective of the crowdsensing system, we now consider the benefits from the perspective of an individual user or a user community, i.e., the crowd.

6.3.2.1 Personal Data

Since crowdsensing lets users collect data, it is a natural choice to collect data about themselves. This is inline with the currently experienced movement known as the *quantified self/me*, also known by other terms like *life logging*, where users collect a variety of data about themselves. Technically, this movement is driven by the various sensors embedded into smartphones or integrated into wearable devices like fitness trackers that can capture personal information. The captured information ranges from raw sensor data, such as heart rate to anything that can be inferred from this data including the user's activities, health status, lifestyle, etc. As a result, crowdsensing applications may contribute in supporting support users belonging to this movement.

6.3.2.2 Communities

Beyond collecting only personal information, crowdsensing allows users sharing common interests to gather into communities. One prominent example is the OpenStreetMap (OSM) community [3], which strives for creating highly accurate maps from crowd-sensed GPS traces completed by crowdsourced map creation from the collected traces or aerial images. This community acts in a purely altruistic way creating highly detailed and precise maps covering any relevant place of the world (see discussion about coverage above).

Another large user community includes drivers sharing the interest to avoid traffic jams, speed traps, etc. For example, the Waze application [170] gathers map data and traffic information from tens of millions of users to provide a navigation service based on crowd-sensed real-time traffic data. The monetary value of the data and user community can be estimated by considering the price that was paid by Google in 2013 to buy Waze: 1.1 billion USD.

6.3.2.3 User Incentives

By contributing to crowdsensing applications, participants can gain several benefits depending on the nature of the applications and the incentive model applied. As detailed in Sect. 6.5.1, applications can reward participants by paying for their contributed data. However, being part of a community and being well recognized within this community for their contributions is often already sufficient. For example, the navigation application Waze [170] has introduced a point system for rewarding users contributing data. Points can be earned by editing the map (adding street names, new road recording, etc.), gas price and road reports, place photos, etc. A user archiving the highest rank is recognized as a "Waze Royalty" showing other users of the community that this is a highly active user.

Introducing gaming aspects and fun is also a means to motivate people to go outside for a walk and exercise, while collecting data on the go. The popular augmented reality game Pokémon Go by Nintendo, where players catch virtual Pokémons located in the physical world, is a good examples showing the possibility to engage millions of users through games.

In summary, participants can benefit from data about themselves and being part of communities aiming at the same goal. Being an active member in a community may further foster social interactions within it. Moreover, they may gain additional revenues when contributing data or benefit from the data themselves within the communities built around the applications. Depending on the applications, it may also allow them to discover new environments or to have fun and compete with others.

6.4 Potential Risks for Crowdsensing Participants

By registering and contributing to crowdsensing applications, participants can be exposed to two main risks: Their privacy can be put at stake and/or they can experience a depletion of their device's resources depending on the underlying design of these applications. In what follows, we describe the associated risks and selected solutions proposed in related work to address these risks.

6.4.1   Threats to Privacy

In absence of privacy-preserving mechanisms, participants of crowdsensing applications can reveal a wide range of information about themselves.

Starting by the registration process, participants may need to provide their real identity and bank information, so that monetary rewards can be paid. Assuming that participants are able to use pseudonyms, their manual search for tasks to be fulfilled in their proximity and the corresponding download can still reveal their location.

Similarly, applications can monitor the participants' location to dynamically distribute sensing tasks to participants closely located to the events of interest.

Moreover, most applications studied in the survey published in [41] annotate the collected sensor readings with time and location information and the current participants' location may also be identified based on the collected sensor readings. For example, pictures, audio samples, and pollution data may include unique features, exposing the participants' whereabouts. Consequently, applications are able to follow the participants' *whereabouts*. Note that additional insights about these threats can be found in the chapter entitled "Location Privacy in Spatial Crowdsourcing" by H. To and C. Shahabi.

In both tasking and sensing steps, the participants' locations may lead to the identification of participants using pseudonyms based on the inference of their home address since participants usually commute from and to this location on a daily basis [103]. Alternatively, a cross-analysis of the participants' mobility patterns could lead to their re-identification based on the uniqueness of their location traces [52].

In addition to the spatiotemporal information about the collected sensor readings, the sensor readings themselves may allow to infer characteristics about the collecting devices and/or participants [107]. Depending on the uniqueness of these characteristics, it may lead to an identification of the participants. For instance, it has been shown that devices can be fingerprinted, i.e., uniquely identified among others, based on an analysis of different sets of collected sensor readings [46, 57, 159]. It is however to be noted that the fingerprinting is usually performed in a lab setting and according to a synthetic scenario.

Besides, participants may be distinguished based on collected accelerometer data when walking [55] or when performing usual daily activities [104]. Additional information about them can be inferred by analyzing the same data. It includes participants' gender [88, 174], height [174], weight [174], as well as current activities [105]. A combination of these information may refine the portrait of the contributing participants and might lead to their identification when the size of the participants' pool is limited.

Accelerometer data can further contribute in revealing participants' keyboard inputs [20, 81] or their current touch actions, such as tapping, scrolling, and zooming, in a browser by leveraging JavaScript [119]. An analysis of the participants' keystroke patterns can also reveal their gender, estimated age, and used hand, thus potentially leading to an identification when combined [9, 18]. Similarly, the characteristics of swipe gestures may contribute in determining the participants' gender [124]. Note that the phone's usage pattern and the installed apps may also provide insights about both participants' gender and age as shown in [7].

6.4.1.1 Threat Modalities

After having highlighted the potential information that can be revealed about the participants, we consider the modalities under which their privacy can be endangered. Like most existing systems, crowdsensing applications can be subject to attacks mounted by *external* attackers.

To gain access to, e.g., the participants' identity or their whereabouts, external attackers can target the application server or communication happening between the different crowdsensing stakeholders. In this case, well-established security solutions can be applied to protect the concerned architecture components from these threats.

Since these threats are not specific to crowdsensing applications, most solutions proposed to protect the participants' privacy adopt an *internal* attacker model, in which the administrators of the crowdsensing applications, end users, or participants may threaten the participants' privacy.

Multiple threat scenarios have been envisaged in this context [31]. The privacy threats can be the result of either passive or active attacks. For example, in the former category, participants reporting their current location at a fine granularity may involuntarily reveal to the crowdsensing administrators the locations of others reporting their location at a coarser granularity based on potential similarities between the associated sensor readings. In the same category, stakeholders may also be honest-but-curious. In this case, they do not launch any active attacks to breach the participants' privacy, but leverage data they have normally access to in order to infer additional information about them.

In contrast, malicious crowdsensing administrators may deliberately distribute selected tasks to specific participants to be able to distinguish and identify them in the case of a selective tasking attack [153] or participants may impersonate others to disclose sensitive data about them [77]. In addition to attacks targeting the whole user base, specific groups of participants can be targeted depending on both the scope and scale of the attack. For example, their selection can be based on the participants' physical proximity to the attacker or be random when considering a remote attack [47, 48].

6.4.1.2 Threat Mitigation

Different solutions have been proposed to address the aforementioned threats to privacy. In what follows, we give an overview of selected solutions targeting different stages of the crowdsensing campaigns. For interested readers, additional details can be found in [31, 41].

Distribution of Sensing Tasks

Before the actual sensing step, insights about the participants can already be disclosed to the campaign administrators when participants search for interesting sensing tasks, e.g., in their physical vicinity or dedicated to a set of particular sensors.

Instead of directly querying the application server, alternative solutions can be implemented to protect the participants' anonymity. For example, tasking beacons including the task details can be broadcasted, so that nearby participants receiving them will not reveal their identity to the application server.

Besides, participants can download the tasks when located in densely populated locations [153] in order to become indistinguishable from participants sharing the same locations. In this case, their anonymity may be ensured during the task distribution, but further interactions within the crowdsensing systems may lead to their identification.

To protect their location, participants can apply privacy-aware routing schemes [95] or reduce the granularity at which their location is transmitted to the application server by applying spatial cloaking. Assuming that participants' groups share the same cloaked location, they become indistinguishable within this region. To build such shared regions, different techniques can be deployed, which mainly differ in the nature of the trusted entities. For example, a distributed and collaborative scheme relying on other participants has been proposed in [98, 99], while a scheme relying on a unique central entity has been introduced in [161]. In both cases, the participants would receive tasks based in their common cloaked regions. To reduce the trust in other entities to the minimum, the network provider can be leveraged, as it knows de facto the participants' location and serves as a broker [157]. By doing so, the network provider allows a distribution of the location-based tasks without revealing the participants' location to the campaign administrators.

In addition to campaign administrators, honest-but-curious participants may also become adversaries when the applications implement a bidding scheme to distribute the tasks between the participants. Assuming that participants are able to see who is participating in a bid, the participants' location or interests can be revealed to other participants. In this case, the privacy-preserving auction model based on the concept of differential privacy [92] can be applied.

Once the tasks have been distributed to the participating devices, different filtering methods can be applied ex ante to reduce the granularity at which both the sensor readings and the associated location are collected.

To allow the participants to control the granularity, dedicated interfaces [40] can be used and completed by picture-based warnings [37], which aim at increasing the participants' awareness about potential threats to their privacy based on their current selected settings.

Spatiotemporal Annotations of Sensor Readings

Post hoc methods can be further leveraged to mitigate the disclosure of location information to the campaign administrators. These methods can be categorized along the following categories: mixing, spatial cloaking, data perturbation, and data aggregation.

Mixing-based solutions build on breaking the link between the devices and the collected sensor readings annotated with spatio-temporal information. They can further be divided between solutions focusing on mixing individual sensor readings and those considering participants' trajectories, i.e., series of individual sensor readings. In the former category, mixing the sensor readings implies exchanging sensor readings between participants. To support these exchanges, servers [153], peer-to-peer routing [162], or ad-hoc communication [36, 39, 143] can be leveraged. The trustworthiness of ad-hoc exchange partners can be evaluated [38] and a minimum trust level defined by the participants using dedicated interfaces [35]. While trajectory-based solutions often follow similar principles, the consideration of consecutive sensor readings introduces additional constraints, such as ensuring similar mobility patterns between mixed trajectories. To cater for these constraints, a trusted server can be used [74, 75]. Reducing the trust to such a server to the minimum can be achieved by implementing peer-based exchanges before the server-based processing step [129] or adopting a fully distributed solution [17]. In the latter solution, the trust in other participants is lowered by alternatively exchanging trajectories including or excluding sensitive locations. Besides, the chronology of the collected sensor readings can be locally modified to additionally break the link between both the collection time and location [25].

Like for the task distribution step, *spatial cloaking* can be applied to reduce the degree of granularity at which the collection location is reported to the application. With spatial cloaking, the participants report a coarser region that includes their original location and can be shared with others instead of the exact collection location. Again, different proposed solutions [5, 24, 60, 160] range from centralized to collaborative and distributed schemes that require to trust either a third party or other participants. In contrast, both solutions [32, 149] build on the concepts of secure multi-party computation, so that participants only reveal their locations to participants sharing them.

In comparison, *data perturbation* preserves the degree of granularity of the location information, but the participating devices apply noise on the collected data. By doing so, the individual sensor readings are protected, but the application server can still compute statistical trends over the submitted data. As a result, the privacy protection depends on the applied noise distribution. Assuming that all participants apply the same noise distribution, malicious participants can infer the original data by launching a brute-force attack and observing the resulting perturbed data when knowing the original data. Therefore, different perturbation schemes [79, 180, 185] have been proposed and build on different noise distributions within the same crowdsensing campaign. To further enhance the privacy protection, the noise distribution can also be adapted to previously collected sensor readings.

An alternative method to protect the collected sensor readings before reporting them to the application server is *data aggregation*. In this case, the sensor readings collected by different participants are merged, so that individual contributions become indistinguishable. Similarly to data perturbation, the individual data are protected against curious application administrators, but the application sever can still obtain aggregated results, such as averaged values, computed over participants' sets belonging to the same aggregation group [62, 63, 109, 183].

The aforementioned privacy-preserving solutions based mixing, spatial cloaking, data perturbation, or data aggregation can be applied before the participants report their collected data to the application server.

Storage and Access of Sensor Readings

Alternatively, participants can retain the control over their sensor readings by installing and managing individual repositories or use cloud-based solutions. In addition to protecting the participants' data against untrusted cloud providers, cryptographic solutions, such as the one proposed in [15], allow users to grant access rights to these data by selecting particular end users or crowdsensing campaigns. To support this access control, different solutions have been proposed taking into account the crowdsensing specificities [22, 23, 28, 29, 128].

In summary, we have highlighted in this section the potential risks for participants of crowdsensing applications in terms of privacy. We have illustrated these risks by different examples, resulting from the collection of sensor readings, their spatiotemporal annotation, or a combination of both. We have discussed how different stakeholders may potentially threaten the participants' privacy and given an overview of different privacy-preserving solutions that can be applied at different stages of the crowdsensing campaign.

6.4.2 Resource Investment

In addition to threats to privacy, the participants' contributions to crowdsensing campaigns require resource investments, which range from personal to technical investments. Participants may first invest time and physical efforts to fulfill sensing tasks. For example, sensing tasks may not be in the participants' direct vicinity, and hence require them to cover an additional distance to be able to execute them. Similarly, sensing tasks may require the participants' involvement, e.g., by manually triggering the data collection process. We have shown in [150] that our 207 participants of a mobile crowdsourcing platform similarly value the efforts required to walk between two waypoints and take a picture. In addition, crowdsensing applications can lead to increased resource consumption for the contributing devices, especially in terms of both battery and data volume consumption.

6.4.2.1 Battery Lifetime

Energy is one of the most critical resources to be conserved since participants will not tolerate a significant decrease in device runtime. Smartphones acting as mobile sensors nodes are typically recharged once per day. Thus, energy is not as critical as for wireless sensor nodes that need to run for years from a single non-rechargeable battery. Still, the different stages of a crowdsensing application can contribute to reduce the battery lifetime of the participants' device. For example, crowdsensing applications may offer a location-based search of new sensing tasks in physical proximity. In this case, the devices' localization based on GPS and/or Wi-Fi can rapidly deplete the battery. Likewise, most of the applications considered in [41] annotate the collected sensor readings with time and location information, hence requiring the activation of the devices' positioning system when collecting data. The data collection in itself also contributes to increased energy consumption, especially when, e.g., a depth-camera used for capturing point clouds is involved and can reduce the device runtime by hours.

Furthermore, the task distribution to the participants and the repeated trans-mission of the collected sensor readings to the application server shrink the available energy budget. To lower the impact of the participants' contributions to crowdsensing applications and hence extend the battery lifetime, energy-efficient sensing is therefore mandatory. Several methods like scheduling sensing tasks to only sense where and when necessary [11, 137, 172, 173], or model-driven sensing [138] learning a model of the sensed phenomenon—e.g., spatio-temporal model of temperature at certain locations and time—and then deriving values from the trained model without sensing have been proposed. Additionally, sensing activities can be scheduled in parallel to other tasks to prevent dedicated wake-ups when the devices are in an idle state [106]. This means that the devices can collect sensor readings during phone calls or the utilization of particular apps. Consequently, the data collection is determined by the participants' or devices' behavior that may hence lead to irregular collection frequencies to the benefit of the devices' lifetime. A similar approach is to piggyback the transmission of both sensing tasks and results during phone calls as introduced in [176].

6.4.2.2 Data Volume

In addition to energy consumption, additional costs for the participants can be incurred by the necessary communication between the devices and the application server. These costs can be a significant overhead for participants with a limited data volume. To reduce these costs to the minimum, a solution is to schedule the communication with the server when a Wi-Fi connection can be established, e.g., when the participants are at home. This solution can, however, only be applied in delay-tolerant application scenarios. To go a step further and potentially reduce the delay between transmissions of sensor readings to the server, a framework is proposed in [165]. This framework optimizes the incurred communication costs for participants with limited data volume by offloading the data to Bluetooth and/or Wi-Fi gateways in the participants' proximity. Additionally, nearby participants having subscribed to an unlimited data volume are leveraged as relays to the application server. In the latter case, acting as relays will impact the energy budget of these participants.

In summary, crowdsensing applications introduce overheads for the participants in terms of time and energy. Additionally, they may reduce the devices' battery lifetime and incur additional costs for participants with limited data volume. These overheads in both isolation and combination may prevent potential participants from joining crowdsensing applications or lead to later opt outs. It is therefore of primary importance to limit resource depletion, especially in terms of battery consumption, to both engage new participants and maintain them in the user base.

6.5 Potential Risks for Crowdsensing Applications

For crowdsensing applications, relying on volunteers to fulfill sensing tasks offers many opportunities as highlighted in Sect. 6.3.1. Simultaneously, this represents a risk, as the applications are dependent on the participants and their contributions.

To be viable, applications should be supported by a sufficient number of participants, who will be able to execute the sensing tasks and collect sensor readings of satisfying quality. In addition to a large user base, this implies that the participants should be located in proximity of the phenomena to be observed and own devices equipped with the appropriate sensors as we discuss in detail in Sect. 6.5.2.

To foster contributions in both the short- and long-term, different factors come into play. For example, the risks for the crowd in terms of *privacy* and *resource depletion* should be reduced to the minimum by, e.g., applying methods discussed in Sects. 6.4.1 and 6.4.2, respectively. Indeed, we have shown in [42] that it is important for potential participants that their privacy is respected when contributing to a participative application. Moreover, *incentives* can be introduced to motivate participants to report sensing readings. In Sect. 6.5.1, we give an overview of existing incentive models.

Having a large user base is, however, not sufficient to ensure the applications' sustainability: Being open systems, crowdsensing applications not only depend on the *contributions' quantity*, but also on their *quality*. Erroneous contributions can result from malfunctioning devices or intentional tampering of the sensor readings by malicious participants. To motivate participants to contribute data of good quality, incentives may play an important role. Nevertheless, we have shown in [150] that tasks associated to high monetary rewards lead to a greater rate of low-quality submissions than less rewarded tasks. Moreover, applications are often running on limited financial budgets, so that solutions aiming at minimizing the participants' reward while maximizing the contributions' quality have been proposed and are presented along with other alternatives to ensure the contributions' quality in [92].

6.5.1 Engagement of Participants

Participants' contributions are the sine qua none condition for the success of crowdsensing applications. This means that potential participants do not only need to install a crowdsensing application, but also continue to use it over a longer time period. The initial engagement of potential participants is eased by the distribution of the crowdsensing applications via the existing app stores. With a forecasted number of 5 million of apps available in the Apple App Store in 2020 [135], crowdsensing applications may however remain unnoticed to most users. To reach the chasm, dissemination is a crucial aspect. Word of mouth can be determinant and contribute to the adoption of applications at a larger scale, eventually leading to build communities around the applications. Communication and marketing may help to increase the visibility of applications, but the participants' decision to install and later use the applications can be influenced by different factors. In what follows, we especially consider existing incentives that may motivate potential participants to engage in crowdsensing applications based on the taxonomy proposed in [87].

6.5.1.1 Monetary Incentives

To motivate users to participate in crowdsensing applications and reward them for their contributions, multiple schemes based on monetary incentives have been proposed. Their common goal is to optimize the price paid by the applications by taking into account different constraints, such as the quality and quantity of the submitted contributions, the participants' physical distribution, their privacy protection, or their long-term engagement.

In most cases, the proposed schemes are based on dynamic pricing strategies. For example, the so-called participants' quality of information is introduced in [91] and is taken into consideration in the design of reverse combinatorial auction models. Following the same line, methods are designed to balance the incentives and the data quality for sensing tasks decomposable into subtasks and needed to be fulfilled within a given timeframe in [163, 177], respectively. In [93], the authors propose an incentive scheme that computes the rewards not only based on the quality of the reported sensor readings, but also on the resulting impact on the participants' privacy. The proposed incentive scheme is completed by both aggregation and perturbation mechanisms, so that data reliability and participants' privacy are simultaneously addressed.

The aforementioned models assume that the participants contribute the sensor readings as expected. In contrast, the solutions introduced in [89, 123] consider the uncertainty introduced by the open nature of the crowdsensing applications. They integrate the risks that participants may submit the requested data at a later time or not at all, respectively. If the participants do not deliver the data, the incentives can be revoked and redistributed between the remaining participants as proposed in [123]. To further reduce this uncertainty factor, participants' profiles are introduced in [96]. These profiles aim at better understanding the participants' preferences in terms of type of tasks, distance to the tasks, as well as rewards. This allows potential campaign administrators to optimize the distribution of both tasks and rewards.

Alternatively, participants can be rewarded by incentives whose price is known beforehand. This, however, requires the applications to be able to estimate the appropriate rewards a priori, so that the participants will be neither underpaid nor overpaid. On the one hand, underpaid participants may take more time to fulfill tasks or not fulfill them in the worst case. On the other hand, overpaying the participants obviously impacts the campaign budget in the short term, but it may contribute to maintain the participants' motivation to contribute to the applications in the long term. Consequently, the applied pricing models need to be fair with the participants, while ensuring the collection of sensor readings of good quality.

In [150], we have analyzed the impact of newly introduced location-based notifications in an existing and running mobile crowdsourcing application. Using two questionnaire-based studies counting 335 users in total, we have explored the participants' expectations in terms of rewards and the value they attribute to invested resources. We have shown based on the users' answers coupled with an analysis of the application database that rewards impact both the time to the tasks' fulfillment and the quality of the contributions. In this case, the observed quality is, however, lower when the reward is higher. More generally, we have examined the impact of demographics, incentives, and collection conditions on the willingness of 200 anonymous participants to contribute to crowdsensing applications in [33]. Our results show that young participants sharing information online would be more willing to continuously report sensor readings to sensing campaigns initiated by academic institutions for a monthly reward of 50 euros on average.

6.5.1.2 Non-Monetary Incentives

Depending on the nature of the crowdsensing applications, non-monetary incentives may also foster users' contributions. In this case, other factors like altruism and competitiveness [146] may come into play. Both monetary and non-monetary incentives can be combined to achieve a short- and long-term participants' involvement. To be able to rely on altruism only, applications should usually have a goal to which multiple potential participants can adhere and which is beneficial to the community as a whole.

The ideal is to build communities around these applications, so that participants not only contribute data, but also gain insights about the observed phenomena and are able to socially exchange with others. Campaigns organized around communities can further leverage participants located in physical proximity to fulfill tasks requiring multiple stages as proposed in [27]. In community-based campaigns, reputation and competition may be an additional factor to motivate users to participate by displaying rankings of participants based on, e.g., the total number of collected sensor readings, their quality, the number of fulfilled tasks, or the covered area. Showing the results of others may encourage participants to overbid them by contributing more or better data. To reward contributing participants, virtual rewards in the form of, e.g., badges can also be distributed. The badges attest that the participants have reached given contributor levels. Alternatively, participants submitting sensor readings can be allowed to access advanced features in the application as proposed in [116].

To benefit from existing communities, we have proposed the concept of Cached-Sensing in [34] with which users can both create and execute sensing tasks. The sensing tasks are written on NFC tags hidden by their creator. As soon as the creator activates a new task, it becomes available to other users, who can search for it using the coordinates and information provided by the creator. Users who find hidden tags, read and execute the associated sensing tasks with their device. For each successfully fulfilled task, a new logbook entry is created for the corresponding user. As a result, CachedSensing shares similarities with existing location-based games, in which the players explore their environment to find different objects. For example, users hide containers with a pen and a logbook in Geocaching, search for QR codes in Munzee, and collaborate to virtually defend particular locations in Ingress. CachedSensing, however, differs from existing games, as it introduces a new component: the sensing tasks.

By merging both crowdsensing and location-based gaming paradigms, our objective is to open sensing tasks to existing gaming communities, as Geocaching.com and Munzee count worldwide over 3 million and 300,000 players, respectively [1, 2]. To this end, we have conducted a questionnaire-based study with 337 anonymous gamers belonging at least to one of the aforementioned communities. The results published in [148] show that our participants would be more ready to contribute to sensing tasks when those would be integrated into a game they are already playing. Virtual badges and associated points have not been perceived in our sample as an efficient incentive to motivate users to contribute data, especially within the Geocaching community. On the contrary, some participants commented that such incentives would even dissuade them to contribute to sensing tasks. Overall, pleasure and social contacts seem to be more efficient motivating factors than competition and rankings in these communities. Similar observations have been made in other online communities, such as computer role-playing gamers and an online tourism community [158, 168]. Moreover, our participants are overall both greedy and altruistic: they would be ready to contribute to crowdsensing if it is fun, but would refuse if it would serve commercial interests.

In summary, different incentive models can be introduced in crowdsensing applications to motivate participants to contribute. Their choice depends on the nature of the application, the involved parties (e.g., scientific community vs. companies), the difficulty and the frequency of the sensing tasks to be executed, the physical distribution of the participants, as well as the available campaign budget. As shown in [148], competition may not reach the expected results in specific communities, such as geocachers, where enjoyment would better foster participants' contributions.

6.5.2   *Quantity and Quality of Contributions*

In addition to motivating participants to contribute, crowdsensing applications need to ensure that the reported data match their requirements in terms of quantity and quality.

While the number of reported sensor readings is closely related to the applied incentive model(s) (see Sect. 6.5.1), crowdsensing applications can apply different strategies to further optimize the distribution of the tasks to participants of interest, so that they only obtain tasks tailored to their current locations or personal interests. By using personalized models such as proposed in [96], the answer rate may increase as participants would be less bothered by uninteresting tasks.

Obtaining a sufficient number of data points is especially crucial in applications requiring a large coverage over a long period of time, such as needed in the construction of noise pollution or air quality maps. To reach these objectives, participants can adopt several roles as proposed in [97], so that the spatiotemporal requirements of the data collection can be fulfilled. In this case, the participants are not only contributing sensor readings, but can also act as relays and/or uploading agents for other participants to ensure timely reporting to the crowdsensing application. Using a real-world deployment involving 85 participants, the relationships between the number of participants and the coverage of specific locations based on their characteristics are analyzed and modeled in [30]. The results confirm that a relatively small set of participants can be sufficient to cover most of the popular locations and provide guidelines for application developers interested in monitoring specific locations.

Due to their open nature, crowdsensing applications may be prone to erroneous contributions, which may result from, e.g., malfunctioning devices, non-compliant sensing conditions, or ill-intentioned participants.

To ensure the quality of the obtained results, the participants can first be selected based on the quality of their last contributions. For example, the k-best participants are identified in [118] based on the expected sensing quality and resulting coverage. In [111], the quality level of the results is defined at the time of the task creation, and hence determines to whom the task will be distributed based on the quality level achieved by the participants in the past. Such quality-aware selection may not only contribute to more accurate results, but also reduce the number of participants to whom the sensing tasks need to be distributed. As a result, this may lead to an overall reduction of the resources invested by the participants and the associated incentives (see Sects. 6.4.2 and 6.5.1). The same goal is pursued in [164], where resources are saved by reducing the number of sensing participants. Spatiotemporal correlation is then applied to obtain missing data with high accuracy. Different techniques, such as semantic hashing, space-efficient filters, or compressing sensing, can be implemented to balance quality and resource consumption [112].

Once the participants have been selected and the sensor readings collected, additional methods can be applied to sort out and eliminate low-quality data. For example, the Expectation Maximization framework introduced in [182] can help campaign administrators in identifying and eliminating duplicates about the same events. In the case of both sparse and redundant data, the solution proposed in [121] can be applied to first reconstruct missing data, before identifying correct contributions. In absence of duplicates, further methods have been proposed to ensure data integrity. Most of them build on attributing a reputation score to each participant. These reputation scores can serve as weight in the computation of summaries over all participants or as threshold defining which sensor readings will be included or excluded from the computed summaries. They also allow for ranking the participants in different groups based on their degree of trustworthiness as proposed in [178]. The reputation scores mostly reflect the quality of participants' contributions, but additional factors, such as timeliness of the results as considered in [145], can also be taken into account in their computation. For example, the Gompertz function is used in [85] and a voting approach supported by trustworthy participants is presented in [139] to assess the quality of the participants' contributions. In addition to detect incorrect contributions, the solution introduced in [26] proposes to correct them using spatio temporal compressive sensing techniques.

In the aforementioned solutions, assessing the participants' trustworthiness requires to link their identity/device with their contributions. By doing so, their identity is hence linked with the spatiotemporal annotation of the sensor readings, thus putting the participants' location privacy at stake (see Sect. 6.4.1). To address both a priori conflicting aspects, namely reputation and privacy, we have first proposed a framework based on periodic pseudonyms and a transfer of reputation between these pseudonyms based on blind signatures in [43, 44]. Our framework is agnostic to the applied reputation algorithm. Further solutions addressing the same challenge have been later proposed in [78, 86, 110, 122, 166, 167, 181].

In summary, applications can minimize the risks of suffering from a lack of data by optimizing the participants' selection based on their personal characteristics. This ex-ante selection can also take into consideration the quality of the reported data, so that the tasks are predominantly distributed to reliable participants. To ensure the quality of the results, post-ante mechanisms should be applied to recover missing data and eliminate erroneous contributions. Participants having contributed data of good quality should be rewarded by a good reputation score, which can be considered in the distribution of both incentives and future tasks.

6.6 Potential Risks for End Users

After having adopted the perspectives of both participants and campaign administrators, we finally consider the perspective of end users. End users can include the participants and the campaign administrators, but also any person or group of persons interested in the collected sensor readings. For example, end users can be scientists interested in monitored phenomena or friends and family of contributing participants.

While end users do not directly contribute data to the crowdsensing applications at the same level as participants, they still play an important role in the crowdsensing ecosystem. Depending on the application scenarios, end users may directly contribute to the sustainability of the crowdsensing applications by, e.g., sponsoring the incentives distributed to the participants. Moreover, large end user bases may indirectly encourage participants to contribute to the applications, as the insights drawn from the collected data may benefit a large community. Like for the participants, crowdsensing applications, hence, need to be attractive for end users.

By consuming data collected by the participants, end users may, however, disclose information to the campaign administrators or the participants about themselves. Their queries to the application server or their subscription to different data streams may reveal their current locations or personal interests. To preserve their privacy against the campaign administrators, tokens can be distributed by the application server to the end users, who can use them directly to query the participants having collected the data as proposed in [59]. Alternatively, the matching between the data collected by the participants and the queries submitted by the end users can be done without revealing them to the campaign administrators by using the solutions introduced in [49–51]. Consequently, the privacy of both participants and end users are simultaneously protected against the campaign administrators.

In summary, the interactions of the end users with the crowdsensing applications may endanger their privacy in absence of protection mechanisms. Since end users contribute to the viability of these applications, their privacy needs to the protected.

6.7 Conclusions

In this chapter, we have shown that a crowd of participants with personal, sensor-equipped devices can deliver "big data" of high volume, velocity, variety, veracity, and value without investing into a dedicated sensor network or the operational cost for maintaining this sensing infrastructure. We have introduced a common system architecture for crowdsensing applications showing the basic building blocks of crowdsensing systems. Through various application examples, we have highlighted that crowdsensing has multiple application areas including commercial and non-commercial (community) applications.

Besides showing the benefits of crowdsensing, we also have identified and discussed its risks and challenges. In order for crowdsensing to be accepted by participants, their privacy must be protected, and resources of personal devices and participants must be preserved. Otherwise, the participants will not participate in these applications, thus posing a serious problem for the adoption of the crowdsensing paradigm at large scale. Moreover, users have to be motivated to participate in crowdsensing, e.g., by giving them suitable incentives. On the other side, the applications need to ensure a minimum quality of the users' contributions. Although novel concepts for better protecting the participants' privacy, increasing the efficiency of crowdsensing, or ensuring both quantity and quality of sensed data are still subject to ongoing research, already existing results in these research areas lead us to the conclusion that the risks and challenges of crowdsensing can be successfully tackled to make crowdsensing a useful and practical concept to provide data to a broad range of applications.

Acknowledgment This chapter has been written when Delphine Reinhardt was working at the University of Bonn and Fraunhofer FKIE in Wachtberg.

References

1. About Munzee. Online: https://www.munzee.com/about/ (accessed in 02.2017).
2. Geocaching. Online: https://www.geocaching.com/press/faq.aspx (accessed in 02.2017).
3. Open Street Map Project. Online: https://www.openstreetmap.org (accessed in 03.2017).
4. T. F. Abdelzaher, Y. Anokwa, P. Boda, J. A. Burke, D. Estrin, L. Guibas, A. Kansal, S. Madden, and J. Reich. Mobiscopes for Human Spaces. *IEEE Pervasive Computing*, 6(2):20–29, 2007.
5. B. Agir, T. G. Papaioannou, R. Narendula, K. Aberer, and J.-P. Hubaux. User-side Adaptive Protection of Location Privacy in Participatory Sensing. *Geoinformatica*, 18(1):165–191, 2014.
6. M. Alzantot and M. Youssef. CrowdInside: Automatic Construction of Indoor Floorplans. In *Proceedings of the 20th International Conference on Advances in Geographic Information Systems (SIGSPATIAL)*, pages 99–108, 2012.
7. I. Andone, K. Blaszkiewicz, M. Eibes, B. Trendafilov, C. Montag, and A. Markowetz. How Age and Gender Affect Smartphone Usage. In *Proceedings of the 2016 ACM International Joint Conference on Pervasive and Ubiquitous Computing (Adjunct UbiComp)*, pages 9–12, 2016.
8. M. Annavaram, N. Medvidovic, U. Mitra, S. Narayanan, G. Sukhatme, Z. Meng, S. Qiu, R. Kumar, G. Thatte, and D. Spruijt-Metz. Multimodal Sensing for Pediatric Obesity Applications. In *Proceedings of International Workshop on Urban, Community, and Social Applications of Networked Sensing Systems (UrbanSense)*, pages 21–25, 2008.
9. M. Antal and G. Nemes. Gender Recognition from Mobile Biometric Data. In *Proceedings of the 11th IEEE International Symposium on Applied Computational Intelligence and Informatics (SACI)*, pages 243–248, 2016.
10. P. Baier, F. Dürr, and K. Rothermel. MapCorrect: Automatic Correction and Validation of Road Maps Using Public Sensing. In *Proceedings of the 36th Annual IEEE Conference on Local Computer Networks (LCN)*, pages 58–66, 2011.
11. P. Baier, F. Dürr, and K. Rothermel. Efficient Distribution of Sensing Queries in Public Sensing Systems. In *Proceedings of the 10th IEEE International Conference on Mobile Ad-hoc and Sensor Systems (MASS)*, pages 272–280, 2013.
12. E. Bales, N. Nikzad, N. Quick, C. Ziftci, K. Patrick, and W. Griswold. Citisense: Mobile Air Quality Sensing for Individuals and Communities Design and Deployment of the Citisense Mobile Air-Quality System. In *Proceedings of the 6th International Conference on Pervasive Computing Technologies for Healthcare (PervasiveHealth)*, pages 155–158, 2012.
13. X. Bao and R. R. Choudhury. MoVi: Mobile Phone based Video Highlights via Collaborative Sensing. In *Proceedings of the 8th ACM International Conference on Mobile Systems, Applications, and Services (MobiSys)*, pages 357–370, 2010.
14. M. Bilandzic, M. Banholzer, D. Peev, V. Georgiev, F. Balagtas-Fernandez, and A. De Luca. Laermometer: A Mobile Noise Mapping Application. In *Proceedings of the 5th ACM Nordic Conference on Human-Computer Interaction (NordiCHI)*, pages 415–418, 2008.

15. D. Biswas and K. Vidyasankar. Privacy Preserving Profiling for Mobile Services. *Procedia Computer Science*, 10(0):569–576, 2012.
16. U. Blanke, G. Troster, T. Franke, and P. Lukowicz. Capturing Crowd Dynamics at Large Scale Events using Participatory GPS-localization. In *Proceedings of the 9th IEEE International Conference on Intelligent Sensors, Sensor Networks and Information Processing (ISSNIP)*, pages 1–7, 2014.
17. I. Boutsis and V. Kalogeraki. Privacy Preservation for Participatory Sensing Data. In *Proceedings of the 11th IEEE International Conference on Pervasive Computing and Communications (PerCom)*, pages 103–113, 2013.
18. A. Buriro, Z. Akhtar, B. Crispo, and F. Del Frari. Age, Gender and Operating-hand Estimation on Smart Mobile Devices. In *Proceedings of the Biometrics Special Interest Group (BIOSIG)*, pages 1–5, 2016.
19. J. Burke, D. Estrin, M. Hansen, A. Parker, N. Ramanathan, S. Reddy, and M. Srivastava. Participatory Sensing. In *Proceedings of the 1st Workshop on World-Sensor-Web (WSW)*, pages 1–5, 2006.
20. L. Cai and H. Chen. TouchLogger: Inferring Keystrokes On Touch Screen From Smartphone Motion. In *Proceedings of the 6th USENIX Conference on Hot Topics in Security (HotSec)*, pages 9–9, 2011.
21. A. T. Campbell, S. B. Eisenman, N. D. Lane, E. Miluzzo, R. A. Peterson, H. Lu, X. Zheng, M. Musolesi, K. Fodor, and G.-S. Ahn. The Rise of People-centric Sensing. *IEEE Internet Computing*, 12(4):12–21, 2008.
22. S. Chakraborty, Z. Charbiwala, H. Choi, K. R. Raghavan, and M. B. Srivastava. Balancing Behavioral Privacy and Information Utility in Sensory Data Flows. *Pervasive and Mobile Computing*, 8(3):331–345, 2012.
23. S. Chakraborty, H. Choi, and M. B. Srivastava. Demystifying Privacy in Sensory Data: A QoI based Approach. In *Proceedings of the 9th Annual IEEE International Conference on Pervasive Computing and Communications (PerCom Workshops)*, pages 38–43, 2011.
24. S. Chakraborty, K. R. Raghavan, M. P. Johnson, and M. B. Srivastava. A Framework for Context-aware Privacy of Sensor Data on Mobile Systems. In *Proceedings of the 14th Workshop on Mobile Computing Systems and Applications (HotMobile)*, pages 11:1–11:6, 2013.
25. X. Chen, X. Wu, X.-Y. Li, Y. He, and Y. Liu. Privacy-preserving High-quality Map Generation with Participatory Sensing. In *Proceedings of the 33th IEEE Conference on Computer Communications (INFOCOM)*, pages 2310–2318, 2014.
26. L. Cheng, J. Niu, L. Kong, C. Luo, Y. Gu, W. He, and S. K. Das. Compressive Sensing based Data Quality Improvement for Crowd-sensing Applications. *Journal of Network and Computer Applications*, 77:123–134, 2017.
27. S. Chessa, A. Corradi, L. Foschini, and M. Girolami. Empowering Mobile Crowdsensing through Social and Ad Hoc Networking. *IEEE Communications Magazine*, 54(7):108–114, 2016.
28. H. Choi, S. Chakraborty, Z. Charbiwala, and M. Srivastava. SensorSafe: A Framework for Privacy-Preserving Management of Personal Sensory Information. In W. Jonker and M. Petković, editors, *Secure Data Management*, volume 6933 of *Lecture Notes in Computer Science*, pages 85–100. Springer Berlin Heidelberg, 2011.
29. H. Choi, S. Chakraborty, and M. B. Srivastava. Design and Evaluation of SensorSafe: A Framework for Achieving Behavioral Privacy in Sharing Personal Sensory Information. In *Proceedings of the 11th IEEE International Conference on Trust, Security and Privacy in Computing and Communications (TrustCom)*, pages 1004–1011, 2012.
30. Y. Chon, N. D. Lane, Y. Kim, F. Zhao, and H. Cha. Understanding the Coverage and Scalability of Place-centric Crowdsensing. In *Proceedings of the ACM International Joint Conference on Pervasive and Ubiquitous Computing (UbiComp)*, pages 3–12, 2013.
31. D. Christin. Privacy in Mobile Participatory Sensing: Current Trends and Future Challenges. *Journal of Systems and Software (JSS)*, 116:57–68, 2016.

32. D. Christin, D. Bub, A. Moerov, and S. Kasem-Madani. A Distributed Privacy-Preserving Mechanism for Mobile Urban Sensing Applications. In *Proceedings of the 10th IEEE International Conference on Intelligent Sensors, Sensor Networks and Information Processing (ISSNIP)*, pages 1–6, 2015.

33. D. Christin, C. Büchner, and N. Leibecke. What's the Value of Your Privacy? Exploring Factors That Influence Privacy-sensitive Contributions to Participatory Sensing Applications. In *Proceedings of the IEEE Workshop on Privacy and Anonymity for the Digital Economy (LCN Workshops)*, pages 946–951, 2013.

34. D. Christin, C. Büttner, and N. Repp. CachedSensing: Exploring and Documenting the Environment as a Treasure Hunt. In *Proceedings of the 7th IEEE International Workshop on Practical Issues in Building Sensor Network Applications (SenseApp, LCN workshop)*, pages 977–985, 2012.

35. D. Christin, F. Engelmann, and M. Hollick. Usable Privacy for Mobile Sensing Applications. In D. Naccache and D. Sauveron, editors, *Information Security Theory and Practice. Securing the Internet of Things*, volume 8501 of *Lecture Notes in Computer Science*, pages 92–107. Springer Berlin Heidelberg, 2014.

36. D. Christin, J. Guillemet, A. Reinhardt, M. Hollick, and S. S. Kanhere. Privacy-preserving Collaborative Path Hiding for Participatory Sensing Applications. In *Proceedings of the 8th IEEE International Conference on Mobile Ad-hoc and Sensor Systems (MASS)*, pages 341–350, 2011.

37. D. Christin, M. Michalak, and M. Hollick. Raising User Awareness about Privacy Threats in Participatory Sensing Applications through Graphical Warnings. In *Proceedings of the 11th International Conference on Advances in Mobile Computing and Multimedia (MoMM)*, pages 445–454, 2013.

38. D. Christin, D. R. Pons-Sorolla, M. Hollick, and S. S. Kanhere. TrustMeter: A Trust Assessment Framework for Collaborative Path Hiding in Participatory Sensing Applications. In *Proceedings of the 9th IEEE International Conference on Intelligent Sensors, Sensor Networks and Information Processing (ISSNIP)*, pages 1–6, 2014.

39. D. Christin, A. Reinhardt, and M. Hollick. On the Efficiency of Privacy-Preserving Path Hiding for Mobile Sensing Applications. In *Proceedings of the 38th IEEE Conference on Local Computer Networks (LCN)*, pages 846–854, 2013.

40. D. Christin, A. Reinhardt, M. Hollick, and K. Trumpold. Exploring User Preferences for Privacy Interfaces in Mobile Sensing Applications. In *Proceedings of 11th ACM International Conference on Mobile and Ubiquitous Multimedia (MUM)*, pages 14:1–14:10, 2012.

41. D. Christin, A. Reinhardt, S. S. Kanhere, and M. Hollick. A Survey on Privacy in Mobile Participatory Sensing Applications. *Journal of Systems and Software*, 84(11):1928–1946, 2011.

42. D. Christin, C. Roßkopf, and M. Hollick. uSafe: A Privacy-aware and Participative Mobile Application for Citizen Safety in Urban Environments. *Pervasive and Mobile Computing (PMC)*, 9(5):695–707, 2013.

43. D. Christin, C. Roßkopf, M. Hollick, L. A. Martucci, and S. S. Kanhere. IncogniSense: An Anonymity-preserving Reputation Framework for Participatory Sensing Applications. In *Proceedings of the 10th IEEE International Conference on Pervasive Computing and Communications (PerCom)*, pages 135–143, 2012.

44. D. Christin, C. Roßkopf, M. Hollick, L. A. Martucci, and S. S. Kanhere. IncogniSense: An Anonymity-preserving Reputation Framework for Participatory Sensing Applications. *Pervasive and Mobile Computing (PMC)*, 9(3):353–371, 2013.

45. Cisco Systems Inc. Cisco Visual Networking Index: Global Mobile Data Traffic Forecast Update, 2016–2021 White Paper. Online: http://www.cisco.com/c/en/us/solutions/collateral/service-provider/visual-networking-index-vni/mobile-white-paper-c11-520862.html (accessed in 05.2017), Feb. 2017.

46. A. Das, N. Borisov, and M. Caesar. Tracking Mobile Web Users Through Motion Sensors: Attacks and Defenses. In *Proceedings of the 23rd Annual Network and Distributed System Security Symposium (NDSS)*, pages 1–15, 2016.

47. E. De Cristofaro and R. Di Pietro. Preserving Query Privacy in Urban Sensing Systems. In *Proceedings of the 13th International Conference on Distributed Computing and Networking (ICDCN)*, pages 218–233, 2012.
48. E. De Cristofaro and R. Di Pietro. Adversaries and Countermeasures in Privacy-Enhanced Urban Sensing Systems. *IEEE Systems Journal*, 7(2):311–322, 2013.
49. E. De Cristofaro and C. Soriente. Short Paper: PEPSI—Privacy-enhanced Participatory Sensing Infrastructure. In *Proceedings of the 4th ACM Conference on Wireless Network Security (WiSec)*, pages 23–28, 2011.
50. E. De Cristofaro and C. Soriente. Extended Capabilities for a Privacy-Enhanced Participatory Sensing Infrastructure (PEPSI). *IEEE Transactions on Information Forensics and Security*, 8(12):2021–2033, 2013.
51. E. De Cristofaro and C. Soriente. Participatory Privacy: Enabling Privacy in Participatory Sensing. *IEEE Network*, 27(1):32–36, 2013.
52. Y.-A. de Montjoye, C. A. Hidalgo, M. Verleysen, and V. D. Blondel. Unique in the Crowd: The Privacy Bounds of Human Mobility. *Scientific reports*, 3(1–5), 2013.
53. L. Deng and L. Cox. LiveCompare: Grocery Bargain Hunting through Participatory Sensing. In *Proceedings of the 10th Workshop on Mobile Computing Systems and Applications (HotMobile)*, pages 1–6, 2009.
54. T. Denning, A. Andrew, R. Chaudhri, C. Hartung, J. Lester, G. Borriello, and G. Duncan. BALANCE: Towards a Usable Pervasive Wellness Application with Accurate Activity Inference. In *Proceedings of the 10th Workshop on Mobile Computing Systems and Applications (HotMobile)*, pages 5:1–5:6, 2009.
55. M. O. Derawi, C. Nickel, P. Bours, and C. Busch. Unobtrusive User-authentication on Mobile Phones using Biometric Gait. In *Proceeding of the 6th IEEE International Conference on Intelligent Information Hiding and Multimedia Signal Processing (IIH-MSP)*, pages 306–311, 2010.
56. S. Devarakonda, P. Sevusu, H. Liu, R. Liu, L. Iftode, and B. Nath. Real-time Air Quality Monitoring through Mobile Sensing in Metropolitan Areas. In *Proceedings of the 2nd ACM SIGKDD International Workshop on Urban Computing*, pages 1–15, 2013.
57. S. Dey, N. Roy, W. Xu, R. R. Choudhury, and S. Nelakuditi. AccelPrint: Imperfections of Accelerometers Make Smartphones Trackable. In *Proceedings of the 21rd Annual Network and Distributed System Security Symposium (NDSS)*, pages 1–16, 2014.
58. E. D'Hondt, M. Stevens, and A. Jacobs. Participatory Noise Mapping Works! An Evaluation of Participatory Sensing as an Alternative to Standard Techniques for Environmental Monitoring. *Pervasive and Mobile Computing*, 9(5):681–694, 2013.
59. T. Dimitriou, I. Krontiris, and A. Sabouri. PEPPeR: A Querier's Privacy Enhancing Protocol for PaRticipatory Sensing. In A. Schmidt, G. Russello, I. Krontiris, and S. Lian, editors, *Security and Privacy in Mobile Information and Communication Systems*, volume 107 of *Lecture Notes of the Institute for Computer Sciences, Social Informatics and Telecommunications Engineering*, pages 93–106. Springer Berlin Heidelberg, 2012.
60. K. Dong, T. Gu, X. Tao, and J. Lu. Privacy Protection in Participatory Sensing Applications Requiring Fine-Grained Locations. In *Proceedings of the 16th IEEE International Conference on Parallel and Distributed Systems (ICPADS)*, pages 9–16, 2010.
61. Y. Dong, S. S. Kanhere, C. Chou, and N. Bulusu. Automatic Collection of Fuel Prices from a Network of Mobile Cameras. In *Proceedings of the 4th IEEE International Conference on Distributed Computing in Sensor Systems (DCOSS)*, pages 140–156, 2008.
62. G. Drosatos, P. S. Efraimidis, I. N. Athanasiadis, E. D'Hondt, and M. Stevens. A Privacy-preserving Cloud Computing System for Creating Participatory Noise Maps. In *Proceedings of the 36th IEEE Annual Computer Software and Applications Conference (COMPSAC)*, pages 581–586, 2012.
63. G. Drosatos, P. S. Efraimidis, I. N. Athanasiadis, M. Stevens, and E. D'Hondt. Privacy-preserving Computation of Participatory Noise Maps in the Cloud. *Journal of Systems and Software*, 92(0):170–183, 2014.

64. S. B. Eisenman and A. T. Campbell. SkiScape Sensing. In *Proceedings of the 4th ACM International Conference on Embedded Networked Sensor Systems (SenSys)*, pages 401–402, 2006.
65. S. B. Eisenman, N. D. Lane, E. Miluzzo, R. A. Peterson, G. Ahn, and A. T. Campbell. MetroSense Project: People-centric Sensing at Scale. In *Proceedings of the 1st Workshop on World-Sensor-Web (WSW)*, pages 6–11, 2006.
66. S. B. Eisenman, E. Miluzzo, N. D. Lane, R. A. Peterson, G.-S. Ahn, and A. T. Campbell. The BikeNet Mobile Sensing System for Cyclist Experience Mapping. In *Proceedings of the 5th ACM International Conference on Embedded Networked Sensor Systems (SenSys)*, pages 87–101, 2007.
67. S. B. Eisenman, E. Miluzzo, N. D. Lane, R. A. Peterson, G.-S. Ahn, and A. T. Campbell. BikeNet: A Mobile Sensing System for Cyclist Experience Mapping. *ACM Transactions on Sensor Networks*, 6(1):1–39, 2009.
68. Endomondo.com. Endomondo. Online: https://www.endomondo.com (accessed in 05.2017).
69. K. Farkas, A. Z. Nagy, T. Tomás, and R. Szabó. Participatory Sensing based Real-time Public Transport Information Service. In *Proceedings of the IEEE International Conference on Pervasive Computing and Communications (PERCOM Workshops)*, pages 141–144. IEEE, 2014.
70. A. Farshad, M. K. Marina, and F. Garcia. Urban WiFi characterization via mobile crowdsensing. In *Proceedings of the IEEE Network Operations and Management Symposium (NOMS)*, pages 1–9, 2014.
71. FitnessKeeper, Inc. RunKeeper - GPS Track Run Walk. Online: https://play.google.com/store/apps/details?id=com.fitnesskeeper.runkeeper.pro&hl=en (accessed in 05.2017).
72. R. K. Ganti, F. Ye, and H. Lei. Mobile Crowdsensing: Current State and Future Challenges. *IEEE Communications Magazine*, 49(11):32–39, 2011.
73. R. Gao, M. Zhao, T. Ye, F. Ye, Y. Wang, K. Bian, T. Wang, and X. Li. Jigsaw: Indoor Floor Plan Reconstruction via Mobile Crowdsensing. In *Proceedings of the 20th ACM Annual International Conference on Mobile Computing and Networking (MobiCom)*, pages 249–260, 2014.
74. S. Gao, J. Ma, W. Shi, and G. Zhan. Towards Location and Trajectory Privacy Protection in Participatory Sensing. In *Proceedings of the 3rd International Conference on Mobile Computing, Applications, and Services (MobiCASE)*, pages 381–386, 2011.
75. S. Gao, J. Ma, W. Shi, G. Zhan, and C. Sun. TrPF: A Trajectory Privacy-Preserving Framework for Participatory Sensing. *IEEE Transactions on Information Forensics and Security*, 8(6):874–887, 2013.
76. S. Gaonkar, J. Li, R. R. Choudhury, L. Cox, and A. Schmidt. Micro-Blog: Sharing and Querying Content through Mobile Phones and Social Participation. In *Proceedings of the 6th ACM International Conference on Mobile Systems, Applications, and Services (MobiSys)*, pages 174–186, 2008.
77. T. Giannetsos, S. Gisdakis, and P. Papadimitratos. Trustworthy People-Centric Sensing: Privacy, Security and User Incentives Road-map. In *Proceedings of the 13th Annual Mediterranean Ad Hoc Networking Workshop (MED-HOC-NET)*, pages 39–46, 2014.
78. S. Gisdakis, T. Giannetsos, and P. Papadimitratos. SPPEAR: Security and Privacy-preserving Architecture for Participatory-sensing Applications. In *Proceedings of the 7th ACM Conference on Security and Privacy in Wireless Mobile Networks (WiSec)*, pages 39–50, 2014.
79. M. Groat, B. Edwards, J. Horey, W. He, and S. Forrest. Enhancing Privacy in Participatory Sensing Applications with Multidimensional Data. In *Proceedings of the 10th IEEE International Conference on Pervasive Computing and Communications (PerCom)*, pages 144–152, 2012.
80. B. Guo, H. Chen, Z. Yu, X. Xie, S. Huangfu, and D. Zhang. FlierMeet: A Mobile Crowdsensing System for Cross-space Public Information Reposting, Tagging, and Sharing. *IEEE Transactions on Mobile Computing*, 14(10):2020–2033, 2015.
81. J. Han, E. Owusu, L. Nguyen, A. Perrig, and J. Zhang. ACComplice: Location Inference using Accelerometers on Smartphones. In *Proceedings of the 4th International Conference on Communication Systems and Networks (COMSNETS)*, pages 1–9, 2012.

82. D. Hasenfratz, O. Saukh, S. Sturzenegger, and L. Thiele. Participatory Air Pollution Monitoring Using Smartphones. In *Proceedings of the 2nd ACM International Workshop on Mobile Sensing*, pages 1–5, 2012.
83. J. Hicks, N. Ramanathan, D. Kim, M. Monibi, J. Selsky, M. Hansen, and D. Estrin. AndWellness: An Open Mobile System for Activity and Experience Sampling. In *Proceedings of the 1st Wireless Health Scientific Conference (WH)*, pages 34–43, 2010.
84. K. Hu, Y. Wang, A. Rahman, and V. Sivaraman. Personalising Pollution Exposure Estimates using Wearable Activity Sensors. In *Proceedings of the IEEE 9th International Conference on Intelligent Sensors, Sensor Networks and Information Processing (ISSNIP)*, pages 1–6, 2014.
85. K. L. Huang, S. S. Kanhere, and W. Hu. Are you Contributing Trustworthy Data?: The Case for a Reputation System in Participatory Sensing. In *Proceedings of the 13th ACM International Conference on Modeling, Analysis, and Simulation of Wireless and Mobile Systems (MSWIM)*, pages 14–22, 2010.
86. K. L. Huang, S. S. Kanhere, and W. Hu. A Privacy-preserving Reputation System for Participatory Sensing. In *Proceedings of the 37th IEEE Conference on Local Computer Networks (LCN)*, pages 10–18, 2012.
87. L. G. Jaimes, I. J. Vergara-Laurens, and A. Raij. A Survey of Incentive Techniques for Mobile Crowd Sensing. *IEEE Internet of Things Journal*, 2(5):370–380, 2015.
88. A. Jain and V. Kanhangad. Investigating Gender Recognition in Smartphones using Accelerometer and Gyroscope Sensor Readings. In *Proceedings of the International Conference on Computational Techniques in Information and Communication Technologies (ICCTICT)*, pages 597–602, 2016.
89. S. Ji, T. Chen, and F. Wu. Crowdsourcing with Trembles: Incentive Mechanisms for Mobile Phones with Ucertain Sensing Time. In *Proceedings of the IEEE International Conference on Communications (ICC)*, pages 3546–3551, 2015.
90. Y. Jiang, Y. Xiang, X. Pan, K. Li, Q. Lv, R. P. Dick, L. Shang, and M. Hannigan. Hallway based Automatic Indoor Floorplan Construction using Room Fingerprints. In *Proceedings of the 2013 ACM International Joint Conference on Pervasive and Ubiquitous Computing (UbiComp)*, pages 315–324, 2013.
91. H. Jin, L. Su, D. Chen, K. Nahrstedt, and J. Xu. Quality of Information Aware Incentive Mechanisms for Mobile Crowd Sensing Systems. In *Proceedings of the 16th ACM International Symposium on Mobile Ad Hoc Networking and Computing (MobiHoc)*, pages 167–176, 2015.
92. H. Jin, L. Su, B. Ding, K. Nahrstedt, and N. Borisov. Enabling Privacy-Preserving Incentives for Mobile Crowd Sensing Systems. In *Proceedings of the 36th IEEE International Conference on Distributed Computing Systems (ICDCS)*, pages 344–353, 2016.
93. H. Jin, L. Su, H. Xiao, and K. Nahrstedt. INCEPTION: Incentivizing Privacy-preserving Data Aggregation for Mobile Crowd Sensing Systems. In *Proceedings of the 17th ACM International Symposium on Mobile Ad Hoc Networking and Computing (MobiHoc)*, pages 341–350, 2016.
94. E. Kanjo, J. Bacon, D. Roberts, and P. Landshoff. MobSens: Making Smart Phones Smarter. *IEEE Pervasive Computing*, 8(4):50–57, 2009.
95. A. Kapadia, D. Kotz, and N. Triandopoulos. Opportunistic Sensing: Security Challenges for the New Paradigm. In *Proceedings of the 1st International Conference on Communication Systems and Networks (COMNETS)*, pages 1–10, 2009.
96. M. Karaliopoulos, I. Koutsopoulos, and M. Titsias. First Learn then Earn: Optimizing Mobile Crowdsensing Campaigns Through Data-driven User Profiling. In *Proceedings of the 17th ACM International Symposium on Mobile Ad Hoc Networking and Computing (MobiHoc)*, pages 271–280, 2016.
97. M. Karaliopoulos, O. Telelis, and I. Koutsopoulos. User Recruitment for Mobile Crowdsensing over Opportunistic Networks. In *Proceedings of the IEEE Conference on Computer Communications (INFOCOM)*, pages 2254–2262, 2015.
98. L. Kazemi and C. Shahabi. A Privacy-aware Framework for Participatory Sensing. *ACM SIGKDD Explorations Newsletter*, 13(1):43–51, 2011.

99. L. Kazemi and C. Shahabi. Towards Preserving Privacy in Participatory Sensing. In *Proceedings of the 9th IEEE International Conference on Pervasive Computing and Communications (PERCOM Workshops)*, pages 328–331, 2011.

100. L. Kazemi and C. Shahabi. GeoCrowd: Enabling Query Answering with Spatial Crowdsourcing. In *Proceedings of the 20th International Conference on Advances in Geographic Information Systems (SIGSPATIAL)*, pages 189–198, 2012.

101. E. Koukoumidis, L.-S. Peh, and M. R. Martonosi. SignalGuru: Leveraging Mobile Phones for Collaborative Traffic Signal Schedule Advisory. In *Proceedings of the 9th ACM International Conference on Mobile Systems, Applications, and Services (MobiSys)*, pages 127–140, 2011.

102. A. Krause, E. Horvitz, A. Kansal, and F. Zhao. Toward Community Sensing. In *Proceedings of the 7th ACM/IEEE International Conference on Information Processing in Sensor Networks (IPSN)*, pages 481–492, 2008.

103. J. Krumm. Inference Attacks on Location Tracks. In *Proceedings of the 5th IEEE International Conference on Pervasive Computing (Pervasive)*, pages 127–143, 2007.

104. J. R. Kwapisz, G. M. Weiss, and S. A. Moore. Cell Phone-based Biometric Identification. In *Proceedings of the 4th IEEE International Conference on Biometrics: Theory Applications and Systems (BTAS)*, pages 1–7, 2010.

105. J. R. Kwapisz, G. M. Weiss, and S. A. Moore. Activity Recognition using Cell Phone Accelerometers. *SIGKDD Explorations Newsletter*, 12:74–82, 2011.

106. N. D. Lane, Y. Chon, L. Zhou, Y. Zhang, F. Li, D. Kim, G. Ding, F. Zhao, and H. Cha. Piggyback CrowdSensing (PCS): Energy Efficient Crowdsourcing of Mobile Sensor Data by Exploiting Smartphone App Opportunities. In *Proceedings of the 11th ACM Conference on Embedded Networked Sensor Systems SenSys*, pages 7:1–7:14, 2013.

107. N. D. Lane, J. Xie, T. Moscibroda, and F. Zhao. On the Feasibility of User De-anonymization from Shared Mobile Sensor Data. In *Proceedings of the 3rd International Workshop on Sensing Applications on Mobile Phones (PhoneSense)*, pages 3:1–3:5, 2012.

108. C. Leonardi, A. Cappellotto, M. Caraviello, B. Lepri, and F. Antonelli. SecondNose: An Air Quality Mobile Crowdsensing System. In *Proceedings of the 8th ACM Nordic Conference on Human-Computer Interaction (NordiCHI)*, pages 1051–1054, 2014.

109. Q. Li and G. Cao. Efficient and Privacy-preserving Data Aggregation in Mobile Sensing. In *Proceedings of the 20th IEEE International Conference on Network Protocols (ICNP)*, pages 1–10, 2012.

110. Q. Li and G. Cao. Providing Privacy-aware Incentives for Mobile Sensing. In *Proceedings of the 11th IEEE International Conference on Pervasive Computing and Communications (PerCom)*, pages 76–84. IEEE, 2013.

111. C. H. Liu, P. Hui, J. W. Branch, C. Bisdikian, and B. Yang. Efficient Network Management for Context-aware Participatory Sensing. In *Proceedings of the 8th Annual IEEE Communications Society Conference on Sensor, Mesh and Ad Hoc Communications and Networks (SECON)*, pages 116–124, 2011.

112. J. Liu, H. Shen, and X. Zhang. A Survey of Mobile Crowdsensing Techniques: A Critical Component for the Internet of Things. In *Proceedings of the 25th IEEE International Conference on Computer Communication and Networks (ICCCN)*, pages 1–6, 2016.

113. H. Lu, D. Frauendorfer, M. Rabbi, M. S. Mast, G. T. Chittaranjan, A. T. Campbell, D. Gatica-Perez, and T. Choudhury. StressSense: Detecting Stress in Unconstrained Acoustic Environments Using Smartphones. In *Proceedings of the 14th ACM International Conference on Ubiquitous Computing (UbiComp)*, pages 351–360, 2012.

114. H. Lu, W. Pan, N. D. Lane, T. Choudhury, and A. T. Campbell. SoundSense: Scalable Sound Sensing for People-centric Applications on Mobile Phones. In *Proceedings of the 7th ACM International Conference on Mobile Systems, Applications, and Services (MobiSys)*, pages 165–178, 2009.

115. T. Ludwig, C. Reuter, T. Siebigteroth, and V. Pipek. CrowdMonitor: Mobile Crowd Sensing for Assessing Physical and Digital Activities of Citizens During Emergencies. In *Proceedings of the 33rd Annual ACM Conference on Human Factors in Computing Systems (CHI)*, pages 4083–4092, 2015.

116. T. Luo and C. K. Tham. Fairness and Social Welfare in Incentivizing Participatory Sensing. In *Proceedings of the 9th Annual IEEE Conference on Sensor, Mesh and Ad Hoc Communications and Networks (SECON)*, pages 425–433, 2012.

117. N. Maisonneuve, M. Stevens, M. E. Niessen, and L. Steels. NoiseTube: Measuring and Mapping Noise Pollution with Mobile Phones. In *Proceedings of the 4th International Symposium on Information Technologies in Environmental Engineering (ITEE)*, pages 215–228, 2009.

118. M. Marjanović, L. Skorin-Kapov, K. Pripužić, A. Antonić, and I. P. Žarko. Energy-aware and Quality-driven Sensor Management for Green Mobile Crowd Sensing. *Journal of Network and Computer Applications*, 59:95–108, 2016.

119. M. Mehrnezhad, E. Toreini, S. F. Shahandashti, and F. Hao. TouchSignatures: Identification of User Touch Actions and PINs based on Mobile Sensor Data via Javascript. *Journal of Information Security and Applications*, 26:23–38, 2016.

120. D. Mendez, A. Perez, M. Labrador, and J. Marron. P-Sense: A Participatory Sensing System for Air Pollution Monitoring and Control. In *Proceedings of the 9th IEEE International Conference on Pervasive Computing and Communications (PERCOM Workshops)*, pages 344–347, 2011.

121. C. Meng, H. Xiao, L. Su, and Y. Cheng. Tackling the Redundancy and Sparsity in Crowd Sensing Applications. In *Proceedings of the 14th ACM Conference on Embedded Network Sensor Systems (SenSys)*, pages 150–163, 2016.

122. A. Michalas and N. Komninos. The Lord of the Sense: A Privacy Preserving Reputation System for Participatory Sensing Applications. In *Proceedings of the 19th IEEE Symposium on Computers and Communication (ISCC)*, pages 1–6, 2014.

123. P. Micholia, M. Karaliopoulos, and I. Koutsopoulos. Mobile Crowdsensing Incentives Under Participation Uncertainty. In *Proceedings of the 3rd ACM Workshop on Mobile Sensing, Computing and Communication (MSCC)*, pages 29–34, 2016.

124. O. Miguel-Hurtado, S. V. Stevenage, C. Bevan, and R. Guest. Predicting Sex as a Soft-Biometrics from Device Interaction Swipe Gestures. *Pattern Recognition Letters*, 79:44–51, 2016.

125. E. Miluzzo, N. D. Lane, K. Fodor, R. A. Peterson, H. Lu, M. Musolesi, S. B. Eisenman, X. Zheng, and A. T. Campbell. Sensing meets Mobile Social Networks: The Design, Implementation and Evaluation of the CenceMe Application. In *Proceedings of the 6th ACM Conference on Embedded Network Sensor Systems (SenSys)*, pages 337–350, 2008.

126. P. Mohan, V. Padmanabhan, and R. Ramjee. Nericell: Rich Monitoring of Road and Traffic Conditions using Mobile Smartphones. In *Proceedings of the 6th ACM Conference on Embedded Network Sensor Systems (SenSys)*, pages 323–336, 2008.

127. M. Mun, S. Reddy, K. Shilton, N. Yau, J. A. Burke, D. Estrin, M. Hansen, E. Howard, R. West, and P. Boda. PEIR, the Personal Environmental Impact Report, as a Platform for Participatory Sensing Systems Research. In *Proceedings of the 7th ACM International Conference on Mobile Systems, Applications, and Services (MobiSys)*, pages 55–68, 2009.

128. M. Y. Mun, D. H. Kim, K. Shilton, D. Estrin, M. Hansen, and R. Govindan. PDVLoc: A Personal Data Vault for Controlled Location Data Sharing. *ACM Transactions Sensor Networks*, 10(4):58:1–58:29, 2014.

129. M. Murshed, A. Iqbal, T. Sabrina, and K. M. Alam. A Subset Coding Based k-Anonymization Technique to Trade-Off Location Privacy and Data Integrity in Participatory Sensing Systems. In *Proceedings of the 10th IEEE International Symposium on Network Computing and Applications (NCA)*, pages 107–114, 2011.

130. M. Musolesi, E. Miluzzo, N. D. Lane, S. B. Eisenman, T. Choudhury, and A. T. Campbell. The Second Life of a Sensor: Integrating Real-world Experience in Virtual Worlds using Mobile Phones. In *Proceedings of the 5th Workshop on Embedded Networked Sensors (HotEmNets)*, pages 1–5, 2008.

131. L. Nachman, A. Baxi, S. Bhattacharya, V. Darera, P. Deshpande, N. Kodalapura, V. Mageshkumar, S. Rath, J. Shahabdeen, and R. Acharya. Jog Falls: A Pervasive Healthcare Platform for Diabetes Management. *Pervasive Computing*, 6030(1):94–111, 2010.

132. E. Niforatos, A. Vourvopoulos, M. Langheinrich, P. Campos, and A. Doria. Atmos: A Hybrid Crowdsourcing Approach to Weather Estimation. In *Proceedings of the ACM International Joint Conference on Pervasive and Ubiquitous Computing (UbiComp, Adjunct Publication)*, pages 135–138, 2014.
133. Nike, Inc. Nike+ Running. Online: https://play.google.com/store/apps/details?id=com.nike.plusgps&hl=en (accessed in 05.2017).
134. E. Paulos, R. Honicky, and E. Goodman. Sensing Atmosphere. In *Proceedings of the Workshop on Sensing on Everyday Mobile Phones in Support of Participatory Research (SenSys Workshop)*, pages 15–16, 2007.
135. S. Perez. App Store to Reach 5 Million Apps by 2020, with Games Leading the Way. Online: https://techcrunch.com/2016/08/10/app-store-to-reach-5-million-apps-by-2020-with-games-leading-the-way, 2016.
136. D. Philipp, P. Baier, C. Dibak, F. Dürr, K. Rothermel, S. Becker, M. Peter, and D. Fritsch. MapGENIE: Grammar-enhanced Indoor Map Construction from Crowd-sourced Data. In *Proceedings of the 12th IEEE International Conference on Pervasive Computing and Communications (PerCom)*, pages 139–147, 2014.
137. D. Philipp, F. Dürr, and K. Rothermel. A Sensor Network Abstraction for Flexible Public Sensing Systems. In *Proceedings of the 8th IEEE International Conference on Mobile Ad-Hoc and Sensor Systems (MASS)*, pages 460–469, 2011.
138. D. Philipp, J. Stachowiak, F. Dürr, and K. Rothermel. Model-Driven Public Sensing in Sparse Networks. In *Proceedings of the 10th International Conference on Mobile and Ubiquitous Systems: Computing, Networking and Services (MobiQuitous)*, pages 17–29, 2013.
139. M. Pouryazdan, B. Kantarci, T. Soyata, and H. Song. Anchor-Assisted and Vote-Based Trustworthiness Assurance in Smart City Crowdsensing. *IEEE Access*, 4:529–541, 2016.
140. B. Predic, Z. Yan, J. Eberle, D. Stojanovic, and K. Aberer. ExposureSense: Integrating Daily Activities with Air Quality using Mobile Participatory Sensing. In *Proceedings of the 11th IEEE International Conference on Pervasive Computing and Communications (PERCOM Workshops)*, pages 303–305, 2013.
141. R. Pryss, M. Reichert, J. Herrmann, B. Langguth, and W. Schlee. Mobile Crowd Sensing in Clinical and Psychological Trials – A Case Study. In *Proceedings of the 28th IEEE International Symposium on Computer-Based Medical Systems (CBMS)*, pages 23–24, 2015.
142. R. Pryss, M. Reichert, B. Langguth, and W. Schlee. Mobile Crowd Sensing Services for Tinnitus Assessment, Therapy, and Research. In *Proceedings of the IEEE International Conference on Mobile Services (MS)*, pages 352–359, 2015.
143. F. Qiu, F. Wu, and G. Chen. SLICER: A Slicing-Based K-Anonymous Privacy Preserving Scheme for Participatory Sensing. In *Proceedings of the 10th IEEE International Conference on Mobile Ad-Hoc and Sensor Systems (MASS)*, pages 113–121, 2013.
144. R. K. Rana, C. T. Chou, S. S. Kanhere, N. Bulusu, and W. Hu. Ear-Phone: An End-to-end Participatory Urban Noise Mapping System. In *Proceedings of the 9th ACM/IEEE International Conference on Information Processing in Sensor Networks (IPSN)*, pages 105–116, 2010.
145. S. Reddy, J. Burke, D. Estrin, M. Hansen, and M. Srivastava. A Framework for Data Quality and Feedback in Participatory Sensing. In *Proceedings of the 5th ACM International Conference on Embedded Networked Sensor Systems (SenSys)*, pages 417–418, 2007.
146. S. Reddy, D. Estrin, M. Hansen, and M. B. Srivastava. Examining Micro-Payments for Participatory Sensing Data Collections. In *Proceedings of the 12th ACM International Conference on Ubiquitous Computing (UbiComp)*, pages 33–36, 2010.
147. S. Reddy, A. Parker, J. Hyman, J. A. Burke, D. Estrin, and M. Hansen. Image Browsing, Processing, and Clustering for Participatory Sensing: Lessons from a DietSense Prototype. In *Proceedings of the 4th Workshop on Embedded Networked Sensors (EmNets)*, pages 13–17, 2007.
148. D. Reinhardt and C. Heinig. Survey-based Exploration of Attitudes to Participatory Sensing Tasks in Location-based Gaming Communities. *Pervasive and Mobile Computing (PMC)*, 27:27–36, 2016.

149. D. Reinhardt and I. Manyugin. OP^4: An OPPortunistic Privacy-Preserving Scheme for Crowdsensing Applications. In *Proceedings of the 41st IEEE Conference on Local Computer Networks (LCN)*, pages 460–468, 2016.
150. D. Reinhardt, M. Michalak, and R. Lokaiczyk. Job Alerts in the Wild: Study of Expectations and Effects of Location-based Notifications in an Existing Mobile Crowdsourcing Application. In *Proceedings of the 13th International Conference on Mobile and Ubiquitous Systems: Computing, Networking and Services (MobiQuitous)*, pages 65–74, 2016.
151. J. Ren, Y. Zhang, K. Zhang, and X. Shen. Exploiting Mobile Crowdsourcing for Pervasive Cloud Services: Challenges and Solutions. *IEEE Communications Magazine*, 53(3):98–105, 2015.
152. K. Shilton. Four Billion Little Brothers?: Privacy, Mobile Phones, and Ubiquitous Data Collection. *Communications of the ACM*, 52(11):48–53, 2009.
153. M. Shin, C. Cornelius, D. Peebles, A. Kapadia, D. Kotz, and N. Triandopoulos. AnonySense: A System for Anonymous Opportunistic Sensing. *Journal of Pervasive and Mobile Computing*, 7(1):16–30, 2010.
154. A. Stopczynski, J. Larsen, S. Lehmann, L. Dynowski, and M. Fuentes. Participatory Bluetooth Sensing: A method for Acquiring Spatio-temporal Data about Participant Mobility and Interactions at Large Scale Events. In *Proceedings of the 11th IEEE International Conference on Pervasive Computing and Communications (PERCOM Workshops)*, pages 242–247, 2013.
155. E. P. Stuntebeck, J. S. Davis, II, G. D. Abowd, and M. Blount. HealthSense: Classification of Health-related Sensor Data through User-assisted Machine Learning. In *Proceedings of the 9th Workshop on Mobile Computing Systems and Applications (HotMobile)*, pages 1–5, 2008.
156. W. Sun, Q. Li, and C.-K. Tham. Wireless Deployed and Participatory Sensing System for Environmental Monitoring. In *Proceedings of the 11th IEEE International Conference on Sensing, Communication, and Networking (SECON)*, pages 158–160, 2014.
157. H. To, G. Ghinita, and C. Shahabi. A Framework for Protecting Worker Location Privacy in Spatial Crowdsourcing. *Proceedings of the Very Large Database Endowment (PVLDB)*, 7(10):919–930, 2014.
158. A. Tychsen, M. Hitchens, and T. Brolund. Motivations for Play in Computer Role-playing Games. In *Proceedings of the ACM Conference on Future Play: Research, Play, Share (Future Play)*, pages 57–64, 2008.
159. T. Van Goethem, W. Scheepers, D. Preuveneers, and W. Joosen. Accelerometer-Based Device Fingerprinting for Multi-factor Mobile Authentication. In *International Symposium on Engineering Secure Software and Systems (ESSoS)*, pages 106–121. Springer, 2016.
160. I. J. Vergara-Laurens, D. Mendez, and M. A. Labrador. Privacy, Quality of Information, and Energy Consumption in Participatory Sensing Systems. In *Proceedings of the 12th IEEE International Conference on Pervasive Computing and Communications (PerCom)*, pages 199–207, 2014.
161. K. Vu, R. Zheng, and J. Gao. Efficient Algorithms for K-anonymous Location Privacy in Participatory Sensing. In *Proceedings of the 31th IEEE Conference on Computer Communications (INFOCOM)*, pages 2399–2407, 2012.
162. C.-J. Wang and W.-S. Ku. Anonymous Sensory Data Collection Approach for Mobile Participatory Sensing. In *Proceedings of the 28th IEEE International Conference on Data Engineering Workshops (ICDEW)*, pages 220–227, 2012.
163. J. Wang, J. Tang, D. Yang, E. Wang, and G. Xue. Quality-Aware and Fine-Grained Incentive Mechanisms for Mobile Crowdsensing. In *Proceedings of the 36th IEEE International Conference on Distributed Computing Systems (ICDCS)*, pages 354–363, 2016.
164. L. Wang, D. Zhang, Y. Wang, C. Chen, X. Han, and A. M'hamed. Sparse Mobile Crowdsensing: Challenges and Opportunities. *IEEE Communications Magazine*, 54(7):161–167, 2016.

165. L. Wang, D. Zhang, and H. Xiong. effSense: Energy-efficient and Cost-effective Data Uploading in Mobile Crowdsensing. In *Proceedings of the ACM Conference on Pervasive and Ubiquitous Computing (UbiComp 'Adjunct)*, pages 1075–1086, 2013.
166. X. O. Wang, W. Cheng, P. Mohapatra, and T. Abdelzaher. ARTSense: Anonymous Reputation and Trust in Participatory Sensing. In *Proceedings of the 32th IEEE Conference on Computer Communications (INFOCOM)*, pages 2517–2525. IEEE, 2013.
167. X. O. Wang, W. Cheng, P. Mohapatra, and T. Abdelzaher. Enabling Reputation and Trust in Privacy-Preserving Mobile Sensing. *IEEE Transactions on Mobile Computing*, 13(12):1–14, 2013.
168. Y. Wang and D. R. Fesenmaier. Assessing Motivation of Contribution in Online Communities: An Empirical Investigation of an Online Travel Community. *Electronic Markets*, 13(1):33–45, 2003.
169. Y. Wang, J. Wang, and X. Zhang. QTime: A Queuing-Time Notification System Based on Participatory Sensing Data. In *Proceedings of the 37th IEEE Annual Computer Software and Applications Conference (COMPSAC)*, pages 770–777, 2013.
170. Waze. Online: https://www.waze.com (accessed in 03.2017).
171. H. Weinschrott, F. Dürr, and K. Rothermel. Efficient Capturing of Environmental Data with Mobile RFID Readers. In *Proceedings of the 10th International Conference on Mobile Data Management (MDM)*, pages 41–51, 2009.
172. H. Weinschrott, F. Dürr, and K. Rothermel. StreamShaper: Coordination Algorithms for Participatory Mobile Urban Sensing. In *Proceedings of the 7th IEEE International Conference on Mobile Ad-hoc and Sensor Systems (MASS)*, pages 195–204, 2010.
173. H. Weinschrott, J. Weißer, F. Dürr, and K. Rothermel. Participatory Sensing Algorithms for Mobile Object Discovery in Urban Areas. In *Proceedings of the 9th Annual IEEE International Conference on Pervasive Computing and Communication (PerCom)*, pages 128–135, 2011.
174. G. M. Weiss and J. W. Lockhart. Identifying User Traits by Mining Smart Phone Accelerometer Data. In *Proceedings of the 5th International Workshop on Knowledge Discovery from Sensor Data (SensorKDD)*, pages 61–69, 2011.
175. F. J. Wu and T. Luo. WiFiScout: A Crowdsensing WiFi Advisory System with Gamification-Based Incentive. In *Proceedings of the IEEE 11th International Conference on Mobile Ad Hoc and Sensor Systems (MASS)*, pages 533–534, 2014.
176. H. Xiong, D. Zhang, L. Wang, J. P. Gibson, and J. Zhu. EEMC: Enabling Energy-Efficient Mobile Crowdsensing with Anonymous Participants. *ACM Transactions on Intelligent Systems and Technology (TIST)*, 6(3):39:1–39:26, 2015.
177. J. Xu, J. Xiang, and D. Yang. Incentive Mechanisms for Time Window Dependent Tasks in Mobile Crowdsensing. *IEEE Transactions on Wireless Communications*, 14(11):6353–6364, 2015.
178. H. Yang, J. Zhang, and P. Roe. Reputation Modelling in Citizen Science for Environmental Acoustic Data Analysis. *Social Network Analysis and Mining*, 3(3):419–435, 2013.
179. P. Zappi, E. Bales, J. H. Park, W. Griswold, and T. Š. Rosing. The Citisense Air Quality Monitoring Mobile Sensor Node. In *Proceedings of the 11th ACM/IEEE Conference on Information Processing in Sensor Networks (IPSN)*, pages 1–5, 2012.
180. F. Zhang, L. He, W. He, and X. Liu. Data Perturbation with State-dependent Noise for Participatory Sensing. In *Proceedings of the 31th IEEE Conference on Computer Communications (INFOCOM)*, pages 2246–2254, 2012.
181. J. Zhang, J. Ma, W. Wang, and Y. Liu. A Novel Privacy Protection Scheme for Participatory Sensing with Incentives. In *Proceedings of the 2nd IEEE International Conference on Cloud Computing and Intelligent Systems (CCIS)*, pages 1017–1021, 2012.
182. J. Zhang and D. Wang. Duplicate Report Detection in Urban Crowdsensing Applications for Smart City. In *Proceedings of the IEEE International Conference on Smart City (SmartCity)*, pages 101–107, 2015.

183. R. Zhang, J. Shi, Y. Zhang, and C. Zhang. Verifiable Privacy-Preserving Aggregation in People-Centric Urban Sensing Systems. *IEEE Journal on Selected Areas in Communications*, 31(9):268–278, 2013.
184. P. Zhou, Y. Zheng, and M. Li. How Long to Wait?: Predicting Bus Arrival Time with Mobile Phone Based Participatory Sensing. In *Proceedings of the 10th ACM International Conference on Mobile Systems, Applications, and Services (MobiSys)*, pages 379–392, 2012.
185. J. Zhu, K.-H. Kim, P. Mohapatra, and P. Congdon. An Adaptive Privacy-preserving Scheme for Location Tracking of a Mobile User. In *Proceedings of the 10th IEEE International Conference on Sensing, Communication, and Networking (SECON)*, pages 140–148. IEEE, 2013.

Chapter 7
Location Privacy in Spatial Crowdsourcing

Hien To and Cyrus Shahabi

Abstract Spatial crowdsourcing (SC) is a new platform that engages individuals in collecting and analyzing environmental, social and other spatiotemporal information. With SC, requesters outsource their spatiotemporal tasks (tasks associated with location and time) to a set of workers, who will perform the tasks by physically traveling to the tasks' locations. However, current solutions require the locations of the workers and/or the tasks to be disclosed to untrusted entities (SC server) for effective assignments of tasks to workers.

This chapter first identifies privacy threats toward both workers and tasks during the two main phases of spatial crowdsourcing, tasking and reporting. *Tasking* is the process of identifying which tasks should be assigned to which workers. This process is handled by a spatial crowdsourcing server (SC server). The latter phase is *reporting*, in which workers travel to the tasks' locations, complete the tasks and upload their reports to the server. The challenge is to enable effective and efficient tasking as well as reporting in SC without disclosing the actual locations of workers (at least until they agree to perform a task) and the tasks themselves (at least to workers who are not assigned to those tasks).

This chapter aims to provide an overview of the state-of-the-art in protecting users' location privacy in spatial crowdsourcing. We provide a comparative study of a diverse set of solutions in terms of task publishing modes (push vs. pull), problem focuses (tasking and reporting), threats (server, requester and worker), and underlying technical approaches (from pseudonymity, cloaking, and perturbation to exchange-based and encryption-based techniques). The strengths and drawbacks of the techniques are highlighted, leading to a discussion of open problems and future work.

H. To (✉) · C. Shahabi (✉)
University of Southern California, Los Angeles, CA, USA
e-mail: hto@usc.edu; shahabi@usc.edu

© Springer Nature Switzerland AG 2018
A. Gkoulalas-Divanis, C. Bettini (eds.), *Handbook of Mobile Data Privacy*,
https://doi.org/10.1007/978-3-319-98161-1_7

167

7.1 Introduction

The increase in computational and communication performance of mobile devices, coupled with the advances in sensor technology, leads to an exponential growth in data collection and sharing by smartphones. Exploiting mobility of such a large volume of potential users, a new mechanism for efficient and scalable data collection has emerged, namely, *spatial crowdsourcing* (SC) [13]. SC has numerous applications in domains such as environmental sensing (iRain [1]), smart cities (TaskRabbit), journalism, and crisis response (MediaQ [15]). With SC, requesters and workers typically register with a crowdsourcing server that acts as a broker between parties, and often also plays a role in how tasks are assigned to workers. A requester issues one or more tasks to the server (i.e., the platform). The server then assigns the task to a worker. We refer to this phase as *tasking* (or task assignment). After tasking, workers travel to the locations of the tasks, perform them and report the results to the server. This phase is referred to as *reporting*.

Both tasking and reporting phases often require workers and requesters to reveal locations of workers and tasks to potentially untrusted entities (server, other workers and other requesters). Several studies (e.g., [5, 13, 14, 30]) focus on effective tasking by maximizing the number of assigned tasks while minimizing workers travel distances, for which they require workers to reveal their locations and requesters to disclose their tasks' locations to the server. Similarly, reporting spatial tasks would enable the server to infer the workers' locations since they must have visited the locations of the tasks. However, disclosing individual locations has serious privacy implications. Leaked locations often lead to a breach of sensitive information such as an individual's health (e.g., presence in a cancer treatment center), alternative lifestyles, political and religious preferences (e.g., presence in a church). Knowing user locations, an adversary can stage a broad spectrum of attacks such as physical surveillance and stalking, and identity theft [25]. Particularly, in [36], the authors show that hackers can stalk users in Waze—a popular SC application—by generating fake events such as accidents. Consequently, mobile users may not agree to engage in spatial crowdsourcing if their privacy is violated; thus, ensuring location privacy is key to the success of SC.

The first step of the tasking phase is task publication. There are two modes of task publication in SC: push (e.g., iRain) vs. pull (e.g., TaskRabbit). With the pull mode, the server publishes the spatial tasks and online workers can choose any spatial task in their vicinity without the need to coordinate with the server. With the push mode, online workers send their locations to the server, which then assigns to every worker his nearby tasks (posted by requesters). Each mode shares similar challenges and has its own unique challenge. The common challenges are that a worker should know a task location only if he plans to perform the task; likewise, only requesters who have tasks performed by the worker should know his location. Furthermore, the unique challenge with the push mode is that the server must match workers to tasks without compromising their privacy. This requires strategies to ensure effective task assignment without revealing locations of tasks and workers. On the other hand, the

Table 7.1 Attacks on SC users

	Tasking	Reporting
Push	[12]	[27]
Pull	[Sect. 7.3.2]	[27, 36], [Sect. 7.3.2]

unique challenge with the pull mode is to enable every worker to request tasks, perform them and subsequently post the results to the server without revealing his location and identity. Finally, providing privacy protection simultaneously both tasking and reporting phases introduces another set of challenges to both push and pull modes.

Among the two modes of task publishing, privacy protection in the push mode is more challenging because tasking in the push mode is more complex than that of the pull mode. Countermeasure studies in the pull mode have been the main focus in the past decade with an emphasis on a special class of SC, named *participatory sensing* (PS). PS usually assumes the pull mode of task publication (workers choose tasks); therefore, the main privacy threats to workers occur during reporting. Meanwhile, the most recent studies in SC have focused on the push mode (server assigns tasks to workers); for this reason, main privacy breaches occur during tasking [12]. Consequently, the existing studies in SC can be classified into two groups: (1) preserving privacy during *reporting in the pull mode* [2, 27, 37], and (2) preserving privacy when *tasking in the push mode* [8, 10, 12, 22, 26, 31, 32, 35, 38].

In this chapter we study the privacy threats to workers and requesters[1] in SC, during both tasking and reporting phases with either push or pull mode. Throughout this chapter we also identify three major drawbacks of the existing studies. First, they solely focus on protecting privacy during either phase of tasking or reporting, but not both. Second, most of these studies ensure privacy for workers only. To elaborate, we perform a set of simple attacks on TaskRabbit to demonstrate that locations of workers and requesters can be learned during both tasking and reporting phases. Third, despite the fact that most studies focus on either reporting in the pull mode or tasking in the push mode, privacy threats to SC users may also occur in other scenarios. Table 7.1 shows that there have been known attacks under the tasking and reporting phases with either the push or pull mode of task publishing. We demonstrate such threats in Sect. 7.3.2 via another set of attacks on TaskRabbit. These observations open some new research questions such as: how do we protect location privacy of both workers and tasks, simultaneously, during both the tasking and reporting phases of SC, and what are the promising privacy techniques to be used?

There have been recent surveys in privacy-preserving participatory sensing [4, 24] and mobile crowdsourcing [21]. Unlike these surveys, which provide an overview of a broad range of related problems, this chapter provides an in-depth study of the privacy challenges and the solutions proposed in the prior studies.

---

[1]Task locations can indirectly reveal requesters' location, i.e., requesters often post tasks in the proximity of their locations.

The remainder of this chapter is organized as follows. In Sect. 7.2 we introduce spatial crowdsourcing and compare it with related concepts. Section 7.3 illustrates potential privacy risks to both workers and requesters. Section 7.4 summarizes existing solutions addressing the privacy concerns in both the tasking and reporting phases of SC. Finally, we present our conclusions and future research directions in Sect. 7.5.

7.2 Spatial Crowdsourcing

In this section we define spatial crowdsourcing and present two modes of task publishing, push vs. pull, with the push mode recently being dominant in the research community. Thereafter, we differentiate SC from the related topic of participatory sensing, which usually assumes the pull mode of task publication.

7.2.1 Generic Framework

Spatial crowdsourcing (SC) [13] is a type of online crowdsourcing where performing tasks requires workers to physically be present at the locations of the tasks, termed *spatial tasks*. A spatial task is a query to be answered at a particular location and must be performed before a deadline. An example of a spatial task is taking a picture of a particular dish in a restaurant. This means that the workers need to physically travel to the location of the restaurant in order to take the picture. A worker is a carrier of a mobile device who will perform spatial tasks for some incentives.

Spatial crowdsourcing has gained popularity in both the research community (e.g., [13, 32]) and industry (e.g., TaskRabbit, Gigwalk). A recent study [34] distinguishes SC from related fields, such as generic crowdsourcing, participatory sensing, volunteered geographic information, and online matching. Research efforts have focused on different aspects of SC, including task assignment, task scheduling, privacy, trust and incentive mechanism.

7.2.2 Task Assignment: The Focus of Spatial Crowdsourcing

The main challenges of spatial crowdsourcing are due to the large-scale, ad hoc and dynamic nature of the workers and tasks. To continuously match thousands of SC campaigns, where each campaign consists of many spatiotemporal tasks with millions of workers, a server must be able to run efficient task assignment (aka tasking). According to [13], there are two types of tasking modes based on how workers are matched to tasks—server-assigned tasks (SAT) and worker-selected tasks (WST)—which are also known as push and pull modes, respectively. Depending on the choice of a particular mode, the focus of privacy protection is either at the tasking or the reporting stage of SC.

With the *pull* mode, the server publicly[2] publishes the spatial tasks, and online workers autonomously choose tasks in their vicinity without coordinating with the server. One advantage of the pull mode is that the workers do not need to reveal their locations to server. However, one drawback of this mode is that the server does not have any control over the allocation of spatial tasks; this may result in some spatial tasks never be assigned, while others are assigned redundantly. Another drawback of the pull mode is that workers choose tasks based on their own objectives (e.g., choosing the k closest spatial tasks to minimize their travel cost), which may not result in a globally optimal assignment. Examples of the pull mode are TaskRabbit and Waze.

With the *push* mode, requesters post tasks that include locations, while online workers send their locations to the server, which assigns tasks to nearby workers. The advantage of this mode is that unlike the pull mode, the server has the big picture and can assign to every worker his nearby tasks while maximizing the overall task assignment. However, the drawback is that locations of both tasks and workers should be sent to the server for effective assignment, which can pose privacy threats. Examples of the push mode include Uber, iRain [30] and MediaQ [15].

Most SC studies assume the push mode and thus emphasize privacy protection during the tasking phase. With the pull mode, the main focus of privacy protection is shifted to the reporting phase, which has been well studied in the context of participatory sensing (e.g., [2, 12, 27, 35, 37]). With participatory sensing, the goal is to exploit the ability of mobile users to collect and share data using their sensor-equipped phones for a given campaign. Most studies on participatory sensing focus on small campaigns with a limited number of workers; hence, they do not have issues of task assignment. However, with SC, the focus is on devising a scalable, generic and multipurpose crowdsourcing framework, similar to Amazon Mechanical Turk, but spatial, where multiple campaigns can be handled simultaneously. Therefore, the main challenge with SC is to devise an efficient approach to assign tasks to workers given the large scale of an environment.

7.3 Privacy Threats

There have been known attacks on SC applications, such as location-based attacks during tasking in the push mode [12] and collusion attacks during reporting in the pull mode [36] (see Table 7.1). Despite the fact that most studies have solely focused on one of the two major threats, privacy risks to SC users may occur in the other

---

[2]Exact geographical coordinates of the tasks may not be published; instead, their cloaked locations or representative names are provided.

scenarios: reporting in the push mode and tasking in the pull mode. In this section we present a threat model which characterizes the *full spectrum of privacy threats to workers and requesters during both tasking and reporting phases with either push or pull mode*. Next, we illustrate the privacy risks on TaskRabbit.

7.3.1 Threat Model

As the privacy threats vary according to the modes of task publishing, we discuss possible threats associated with each mode.

7.3.1.1 Privacy Threats with the Push Mode

With the *push* mode, the server takes as input the perturbed locations of both workers and tasks to perform effective task assignment; hence, there is a serious privacy threat from the server which might become a single point of attack. Figure 7.1a depicts the threat model for the push mode of spatial crowdsourcing. The first row means that locations of workers and tasks are protected from the server at all the time. The role of the server is to create the *assignment links* between the workers and the requesters so that they can establish a direct communication channel among themselves. Each worker-requester pair cooperatively decides whether to accept the assignment from the server. If yes, they send a *consent* message to the server, confirming that the worker will perform the requester's tasks. This agreement is illustrated by the first *reporting link* in Fig. 7.1a. We argue that to preserve location privacy during both tasking and reporting phases, task locations need to be protected from the server. Otherwise, the completion of a task reveals that some workers

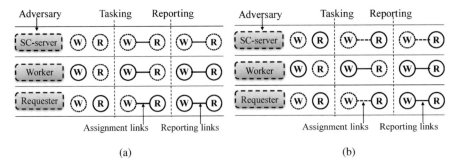

Fig. 7.1 Threat models in spatial crowdsourcing. W and R denote workers and requesters, respectively. The dotted circles surrounding them denote that they are protected from an untrusted entity shown in the first column. After tasking and reporting, the assignment and reporting links between W and R represent the established connections during each phase. The dashed links indicate connections that are oblivious to the corresponding malicious entity. (**a**) Push mode. (**b**) Pull mode

must have visited the task's location. In restrictive privacy settings, workers and requesters can also be malicious to each other. Hence, to ensure minimum disclosure among them, only workers who aim to perform the tasks should know the tasks' locations (see the second row in Fig. 7.1a). Likewise, a requester should only know the workers' locations once her tasks are matched to and then performed by those workers (see the third row in Fig. 7.1a).

We emphasize the minimum disclosure of location information for both workers and tasks. The reason for this is twofold. First, the server knows only the assignment links between workers and tasks. Due to such links, the assigned workers (or tasks) may infer that there exist nearby tasks (or workers). These disclosures are unavoidable in the push mode of SC. Second, the disclosure of workers' locations to their corresponding requester is inevitable at the reporting phase per definition of SC. It is worth mentioning that this threat model is restrictive; hence, weaker variants exist. For example, most existing studies in the push mode assume that workers are trusted [10, 12, 22] and task locations are public [8, 26, 32, 35, 38].

7.3.1.2 Privacy Threats with the Pull Mode

With the *pull* mode, despite the fact that workers do not need to send their locations to the server, the locations can still be learned during both tasking and reporting phases. As long as a worker connects to the server to either *request* some tasks or *report* results, he may reveal to the server patterns of where and when the connections were made and what kind of tasks he wants to perform. Consequently, in [27], the authors show that linking multiple requests or reports of the worker may allow an adversary to trace him since the worker's location information can be tracked through several stationary connection points (e.g., cell towers). In addition, the worker's location trace can be inferred by both the server and requesters since he must be in the vicinity of the tasks in order to perform them. Figure 7.1b depicts the proposed threat model for the pull mode. To preserve privacy and identity of the workers from the server, both assignment links and reporting links should be secure during tasking and reporting phases, respectively. This is because if the connections are discovered by the server, which already knows the locations of tasks, the server learns the locations of workers since they must have visited the locations of the performed tasks. Hence, the workers must request tasks without revealing their identity to the server; once the tasks are performed, the workers must also disassociate their connections with the performed tasks while uploading task content to the server. Similar to the push mode, both workers and requesters themselves can be hostile to one another. Thus, the privacy threats from workers and requesters (rows 2 and 3 in Fig. 7.1a) are similar to those in the push mode (rows 2 and 3 in Fig. 7.1b), except the difference in the assignment links of the two second rows. The reason for this is that the requester is oblivious to the requests between the worker and the server during tasking.

Fig. 7.2 Screenshots of TaskRabbit web application from worker Bob. (**a**) Task locations. (**b**) Task price. (**c**) Task status. (**d**) Performed tasks

7.3.2 Case Study of TaskRabbit

We show that an adversary can perform harmful attacks on a typical SC application without much effort. TaskRabbit is a pull-based[3] online and mobile marketplace that matches workers with requesters, allowing requesters to find immediate help with everyday tasks including, but not limited to, cleaning, moving, and delivery. In the following we discuss the aforementioned threats to TaskRabbit users. Note that the following attacks on TaskRabbit.com were conducted in October 2014; the website has been updated since then.

We first show the breach of task location during tasking. We signed up as a worker account and searched for delivery tasks in Los Angeles; 2381 spatial tasks were found. We obtained various information about a particular task by clicking on it, such as description, price, task status and cloaked locations. Although each location is cloaked in a circle with a radius of half a km[4] (Fig. 7.2a) to protect task locations from workers, the actual drop-off and pick-up locations were mentioned in the task description, i.e., *"Please pick up a box of mini-muffins from (S) promptly at 8 am on Tues, 9/4, and drive them straight to me at (D)."* It is also worth noting that task requests often contain sensitive information, such as health status of the requesters. An example of a sensitive task is one with title *"super easy task deliver a bag to the doorstep of a sick friend."* Nonetheless, these privacy risks are due to the disclosure of task content, which is beyond the scope of this study.

We then show the leak of worker location during tasking and reporting. To gain a competitive advantage, a worker may wish to not disclose locations of his visits to other workers and requesters. The task status (Fig. 7.2c) infers that the worker, referred to as Bob, was at the pick-up and drop-off locations of the task during the 1-h period between his assigned time and his completed time. The risk of precisely inferring Bob's locations is even higher for time-sensitive tasks such as delivery and help at home, which requires him to meet requesters in-person

---

[3]We present the privacy threats to a pull-based SC system only; however, some of these privacy threats also occur in push-based SC such as iRain.

[4]We obtained this information via JavaScript code.

Table 7.2 Three tasks requested by requester Alice

Task description	Corresponding JavaScript
Quick post-party dishwashing clean up needed	"radius" : "0.5", "geo_center" : {"lat" : "33.xxxxxx", "lng" : "-118.xxxxxx"}
Take down light Christmas decorations	"radius" : "0.5", "geo_center" : {"lat" : "33.xxxxxx", "lng" : "-118.xxxxxx"}
Put up 20 yard sale signs in Mid-Wilshire area	"radius" : "0.5", "geo_center" : {"lat" : "33.xxxxxx", "lng" : "-118.xxxxxx"}

We replaced six digits after the decimal point of "geo_center" by 'x' to protect the privacy of the requester

at a specific place and time. This inference attack shows that TaskRabbit does not guarantee privacy protection for the pull mode in Sect. 7.3.1, which says that Bob's locations are private to the server and only requesters who have their tasks performed by Bob should know his locations. In addition, one can also see much more information about Bob, including his previously performed tasks (Fig. 7.2d) and all reviews from the requesters who hired him. These associations between Bob and his performed tasks indicate that the assignment links and reporting links are known to the server.

Among Bob's requesters, we randomly picked one named Alice. We further show that her home location can be learned by tracking her task requests. We searched for household tasks that Alice requested in the past; three of them are shown in Table 7.2. These tasks were in the proximity of each other and likely situated at her home. Our hypothesis is that the tasks' locations were randomly cloaked such that the cloaking regions covered the actual location of the tasks. The location must be in the overlapped area using triangulation. We validated our hypothesis by confirming that the location of another task, whose location was known, is within the overlapped region. This attack suggests that the more task requests are posted, the more accurately their locations can be learned. This simple attack is against the threat model, which states that the locations of Alice's tasks should only be revealed to the workers who performed her tasks.

7.4 Privacy Countermeasures

In this section we survey some state-of-the-art approaches addressing the privacy issues in spatial crowdsourcing. We first categorize the studies into two groups: *tasking in the push mode* and *reporting in the pull mode*. Subsequently, each subgroup is further classified according to the applied techniques. Within each subgroup we identify one key paper shown in boldface to be presented in depth while follow-up studies are briefly discussed. An overview of these studies is presented in Table 7.3. The table shows that the studies solely focus on location privacy of workers and assume that the locations and content of tasks are public.

Table 7.3 Overview of problem focuses (Re: reporting, Ta: tasking); privacy techniques used (Ps: pseudonym, Cl: cloaking, Pt: perturbation, Ex: exchange-based, En: encryption-based); threats (W: worker, T: requester, S: server); trusted third party (TTP); optimization type (ST: single task, MT: multiple tasks). x and (x) represent primary and secondary aspects, respectively

Paper	Phase		Techniques					Protection		Threats			TTP		Opt. type	
	Re	Ta	Ps	Cl	Pt	En	Ex	W	T	W	R	S	Yes	No	ST	MT
Shin et al. 2011 [27]	x	x	x	(x)		(x)		x	N/A		N/A	x	x		x	
Boutsis et al. 2013 [2]	x		(x)				x	x	N/A	(x)	N/A	x		x		x
Zhang et al. 2016 [37]	x						x	x	N/A		N/A	x		x		x
Kazemi et al. 2011 [12]		x	(x)	x				x			(x)	x	x			x
Vu et al. 2012 [35]		x	x		(x)			x		(x)	(x)	x	x			x
Sun et al. 2017 [28]		x	x					x				x	x			x
Pham et al. 2017 [20]		x	x					x	x			x	x			x
To et al. 2014 [32]		x		x				x		(x)	(x)	x	x		x	
Gong et al. 2015 [8]		x		x				x		(x)	(x)	x	x		x	
Zhang et al. 2015 [38]		x		x				x		(x)	(x)	x	x		x	
To et al. 2016 [31]		x		x				x		(x)	(x)	x	x		x	
Pournajaf et al. 2014 [22]		x	x					x			(x)	x	x			x
Hu et al. 2015 [10]		x	x					x			(x)	x	x			x
Shen et al. 2016 [26]		x			x			x		(x)	(x)	x		x	x	
Liu et al. 2017 [17]		x			x			x				x		x	x	
Liu et al. 2017 [16]		x			x			x	x			x		x	x	

Moreover, the server is regarded as a primary threat in all studies, while some consider workers and requesters as secondary adversaries. We also notice that the most recent studies focus on the push mode, which requires privacy protection during tasking. This problem is considerably more challenging when compared to the problem of privacy-preserving reporting in the pull mode.

7.4.1   Protection in the Pull Mode

Privacy protection in the pull mode has been studied in the context of participatory sensing. In this section we highlight recent studies that often focus on the reporting phase of the pull mode. They use either *pseudonymity* [27] or *exchange-based* techniques [2, 37]. The pseudonymity method disassociates the connections between one's uploaded data and his/her identity while the latter exchanges workers' crowsourced data and location information before uploading them to a server so that the server is uncertain about locations of individual workers.

7.4.1.1   Pseudonymity Techniques

Shin et al. [27] propose a privacy-preserving framework for the pull mode as illustrated in Fig. 7.3. A requester submits a task to a *registration authority* (RA) that will verify the task before sending it to a *task service* (TS). Also, a worker connects to TS through an anonymizing network such as Tor to request new tasks, referred to as a *task subset*. After receiving the requested tasks, the worker chooses which tasks to accept. He then performs the tasks and uploads the corresponding task reports to a *report service* (RS) via an *anonymous service* (AS). In this framework, RA and AS are trusted while TS, RS and requesters can be hostile. TS and RS can be considered as services performed by the server.

This study [27] provides privacy protection in both tasking and reporting phases. During tasking, the role of the anonymizing network is to disassociate the worker and his requested tasks, depicted by the first and the third assignment links in Fig. 7.1b. To preserve privacy during reporting, a worker typically sends his task report to RS via AS, which routes the report through multiple servers so that the server (i.e., TS and RS) cannot associate multiple locations (i.e., IP addresses) with

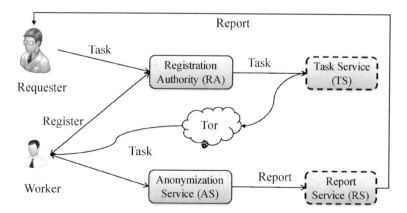

Fig. 7.3 A framework for privacy protection during tasking and reporting in the pull mode. Dashed entities are malicious, while others are trusted

the identity of the same worker. Consequently, the server is oblivious to the first reporting link in Fig. 7.1b. More recently, there has been closely related work in participatory sensing that enables workers to hide their locations and data ownership by passing the collected data through a random neighboring worker multiple times before uploading the data to the server [11].

7.4.1.2 Exchange-Based Techniques

Pseudonymity techniques are ad hoc and do not provide quantifiable privacy protection. For more sensitive tasks that require strong privacy guarantee, k-anonymity [29] is used in [27] to ensure that each report is anonymized with $k - 1$ reports generated by other workers with similar sensitive information. However, such techniques may not be applicable to SC because the worker location is part of the report. To address such a problem, Boutsis and Kalogeraki [2] propose the exchange-based technique to obscure the workers by exchanging their reports between them before disclosing the sensitive information to an untrusted server (i.e., server). Such a technique can be used as AS in Fig. 7.3, aiming to protect the first reporting link in Fig. 7.1b from the server.

To provide a quantifiable privacy guarantee, in [2] the authors use location entropy as the measure of privacy or the attacker's uncertainty. The study aims to make all workers' trajectories as equiprobable to contain sensitive locations by maximizing the location entropy of an individual's trajectories to be defined later. To maximize the location entropy, trajectories with sensitive locations are distributed among multiple workers. Particularly, each worker's mobile phone identifies the k most frequently visited locations as *sensitive data* from a *local* trajectory database. A trajectory is selected for exchange if removing the trajectory increases the entropy of the database, computed as follows.

$$H_i = \sum_{loc_{ij} \in L} Pr(loc_{ij}) log(Pr(loc_{ij})))$$

where L is the set of locations and $Pr(loc_{ij})$ is the fraction of total visits to location j that belongs to user i. Consequently, an attacker will not be able to identify sensitive locations or identities of the workers.

For each worker, the trajectories that contain locations with high frequency are exchanged with other workers since removing high-frequency trajectories (trajectories that contain sensitive locations) makes the frequency of the locations in L more homogeneous and thus increases the entropy. Furthermore, as other workers may not be trustful, not only the set of high-frequency trajectories are exchanged but also another set of trajectories that do not contain the sensitive locations. This guarantees that neighboring workers are not able to associate the worker with his sensitive data. Consequently, both frequent and non-frequent trajectories are selected and forwarded to individual workers so that no worker can be certain about the sensitivity of any trajectory.

A drawback of computing entropy locally is that the exchange decisions can be suboptimal due to the lack of a global view of all workers. This is because individual workers try to maximize their own entropy regardless of each other, which goes contrary to the fact that exchanging trajectories alters the location entropy of multiple workers. Thus, the exchange-based technique should consider the entropy with respect to all workers as opposed to individual workers. Therefore, Zhang et al. [37] introduce a similar framework, but here workers coordinate with each other to exchange their sensing data, including locations before uploading to the server. As a result, all sensitive locations are equally likely visited by any worker so that the actual trajectory of each worker cannot be learned. However, unlike [2] where entropy is computed for a single worker, here entropy is calculated for all workers.

Although the exchange-based technique is simple and does not rely on a trusted server, the actual location information is still uploaded to the server. Therefore, this approach is vulnerable to background knowledge attack. For instance, if the server knows that only worker w_i visits a particular location where a report was uploaded, the server is certain that w_i actually made the report.

7.4.2 Protection in the Push Mode

While preserving privacy during reporting in the pull mode has been largely studied in the context of participatory sensing (a recent survey can be found in [4]), recent SC studies focus on the more challenging phase of tasking. These studies generally assume the push mode. We emphasize that focusing on the tasking step in the push mode is the correct approach, given that SC workers have to physically travel to the task location. The completion of a task discloses the fact that some worker must have been at that location, and this is unavoidable in SC. Focusing on tasking also makes sense from a disclosure volume standpoint. During the assignment, all workers are candidates for participation; therefore, locations of all workers are exposed, absent a privacy-preserving mechanism. Nevertheless, after task request dissemination, only a few workers will participate in task completion, and *only if they give their explicit consent* (see the threat model for the push mode in Sect. 7.3.1).

Various techniques have been proposed to protect location privacy of workers during task assignment in SC, including *cloaking* (hide the accurate location in a cloaked region) [10, 13, 22, 35], *perturbation* (distort the actual location information by adding artificial noise) [8, 31, 32, 38] and *encryption* [26, 27].

7.4.2.1 Cloaking Techniques

The studies in this category generally implement spatial k-anonymity by generating a *cloaking region* (CR) for each worker, which includes $k - 1$ other workers. To guarantee strong privacy protection, peer-to-peer spatial k-anonymity [3] has been adopted in these studies. In the following we first present a simplified version of tasking without constraints. Next, we survey some recent studies that consider real-world constraints, such as the travel budget of each worker and a worker's willingness to perform tasks.

Task Assignment Without Constraints

In [27], each worker requests a *task subset* of size p at a time; however, choosing an appropriate value of p is not trivial. Large p may lead to not only high communication overhead between workers and TS, but tasks are also *unnecessarily* disclosed to the workers. In contrast, small p may result in some tasks that will never be accepted by any worker. One reason for this is that a worker can browse far-away tasks that he cannot complete before the tasks' deadlines. This redundant disclosure incurs additional privacy threats to the requesters of those tasks.

In order to minimize such disclosure, Kazemi and Shahabi [12] propose a privacy framework that enables each worker w_i to query the server for a set of nearby spatial tasks. Particularly, the server needs to distribute a set of spatial tasks to workers such that each worker is assigned a subset of tasks that are closer to him than to any other worker. Without privacy protection, the server can construct a Voronoi diagram of the workers, including a set of cells where each cell belongs to a worker, and any spatial task in the cell is closer to the worker than to any other worker. Once the server computes the Voronoi diagram of the workers, it forwards to each worker all the spatial tasks lying inside the corresponding cell. However, in such a scenario, an adversary may infer the worker's identity by associating the query to query location (i.e., the location from which the query is issued. This is referred to as *location-based attack*. Consequently, the framework aims to protect worker identity from location-based attacks by disassociating a query from the query location.[5] The framework named PiRi (partial-inclusivity and range independence) has both *query formation* and *query selection*.

Query Formation

In the query formation step, each worker w_i computes his Voronoi cell by communicating with his neighboring peers [3]. The worker forms his CR, where his location is blurred among $k - 1$ other peers (with $k = 3$, the solid-lined rectangle in Fig. 7.4a). The worker can send the CR along with the radius r (i.e., the smallest enclosing circle of w_i's Voronoi cell) to the server to retrieve all the tasks which lay inside his Voronoi cell. However, the range query is dependent on the size of the worker's Voronoi cell (range dependency), which is a potential for information leaks. Considering an extreme scenario where the server knows the

---

[5]However, this study assumes that workers trust one another. Hence, a more recent study [35] solves a similar problem as in [12] without the assumption of trusted workers.

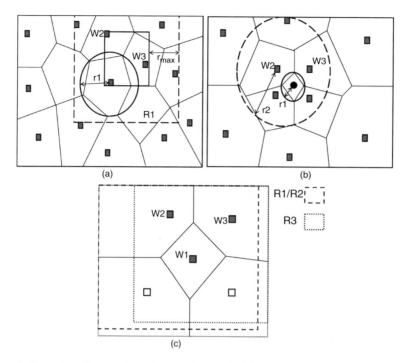

Fig. 7.4 Examples of range dependency and all-inclusivity. (**a**) Query formation. (**b**) Range dependency leak. (**c**) All-inclusivity leak

workers' locations, it also knows their Voronoi cells and therefore the radius r for each of them. Consequently, the server can easily identify the query issuer (i.e, the set of all workers in the CR with radius r). Figure 7.4b depicts such a scenario, where w_1 (black-filled circle) cloaks himself with w_2, and sends the CR along with radius r_1 to the server (see the size of r_1 as compared to r_2). The server, knowing the location of the workers, and hence their Voronoi cells (i.e., r_1, and r_2), relates the query with radius r_1 to its query location (i.e., the location of a worker with the Voronoi cell of the same radius).

In order to avoid the range dependency leak, each worker w_i should cloak not only his location but also his range query among $k - 1$ other peers. In other words, instead of forming his range query with radius r_i, the worker forms his query with radius r_{max}–the maximum radius among all the k peers inside the CR. This guarantees the k-anonymity at all times. In Fig. 7.4a, R_1 (the dotted line rectangle) shows the query region formed by r_{max}.

Query Selection

Once all workers have formed their query regions, they can send them out to the server. However, the server can utilize the gathered information (i.e., query regions) from all workers to attack the system (all-inclusivity leak). Figure 7.4c illustrates such scenario, in which workers $w_{1..3}$ participate in the system. The figure shows that w_1 cloaks himself with w_2. Similarly, w_2 forms a cloaked region with w_1. Subsequently, both w_1 and w_2 form identical query regions. The figure also depicts that w_3 cloaks himself with w_1. Accordingly, the server can easily identify w_3 by relating it to the query region R_3, since w_3 appears only once (i.e., R_3) in all the three submitted query regions to the server. This indicates that the more workers submit queries to the server, the more information the server has to infer the workers' identities. To prevent this leak, the authors attempt to minimize the number of queries submitted to the server while assigning the nearby tasks to *every* single worker.

Since there is a large overlap among the query regions of the workers, a worker can share his result received from the server with all the peers whose Voronoi cells lay completely inside his query region. The problem is how to select the group of representative workers, formally stated as follows. Given a set of workers W, and a set of spatial tasks T, let R and V be the set of query regions and Voronoi cells for the set W, respectively, where R_i corresponds to the query region for worker w_i, and V_i is the Voronoi cell for w_i. The problem is to find a set $C \subseteq R$ that covers the entire set V with minimum cardinality. This problem is shown to be NP-hard by reduction from the minimum set cover problem [12]. One well-known approach for solving the set cover problem is a greedy algorithm that picks a representative worker whose query region covers the largest number of uncovered Voronoi cells from V. However, this approach is applicable only in a centralized setting, where a global knowledge of the environment is available. To address this issue, the greedy heuristic is extended to support the distributed environment. Particularly, a voting mechanism is devised to select the set of representative workers, whose CRs are sent out to the server. These query results will later be shared with the rest of the workers. This step has been shown to prevent the all-inclusivity leak [12].

Task Assignment with Constraints

In [12, 35], spatial tasks are distributed to the corresponding nearest workers. This objective may not necessarily fit SC applications as workers often have various constraints that need to be considered. For example, they may be willing to perform tasks that are far away, but within their daily travel routes. To capture such constraints, each worker w_i has a cloaked area a_i and a limited travel budget b_i, which denotes the maximum distance he is willing to travel [22]. Given the cloaking regions of a set of workers, the objective of the server is to match a set of spatial tasks to the workers such that task assignment is maximized while satisfying the travel budget constraint of each worker.

As *travel cost* (often measured by the distance between tasks and assigned workers) is an important performance metric in SC, in the following we first present

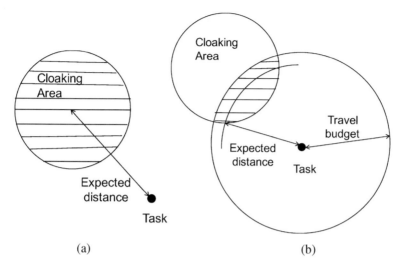

Fig. 7.5 Distance estimation methods. (**a**) Centroid-point method. (**b**) Expected probabilistic method

two methods for estimating the travel cost from the cloaked areas of the workers. Thereafter, we present the problem of *spatial task assignment with cloaked locations* (STAC) [22].

Distance Estimation

Given the cloaked area a_i of the workers, STAC proposes two methods for estimating the expected distances between pairs of workers w_i and tasks t_j, named $\hat{d}_{i,j}$. The baseline method approximates the worker location as the centroid of his cloaking area as depicted in Fig. 7.5a. Another method uses the travel budget of the worker to prune the cloaking area (i.e., the dashed area in Fig. 7.5b), resulting in a shrunk area that contains only accessible locations of the worker. Consequently, $\hat{d}_{i,j}$ is estimated by the distance between the task location and shrunk areas.

Next, we present a two-phase optimization approach to STAC. The first phase, denoted as G-STAC, globally optimizes task assignment using cloaked locations of the workers. The second phase, referred to as L-STAC, locally optimizes the assignment of individual workers using their own exact locations.

Global Optimization

Given a set of workers and a set of spatial tasks, G-STAC aims to achieve a particular goal of task coverage with the minimum travel cost. G-STAC is formally defined as follows.

$$min\ TC = \sum_{i \in W} \sum_{j \in T} \hat{d}_{i,j} x_{i,j}$$

$$s.t.\ TU = \sum_{j \in T} \frac{\sum_{i \in W} x_{i,j}}{k_j} \geq gm$$

$$\sum_{j \in T} \hat{d}_{i,j} x_{i,j} \leq b_i$$

where task cost (TC) is the total distance traveled by all workers, while task coverage or utility (TU) is the total covered fraction of tasks. $\hat{d}_{i,j}$ is the estimated distance between worker i and task j, $x_{i,j} = 1$ means worker i is assigned to task j, otherwise $x_{i,j} = 0$, k_j is the required coverage of t_j (i.e., the number of workers to perform t_j) while $g \in (0, 1]$ indicates the required fraction of coverage for a task. The last constraint guarantees that w_i's travel distance is within his budget b_i.

G-STAC is shown to be NP-hard by reduction from the minimum set cover problem. Therefore, a greedy algorithm is proposed that iteratively selects the most cost-effective worker-task pair and updates TU until either the coverage goal is achieved or the travel budgets of all workers are spent. A worker-task pair is cost-effective if the ratio of expected distance to the expected coverage contributed by this worker is small.

Local Optimization

The output of G-STAC is the best mapping of tasks to workers, which is sent to workers as suggested assignments. However, a worker may be assigned tasks whose locations exceed his travel budget, or nearby tasks are not assigned to him because their distance has been estimated as being farther away. Thus, the local refinement phase (L-STAC) is performed by individual workers' devices for more coverage and lower travel cost. A caveat is that selecting the closest tasks for each worker may result in over-coverage for some tasks, while the others remain unperformed. Consequently, in addition to minimizing the travel cost, L-STAC also tries to minimize the change in the local optimization when compared to the global optimization. L-STAC is formally defined as follows.

$$min\ TC_i = \sum_{j \in T} d_{i,j} y_{i,j}$$

$$s.t.\ |y_i - x_i| < \epsilon$$

$$\sum_{j \in T} \frac{y_{i,j}}{k_j} \geq \sum_{j \in T} \frac{x_{i,j}}{k_j}$$

$$\sum_{j \in T} d_{i,j} y_{i,j} \leq b_i$$

where for each worker w_i, x_i and y_i are the binary assignment vectors of the global and local phases of STAC, respectively. The first constraint, $|y_i - x_i|$, is the Hamming distance between x_i and y_i, which is bounded by a threshold ϵ aiming to keep minimum changes in the local assignment. The second constraint ensures that w_i's contribution to the task coverage is not decreased when compared to his contribution in the global phase. In the same fashion, L-STAC is NP-hard by reduction from the minimum set cover problem; thus, another greedy algorithm has been proposed to solve L-STAC.

Recently, Hu et al. [10] extended the travel budget constraint in [22] to a *spatial region*, represented by a rectangle R, within which the worker is willing to travel. Similar to [12], workers employ the peer-to-peer cloaking technique [3] to cloak their locations among $k - 1$ other workers. Also, each worker's cloaking area must contain his spatial region R, otherwise the cloaking area is extended to cover R. Observing that workers' cloaking areas often contain multiple spatial regions of other workers, to reduce the communication overhead, only some cloaking areas that could cover all the workers' spatial regions will be sent to the server. This technique limits the disclosure of information when compared to sending all the workers' cloaking areas to the server [22].

The cloaking techniques used in [10, 22] are intuitive; nevertheless, their privacy guarantee is weak. Such obfuscation-based techniques do not provide rigorous privacy protection and are prone to homogeneity attack [18] when all k workers are at the same location. Also, the value k needs to be specified to guarantee the desired level of privacy protection. Unfortunately, choosing an appropriate k value can be difficult because k-anonymity does not consider the frequency of user visits. To elaborate, a location may be visited by many workers—those who have a dominant contribution to the location (i.e., home or office) are most likely to be the subject of attack. Consequently, one with a background knowledge of who visits the location the most can easily perform such an attack.

7.4.2.2 Perturbation Techniques

Methods in this category use differential privacy (DP) to protect workers' locations during task assignment [8, 31–33, 38], which overcomes the aforementioned issues of the obfuscation technique. DP has emerged as the de facto standard with strong protection guarantees rooted in statistical analysis. It provides a *semantic* privacy model as opposed to a *syntactic* model in other sanitization techniques (e.g., k-anonymity, l-diversity). DP has been adopted by major industries for various tasks without compromising individual privacy, e.g., discovering users' usage patterns with Apple [9] or crowdsourcing statistics from end-user client software with Google [7]. DP ensures that an adversary is not able to reliably learn from the published sanitized data whether or not a particular individual is present in the original data, regardless of the adversary's prior knowledge.

The authors in [32] propose system model, privacy model and performance metrics, followed by two main steps that preserve privacy and identity of workers: *sanitization* of workers' locations and *task assignment* on the sanitized data.

System Model

To protect location privacy of workers participating in spatial tasks, the server must only have access to data sanitized according to ϵ-*differential privacy* [6] (ϵ is privacy loss or privacy budget). Figure 7.6a shows the system architecture. Workers send their locations (Step 0) to a trusted *cellular service provider* (CSP) which collects updates and releases a *private spatial decomposition* (PSD) according to privacy budget ϵ mutually agreed upon with the workers. The PSD is accessed by the server (Step 1), which also receives tasks from a number of requesters (Step 2). When the server receives a task t, it queries the PSD to determine a *geocast region (GR)* that encloses with high probability workers close to t. Next, the server initiates a *geocast* communication [19] process (Step 3) to disseminate t to all workers within *GR*. According to DP, sanitizing a dataset requires the creation of fake locations in the PSD. If the server is allowed to directly contact workers, then failure to establish

Fig. 7.6 Differentially private framework for spatial crowdsourcing. (**a**) System architecture. (**b**) Worker PSD using adaptive grid

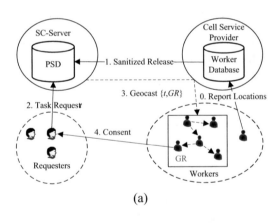

(a)

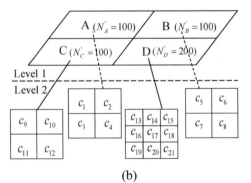

(b)

a communication channel would breach privacy, as the server is able to distinguish fake workers from real ones. Using geocast is a unique feature of the framework which is necessary to achieve privacy protection. Geocast can be performed either with the help of the CSP infrastructure, or through a mobile ad hoc network where the CSP contacts a single worker in the *GR*, and then the message is disseminated on a hop-by-hop basis to the entire *GR*. The latter approach keeps CSP overhead low and can reduce operation costs for workers. Upon receiving request t, a worker w_i decides whether to perform the task or not. If yes (Step 4), she sends a *consent* message to the server (or requesters) confirming w_i's availability. If w_i is not willing to participate in the task, then no consent is sent, and no information about the worker is disclosed.

Privacy Model and Assumptions

The objective of the framework is to protect both the *location* and the *identity* of workers *during task assignment*. Once a worker consents to a task, the worker herself may directly disclose information to the task requester (e.g., to enable a communication channel between worker and requester). However, such additional disclosure is outside the scope of this work, as each worker has the right to disclose his or her individual information. Instead, the focus of the framework is on what happens prior to consent, when worker location and identity must be protected from *both* task requesters and the server. This privacy model is a weaker version of the restrictive model in Fig. 7.1a since task locations are public.

Workers cannot trust the server, especially as there may be many such entities with diverse backgrounds, e.g., private companies, non-profits, government organizations, academic institutions. On the other hand, the CSP already has a signed agreement with workers through the service contract, so there is already a trust relationship established, as well as mutually-agreed upon rules for data disclosure. Furthermore, the CSP already knows where subscribers are, e.g., using cell tower triangulation, so worker location reporting does not introduce additional disclosure. In addition, having the CSP expose a PSD release of the user location dataset can benefit applications beyond crowdsourcing. For instance, the PSD can be shared with law enforcement agencies for public safety, or with commercial organizations to increase the revenue of the CSP. Therefore, there is sufficient motivation for the CSP to provide such a location sanitization service.

However, the CSP has no expertise, and perhaps no financial interest, to host an SC service, which needs to deal with a diverse set of issues such as interacting with various task requester categories, managing profiles (e.g., some workers may only volunteer for environmental tasks), etc. The role of the CSP is to aggregate locations from subscribed workers, transform them according to DP, and release the data in sanitized form to one or more servers for assignment. As multiple servers can use the same PSD, it is practical for the CSP to provide PSDs for a small fee, e.g., a percentage of the workers' payment, or a tax incentive in the case of a public-interest SC application.

Design Goals and Performance Metrics

Protecting worker location significantly complicates task assignment and may reduce the effectiveness and efficiency of worker-task matching. Due to the nature of DP, it is possible for a region to contain no workers, even if the PSD shows a positive count. Therefore, no workers (or an insufficient number thereof) may be notified of the task request, and the task may not be completed. Alternatively, the GR may comprise workers who are a long distance away from the task location, whereas nearer workers are not included. Finally, in the non-private case, only one selected worker, whose location and identity is known, is notified of the task request. With location protection, redundant messages need to be sent, increasing overhead. We focus on the following performance metrics:

- *Assignment success rate (ASR).* Due to PSD data uncertainty, the server may incorrectly assign workers to tasks (e.g., no worker is reached, or task is too far and workers do not accept it). *ASR* measures the ratio of tasks accepted by a worker to the total number of task requests.
- *Worker travel distance (WTD).* The server is no longer able to accurately evaluate worker-task distance, hence workers may have to travel long distances to tasks. The challenge is to keep the worker travel distance low, even when exact worker locations are not known.
- *System overhead.* Dealing with imprecise locations increases the complexity of assignment, which poses scalability problems. A significant metric to measure overhead is the *average number of notified workers (ANW).* This number affects both the *communication overhead* required to geocast task requests, as well as the *computation overhead* of the matching algorithm, which depends on how many workers need to be notified of a task request.

Sanitization of Workers' Locations Using Adaptive Grid

The first step in the proposed framework consists of building a PSD (at the CSP side) to be used later for task assignment at the server. Building the PSD is an essential step because it determines how accurate the released data is, which in turn affects *ASR*, *WTD* and *ANW*. Worker location data are sanitized at the CSP using a PSD, named *adaptive grid* (AG) [23]. PSD is a sanitized spatial index, where each index node contains a noisy count of the workers rooted at that node. Figure 7.6b shows a snapshot of an adaptive grid with four level-1 cells A, B, C, D. Constructing a differentially private AG requires two steps. First, the noisy counts N' of A, B, C, D are computed by adding calibrated random Laplace noise [6]. Second, based on the noisy counts, level-1 cells are further split into level-2 cells. Cell D, which has a higher noisy count of 200 is partitioned according to a 3×3 grid, while the granularity for other cells is 2×2. Thereafter, AG adds to each level-2 cell (c_i,

$i = 1 \ldots 21$) calibrated random Laplace noise. Finally, their corresponding noisy counts n_{c_i} are published together with the structure of the AG.

Although AG yields small errors for general spatial queries, it is not directly applicable to SC due to its rigidity in choosing parameters. Specifically, the granularity m_2 of the level-2 grid is too coarse, leading to large geocast areas and high communication overhead. Thus, the AG method is extended to address the specific requirements of the SC framework. Particularly, a heuristic is proposed to increase the granularity m_2 in order to decrease overhead, but only to the point where there is at least one worker in a cell [32].

Task Assignment on Sanitized Data

On top of the noisy data, to ensure that task assignment has a high success rate, analytical models that consider task completion rate, worker travel distance and system overhead are developed. When a request for a task t is posted, the server queries the PSD and determines a geocast region GR where the task is disseminated. The goal is to obtain a high success rate for task assignment, while at the same time reducing the worker travel distance WTD and request dissemination overhead ANW.

Acceptance Rate and Analytical Utility Model

Travel distance is critical in SC, as workers need to physically visit the task locations. A worker is more willing to accept nearby tasks [13], so acceptance rate is modeled as a decreasing function of travel distance. Also, we denote by *acceptance rate (AR)* the probability $p^a (1 \leq p^a \leq 1)$ that a worker agrees to complete a task for which he has received a request. Thereafter, an analytical *utility* model is developed that allows the server to quantify the probability that a task request disseminated in a certain GR is accepted by a worker. Intuitively, the utility depends on the AR and on the worker count $\bar{w}$ estimated to be enclosed within the GR. A server will typically establish an *expected utility* threshold EU which is the targeted success rate for a task (this is a system goal, rather than an outcome). Generally, EU is considerably larger than an individual worker's p^a, so the GR must contain multiple workers.

We define X as a random variable for the event that a worker accepts a received task: $P(X = True) = p^a$ and $P(X = False) = 1 - p^a$. Assuming w independent workers, $X \sim Binomial(w, p^a)$. We define the *utility* of a geocast region covering w workers as:

$$U = 1 - (1 - p^a)^w \qquad (7.1)$$

U measures the probability that at least one worker accepts the task. The utility definition can be extended to the case of redundant task assignment, where multiple workers are required to complete a task [31].

Geocast Region Construction

The third step in the framework is the construction and dissemination of GR. By the nature of the DP protection model, fake entries may need to be created in the worker PSD. Thus the server cannot directly contact workers, not even if pseudonyms are used, as establishing a network connection to an entity would allow the server to learn whether an entry is real or not, and this breach privacy. To address this challenge, the geocast mechanism was introduced for the task request dissemination. Geocast is a routing and addressing method, which is used to deliver information to all devices situated within a geographical area. Once a PSD partition is identified by the analytical model outlined above, the task request is geocast to all the workers within that partition.

Particularly, given task t, the GR construction algorithm must balance two conflicting requirements: determine a region that (1) contains sufficient workers such that task t is accepted with high probability, and (2) the size of the geocast region must be small. The input to the algorithm is task t as well as the worker PSD, consisting of the two-level AG with a noisy worker count for each grid cell. The algorithm chooses as initial GR the level-2 cell that covers the task, and determines its U value. As long as utility is lower than threshold EU, it expands the GR by adding neighboring cells. Cells are added one at a time, based on their estimated increase in GR utility. Following the task localness property, we take into account the distance of each candidate neighboring cell to the location of t, and give priority to closer cells. The algorithm stops either when the utility of the obtained GR exceeds threshold EU, or when the size of GR is larger than a particular threshold; hence, utility can no longer be increased. The GR construction algorithm is a greedy heuristic, as it always chooses the candidate cell that produces the highest utility increase at each step. The experimental results show that workers' location privacy is protected without compromising performance, and the extra travel cost is tolerable—a 20% increase when compared to the non-private case.

Next, we present various extensions of the worker PSD, followed by an approach toward PSD for moving workers.

Extensions and Enhancements of Worker PSD

There have been recent studies that adopt the privacy model used in [32], assuming a trusted CSP and differentially private location sanitization. Particularly, Gong et al. [8] propose a framework that can protect the workers' location privacy when allocating tasks to the workers. Similar to [32], they develop analytical models and task allocation strategies that balance privacy, utility, and system overhead. In [8], the CSP not only aggregates workers' locations but also their reputation information, which is used to provide quality control over the reports. Consequently, a new structure called reputation-based PSD is proposed to partition the space based on both reputation and location information.

Another work studies reward-based spatial crowdsourcing that enables task assignment with optimized reward allocation (Zhang et al. [38]). The authors also reuse the privacy framework introduced in [32], in which the server and workers are connected by a trusted CSP. However, unlike [32] that uses the adaptive grid to releases a sanitized location view to the server, this study constructs a contour plot to represent the spatial distribution of workers aiming to introduce less noise than the prior technique. The contour plot is used to perform task assignment. The objective of task assignment is to find the minimum radius r to ensure that the *ASR* of a task is equal to *expected utility* threshold *EU*, i.e., the probability that at least one worker performs the task is no less than the threshold.

Protection for Dynamic Workers' Locations

Previous perturbation techniques [8, 32, 38] assume a static scenario where workers' locations do not change. However, SC systems receive continuous requests for task assignment. Hence, it is important to keep track of the whereabouts of *moving* workers and to release a *sequence* of worker PSDs that allow effective spatial task assignment over multiple timestamps. The challenge is that as workers move, new snapshots of sanitized worker locations must be disclosed to maintain task assignment effectiveness. However, access to sequential releases gives an adversary more powerful attack opportunities. To counter such threats, differential privacy requires more noise injection, which in the worst case may reach amounts that are proportional to the length of the released location history (i.e., the number of disclosed snapshots). Clearly, such large noise would render the data useless, since SC is likely to be a continuously offered service in practice. A recent study [31] extends [32] to address the challenge of moving workers by investigating privacy budget allocation techniques across consecutive releases, and employing post-processing techniques based on Kalman filters to reduce the inaccuracy introduced by addition of noise.

7.4.2.3 Encryption Techniques

In this section we discuss studies that use encryption-based approaches. In [27] the identity and location (i.e., IP address) of workers are hidden from TS through multiple Tor relays using Onion encryption. However, Tor does not try to protect against an attacker who can see or measure both traffic going into the Tor network and also traffic coming out of the Tor network—for example, the end-to-end timing correlation attack. Thus, to prevent TS from performing a timing attack by linking multiple task requests, the workers connect to TS at random intervals. Furthermore, during tasking the workers make sure that TS does not tamper the task request from RS; otherwise, the workers can report TS as fraudulent to RS.

Shin et al. [27], however, focus on the pull mode, which likely results in suboptimal task assignment. Therefore, a recent study [26] proposes a secure task-assignment protocol to protect worker location privacy in the push mode. The privacy framework used in [26] is similar to [32] (Fig. 7.6a), except the CSP is replaced by a *privacy service provider* (PSP)—a semi-honest (i.e., honest-but-curious) third party to provide privacy functionality and collect encrypted data from workers, including encrypted location reports. With the framework, the server needs to perform worker-task matching in the encrypted domain. Particularly, given a task, the server communicates with PSP in the encrypted domain to find the worker with minimum travel cost to the task. The travel cost is evaluated in terms of worker-task distance and the degree of interest of the worker to the task.

The advantage of the proposed protocol is twofold. The framework is not relying on a trusted-third-party and is robust to semi-honest adversaries. Also, the privacy guarantees hold for moving workers. However, when compared to the cloaking and perturbation techniques, cryptographic-based approaches may incur higher computation overhead. In addition, the semi-honest adversary model is restrictive in terms of privacy protection and may not always hold in the real-world SC applications. That is, server and PSP may not follow the specified protocol, or requesters can be malicious. Thus, a stronger privacy protocol that is resilient to malicious adversary model needs to be developed.

7.5   Conclusion and Future Directions

With the popularity of mobile devices, spatial crowdsourcing is rising as a frame-work that enables human workers to solve tasks in the physical world. With spatial crowdsourcing, requesters outsource a set of spatiotemporal tasks to a set of workers, i.e., individuals with mobile devices that perform the tasks by physically traveling to the specified locations of interest. However, current solutions require a worker to disclose his location to the server and/or to other requesters even before accepting a task—or a requester to disclose his tasks' locations, which can be used to infer his own location, to untrusted entities. In this chapter we identified the privacy threats to both workers and requesters in the two main phases of crowdsourcing: task assignment and task reporting.

We surveyed some of the most notable solutions proposing various privacy techniques, ranging from pseudonym, cloaking, perturbation to exchange-based and encryption-based approaches. These studies have shown encouraging results in protecting the privacy of both workers or requesters in spatial crowdsourcing. However, protecting the privacy of workers and requesters simultaneously using rigorous privacy guarantees such as differential privacy is still an open problem. Another promising direction is to consider powerful adversaries with knowledge about temporal correlations of a moving user's locations or collusion between workers and the server; for example, some workers may work for the SC company or the company may use driverless cars.

References

1. iRain: new mobile app to promote citizen-science and support water management: http://en.unesco.org/news/irain-new-mobile-app-promote-citizen-science-and-support-water-management, 2016.
2. I. Boutsis and V. Kalogeraki. Privacy preservation for participatory sensing data. In *2013 IEEE International Conference on Pervasive Computing and Communications (PerCom)*, pages 103–113. IEEE, mar 2013.
3. C.-Y. Chow, M. F. Mokbel, and X. Liu. Spatial cloaking for anonymous location-based services in mobile peer-to-peer environments. *GeoInformatica*, 15(2):351–380, 2011.
4. D. Christin. Privacy in mobile participatory sensing: Current trends and future challenges. In *Journal of Systems and Software*, volume 116, pages 57–68, 2016.
5. D. Deng, C. Shahabi, and U. Demiryurek. Maximizing the number of worker's self-selected tasks in spatial crowdsourcing. In *Proc. 21st ACM SIGSPATIAL Int. Conf. Adv. Geogr. Inf. Syst. - SIGSPATIAL'13*, pages 314–323, New York, New York, USA, 2013. ACM Press.
6. C. Dwork. Differential privacy. In *Automata, languages and programming*, pages 1–12. Springer, 2006.
7. Ú. Erlingsson, V. Pihur, and A. Korolova. RAPPOR: Randomized aggregatable privacy-preserving ordinal response. In *SIGSAC*, pages 1054–1067. ACM, 2014.
8. Y. Gong, C. Zhang, Y. Fang, and J. Sun. Protecting Location Privacy for Task Allocation in Ad Hoc Mobile Cloud Computing. In *IEEE Transactions on Emerging Topics in Computing*, pages 1–1, 2015.
9. A. Greenberg. Apple's "differential privacy' is about collecting your data - but not your data. https://www.wired.com/2016/06/apples-differential-privacy-collecting-data/, 2016.
10. J. Hu, L. Huang, L. Li, M. Qi, and W. Yang. Protecting Location Privacy in Spatial Crowdsourcing. In *Asia-Pacific Web Conference*, pages 113–124. Springer International Publishing, 2015.
11. L. Hu and C. Shahabi. Privacy assurance in mobile sensing networks: Go beyond trusted servers. In *2010 8th IEEE International Conference on Pervasive Computing and Communications Workshops, PERCOM Workshops 2010*, pages 613–619, 2010.
12. L. Kazemi and C. Shahabi. A privacy-aware framework for participatory sensing. In *ACM SIGKDD Explorations Newsletter*, volume 13, page 43. ACM, aug 2011.
13. L. Kazemi and C. Shahabi. GeoCrowd: Enabling Query Answering with Spatial Crowdsourcing. In *Proc. 20th Int. Conf. Adv. Geogr. Inf. Syst. - SIGSPATIAL '12*, number c, page 189, 2012.
14. L. Kazemi, C. Shahabi, and L. Chen. GeoTruCrowd: Trustworthy Query Answering with Spatial Crowdsourcing. In *Proc. 21st ACM SIGSPATIAL Int. Conf. Adv. Geogr. Inf. Syst. - SIGSPATIAL'13*, pages 304–313, 2013.
15. S. H. Kim, Y. Lu, G. Constantinou, C. Shahabi, G. Wang, and R. Zimmermann. Mediaq: mobile multimedia management system. In *Proceedings of the 5th ACM Multimedia Systems Conference*, pages 224–235. ACM, 2014.
16. A. Liu, Z.-X. Li, G.-F. Liu, K. Zheng, M. Zhang, Q. Li, and X. Zhang. Privacy-preserving task assignment in spatial crowdsourcing. *Journal of Computer Science and Technology*, 32(5):905–918, 2017.
17. B. Liu, L. Chen, X. Zhu, Y. Zhang, C. Zhang, and W. Qiu. Protecting location privacy in spatial crowdsourcing using encrypted data. In *EDBT*, pages 478–481, 2017.
18. A. Machanavajjhala, D. Kifer, J. Gehrke, and M. Venkitasubramaniam. l-diversity: Privacy beyond k-anonymity. *ACM Transactions on Knowledge Discovery from Data (TKDD)*, 1(1):3, 2007.
19. J. C. Navas and T. Imielinski. Geocast: geographic addressing and routing. In *Proceedings of the 3rd annual ACM/IEEE international conference on Mobile computing and networking*, pages 66–76. ACM, 1997.

20. A. Pham, I. Dacosta, B. Jacot-Guillarmod, K. Huguenin, T. Hajar, F. Tramèr, V. Gligor, and J.-P. Hubaux. Privateride: A privacy-enhanced ride-hailing service. *Proceedings on Privacy Enhancing Technologies*, 2017(2):38–56, 2017.
21. L. Pournajaf, D. A. Garcia-ulloa, L. Xiong, and V. Sunderam. Participant Privacy in Mobile Crowd Sensing Task Management : A Survey of Methods and Challenges. In *SIGMOD Record*, volume 44, pages 23–34. ACM, may 2015.
22. L. Pournajaf, L. Xiong, V. Sunderam, and S. Goryczka. Spatial task assignment for crowd sensing with cloaked locations. In *Proceedings - IEEE International Conference on Mobile Data Management*, volume 1, pages 73–82. IEEE, jul 2014.
23. W. Qardaji, W. Yang, and N. Li. Differentially private grids for geospatial data. In *2013 IEEE 29th International Conference on Data Engineering (ICDE)*, pages 757–768. IEEE, 2013.
24. D. Reinhardt and F. Dürr. Opportunities and risks of delegating sensing tasks to the crowd. In *Handbook on Mobile Data Privacy*. Springer, 2017.
25. J. Scheck. Stalkers exploit cellphone GPS. http://www.wsj.com, 2010.
26. Y. Shen, L. Huang, L. Li, X. Lu, S. Wang, and W. Yang. Towards preserving worker location privacy in spatial crowdsourcing. In *2015 IEEE Global Communications Conference, GLOBECOM 2015*, 2016.
27. M. Shin, C. Cornelius, D. Peebles, A. Kapadia, D. Kotz, and N. Triandopoulos. AnonySense: A system for anonymous opportunistic sensing. *Pervasive and Mobile Computing*, 7(1):16–30, 2011.
28. Y. Sun, A. Liu, Z. Li, G. Liu, L. Zhao, and K. Zheng. Anonymity-based privacy-preserving task assignment in spatial crowdsourcing. In *International Conference on Web Information Systems Engineering*, pages 263–277. Springer, 2017.
29. L. Sweeney. k-anonymity: A model for protecting privacy. *International Journal of Uncertainty, Fuzziness and Knowledge-Based Systems*, 10(05):557–570, 2002.
30. H. To, L. Fan, L. Tran, and C. Shahabi. Real-time task assignment in hyperlocal spatial crowdsourcing under budget constraints. In *2016 IEEE Int. Conf. Pervasive Comput. Commun. PerCom 2016*, pages 1–8. IEEE, mar 2016.
31. H. To, G. Ghinita, L. Fan, and C. Shahabi. Differentially Private Location Protection for Worker Datasets in Spatial Crowdsourcing. In *IEEE Transactions on Mobile Computing*, volume PP, pages 1–1, 2016.
32. H. To, G. Ghinita, and C. Shahabi. A framework for protecting worker location privacy in spatial crowdsourcing. In *Proceedings of the VLDB Endowment*, volume 7, pages 919–930. VLDB Endowment, jun 2014.
33. H. To, G. Ghinita, and C. Shahabi. PrivGeoCrowd: A toolbox for studying private spatial Crowdsourcing. In *Proceedings - International Conference on Data Engineering*, volume 2015-May, pages 1404–1407. IEEE, apr 2015.
34. H. To, C. Shahabi, and L. Kazemi. A server-assigned spatial crowdsourcing framework. *ACM Transactions on Spatial Algorithms and Systems*, 1(1):2, 2015.
35. K. Vu, R. Zheng, and J. Gao. Efficient algorithms for K-anonymous location privacy in participatory sensing. In *Proceedings - IEEE INFOCOM*, pages 2399–2407, 2012.
36. G. Wang, B. Wang, T. Wang, A. Nika, H. Zheng, and B. Y. Zhao. Defending against Sybil Devices in Crowdsourced Mapping Services. *Proceedings of the 14th Annual International Conference on Mobile Systems, Applications, and Services - MobiSys '16*, pages 179–191, 2016.
37. B. Zhang, C. H. Liu, J. Lu, Z. Song, Z. Ren, J. Ma, and W. Wang. Privacy-preserving QoI-aware participant coordination for mobile crowdsourcing. In *Computer Networks*, volume 101, pages 29–41, 2016.
38. L. Zhang, X. Lu, P. Xiong, and T. Zhu. A Differentially Private Method for Reward-Based Spatial Crowdsourcing. In *International Conference on Applications and Techniques in Information Security*, pages 153–164. Springer Berlin Heidelberg, 2015.

Chapter 8
Privacy in Geospatial Applications and Location-Based Social Networks

Igor Bilogrevic

Abstract The use of location data has greatly benefited from the availability of location-based services, the popularity of social networks, and the accessibility of public location data sets. However, in addition to providing users with the ability to obtain accurate driving directions or the convenience of geo-tagging friends and pictures, location is also a very sensitive type of data, as attested by more than a decade of research on different aspects of privacy related to location data.

In this chapter, we focus on two domains that rely on location data as their core component: Geospatial applications (such as thematic maps and crowdsourced geo-information) and location-based social networks. We discuss the increasing relevance of geospatial applications to the current location-aware services, and we describe relevant concepts such as volunteered geographic information, geo-surveillance and how they relate to privacy. Then, we focus on a subcategory of geospatial applications, location-based social networks, and we introduce the different entities (such as users, services and providers) that are involved in such networks, and we characterize their role and interactions. We present the main privacy challenges and we discuss the approaches that have been proposed to mitigate privacy risks in location-based social networks. Finally, we conclude with a discussion of open research questions and promising directions that will contribute to improve privacy for users of location-based social networks.

8.1 Introduction

The rate at which new online data is being generated is unprecedented. It is believed that 90% of all of the online data has been produced over the past 2 years [127]. Such data is used in various domains, including healthcare, research, agriculture, logistics, urban design, energy, retailing, crime reduction and business

I. Bilogrevic (✉)
Google, Zurich, Switzerland
e-mail: ibilogrevic@google.com

© Springer Nature Switzerland AG 2018
A. Gkoulalas-Divanis, C. Bettini (eds.), *Handbook of Mobile Data Privacy*,
https://doi.org/10.1007/978-3-319-98161-1_8

operations [133]. In particular, location data is extremely useful for transportation, mapping, urban design, environmental monitoring and advertisement. For instance, mobility patterns of hundreds of millions of users have been mined in order to analyse the Chinese economy [55]; in another case, location data from cell-phone users, as well as buses and taxi drivers, has been used to better understand city dynamics and environmental issues [90]; similarly, location information has been mapped to crime statistics [71] and used for poverty prediction [62]. In yet another instance, location data was used for disaster relief and coordination [111, 146].

Location is one among several aspects of a person's context, such as the time, the activity, the objects or the people in proximity of a person. In order to infer the context, people use their senses. Similarly, mobile devices require sensors to determine their context, and often also communication with third-party service providers and other devices. By being aware of their context, mobile devices can provide users with a multitude of services that enrich their experience and simplify their everyday activities. For example, location awareness enables devices to provide relevant and timely driving and walking directions, or to obtain local weather forecasts. In addition to services that use location as their core functionality, more recently location data became very relevant for online social networks, by enabling users to share their locations with their social circles, by adding location information to shared media (i.e., geo-tagging) or co-presence with other people.

Location-based services are extremely popular. In the U.S., 90% of smartphone owners reported using their devices to obtain information related to their location [98]. Similarly, one of the largest (in terms of number of registered users) online social networks that uses location data has reported having surpassed one billion monthly active users [35]. In addition to being very popular among users, location data is often processed by service providers in order to enhance their services; a recent report stated that location is among the top-3 identity-related data sources used for personalization [126]. Therefore, location data is not only valuable to the users, but also to the service providers and third parties, as they frequently use it in order to drive their revenues.

In addition to being valuable, location is also a sensitive type of data [10, 79], as it can be used to reveal aspects of one's life that go beyond the location itself. Research has shown that location traces can be used to infer one's home/work places [48, 56], political affiliations [65], activities [140], interests [94] and social networks [9, 83]. Hence, being able to control the access to and flow of location data is of paramount importance for the users. Currently both Google and Facebook, two of the largest online service providers, allow their users to manage privacy settings and controls, enabling them to decide who can see their information and how it is used to personalize online services [34, 46]. For example, Google enables its users to see, correct and delete location data about them. Similarly, Facebook allows its users to decide how location check-ins and other social features (such as friend geo-tagging) work, by limiting and removing location tags [33].

In this chapter, we discuss privacy issues for two popular use-cases of location data on mobile devices: (1) geospatial applications (such as crowdsourced mapping, urban design, crisis and poverty thematic maps) and (2) location-based social networks (such as proximity-based friend finders, online dating, social and media geotagging, as well as event planning). We begin by discussing the increasing relevance of the geospatial applications in Sect. 8.2, which have paved the way for the current location-based services. We cover topics such as crowdsourced geographic data, geo-surveillance and their relevance to privacy. Afterwards, in Sect. 8.3 we focus on a subcategory of geospatial applications, i.e., location-based social networks (LBSNs), where we discuss their different entities and their roles. For instance, users may be concerned with what other users know or can learn about them, but they can also worry about how service providers and other third parties are using their data. Next, we present the main privacy challenges and we discuss the approaches that have been proposed to mitigate the privacy risks, by surveying solutions from both the engineering field as well as the Human-Computer Interaction (HCI) domain. It is crucial to consider these related but separate aspects, as Privacy-Enhancing Technologies (PETs) are most effective when they are intuitive and bring benefits to users [10].

8.2 Privacy in Geospatial Applications

One of the most ancient instances of geospatial applications is cartography, which can be defined as the science of creating maps.[1] Although the first examples of maps were used to describe the stars rather than Earth's surface [92], modern maps are able to capture and summarize a plethora of information about the surface of our planet and its inhabitants, such as the road networks, ocean dynamics, environmental aspects related to natural disasters and thematic maps of economic indicators. For instance, road maps have been widely used to help people decide on the optimal way to reach their destinations, whereas thematic maps—which associate a specific type of information, such as poverty or crime levels, with a geographic region[2]—are routinely employed as tools to inform and guide policy and political efforts [71].

The increase in availability of different types of maps has benefited from a wider accessibility of public geographic information and geodemographic databases [25]. For example, several countries make census data publicly available to download and use.[3] In the U.S., such data contains anonymized information, at a block-level resolution, about citizens' incomes, education levels, housing and general demographics, including ethnicity, gender, age and sex.[4] In addition to census data, some countries have started releasing geo-referenced statistics related to public safety aspects, such as crime rates. In the U.S. and U.K., for instance, police

---

[1] http://www.merriam-webster.com/dictionary/cartography, last retrieved Dec. 4, 2016.

[2] https://www.census.gov/geo/maps-data/maps/thematic.html, last retrieved Dec. 4, 2016.

[3] http://unstats.un.org/unsd/demographic/sources/census/wphc/default.htm, last retrieved Dec. 4, 2016.

[4] http://www.census.gov/data/data-tools.html, last retrieved Dec. 4, 2016.

departments have been releasing such data on interactive websites as of 1999 and 2005 [71], respectively. In Sect. 8.2.2 we discuss in more detail the role of thematic maps and the inherent privacy issues.

Technological advances have undoubtedly helped to expand the accessibility of geo-referenced data, which has evolved in terms of both quantity and quality of the information it conveys. Currently, high-resolution satellite imagery can be accessed online for free from both governmental sources[5] and private companies such as Google,[6] Microsoft[7] and Esri.[8] With the advent of Web 2.0 and the mobile revolution of the past decade, which dramatically changed the way Internet users exchange information, interact and generate online content, the creation and curation of geographic data was no longer limited to the subject experts (such as geographers and cartographers). In fact, more and more people without a formal training in any of those fields started contributing geographic information through open access platforms [88], such as OpenStreetMap[9] and Wikimapia.[10] In Sect. 8.2.1 we discuss the benefits and disadvantages, from a privacy standpoint, of crowdsourced geospatial systems for both users and service providers.

So far, we have described how technological advances—amount of publicly-accessible data, technological advances and crowdsourced contributions—have increased both the coverage and detail of cartography in the past decades. By changing the way people interact with and search for geo-referenced data, such an evolution has also altered another important dimension for both offline and online users, which is privacy. In fact, each of the three aforementioned advances have had a distinct and yet complementary effect on the erosion of user privacy. First, the increase in the availability of geo-referenced data has potentially exposed demographic and social elements, such as gender, income and housing, to anyone with an Internet connection, anywhere in the world. In the era of big data, such abundance and availability has made it possible for researchers to develop algorithms that combine different sources of geo-referenced data to predict socio-economic, environmental and safety-related outcomes with high accuracy [62, 69, 71, 88]. Second, the increase in quality of the data that is collected (through, for example, high-resolution satellite imagery, widespread use of mobile devices and ZIP-code-level statistics) has amplified the effect on the erosion of privacy by pinpointing more accurately the spaces and places in which people live and interact. Third, if on the one hand citizen-contributed geographic information has dramatically increased the speed and coverage of geographic and sociographic data, it also added more uncertainty in the veracity of such data—especially in regions where more traditional data collection methods, such as surveys, are scarce and rare [62].

---

[5] http://earthexplorer.usgs.gov/, last retrieved Dec. 4, 2016.

[6] https://www.google.com/earth/, last retrieved Dec. 4, 2016.

[7] https://www.bing.com/maps/, last retrieved Dec. 4, 2016.

[8] https://www.arcgis.com/features/index.html, last retrieved Dec. 4, 2016.

[9] https://www.openstreetmap.org/, last retrieved Dec. 4, 2016.

[10] http://wikimapia.org/, last retrieved Dec. 4, 2016.

In the next subsections, we discuss privacy in geospatial applications from three different but related perspectives. First, in Sect. 8.2.1 we focus on the crowdsourcing aspect, by elaborating the ways in which such data is collected and how it could impact both the users that contributed it, as well as those it pertains to. Then, in Sect. 8.2.2 we discuss aspects related to surveillance and privacy, two elements that are increasingly relevant to users due to the increase in quantity and quality of geo-referenced data and big-data processing algorithms. In particular, we cover governmental surveillance and the privacy of socioeconomic and environmental factors, such as poverty.

8.2.1   Volunteered and Contributed Geographic Information

The Web 2.0 has made it possible for online users to generate and curate content on the Internet at an unprecedented scale. Geographic and geo-referenced data are two very popular types of data that have benefited from such a technology. Online social networks such as Facebook and Twitter have more than one billion mobile daily active users [36], and many of those users routinely share their exact location with other users of these services [15], by means of geo-tagged media content, check-ins to places and geo-referenced posts and tags (more about this in Sect. 8.3). In addition to contributing location information to online social networks, users are also voluntarily adding, updating and deleting geographic information from other types of platforms, such as online mapping ones. One notable example of such a platform is OpenStreetMap, where maps are "created by people like you and free to use under an open license."[11]

In both of these scenarios (social networks and online mapping), users are contributing geographic or geo-referenced data to a service. When users choose to add a geographic reference to a picture they post on a social network, they are aware that they are sharing location data with other users. Similarly, when a contributor on OpenStreetMap adds a new Point of Interest (POI) to a place, she or he knows that it is her or his responsibility to be as accurate and truthful as possible. In addition to such explicit choices to either attach location data or to contribute geographic information, there are more implicit ways in which users of online services are contributing geographic information, sometimes without even being aware of it. For instance, mobile apps that require access to location information are able to infer the coarse position even if users do not grant such access, simply due to the way IP addresses are shared by users or assigned by network operators [131].

Volunteered geographic information (VGI) is an expression first formulated by Goodchild [11, 45] in order to define the practice of generating geographic information by those who are not trained in geospatial data collection and analysis, and whose information may not be as accurate as those generated by official agencies.

---

[11] http://www.openstreetmap.org/, last retrieved Dec. 4, 2016.

More recently, geographers have started distinguishing between "volunteered" and "contributed" geographic information (CGI) [50, 66]. According to Harvey [50], one can define the two expressions in the following way:

Definition 8.1 *Volunteered geographic information*, or VGI, is crowd-sourced information with clarity about purposes and abilities to control collection and reuse. VGI refers to geographic information collected with the knowledge and explicit decision of a person.

Definition 8.2 *Contributed geographic information*, or CGI, refers to geographic information that has been collected without the immediate knowledge and explicit decision of a person using mobile technology that records location.

The difference between VGI and CGI relies in the way data is collected from the users: if it is an "opt-in" approach, then the data is volunteered, whereas if it is an "opt-out" approach, the data is contributed. Such a distinction is fundamental in order to better understand the differences in data quality and biases that could derive as a result of crowdsourced geographic data.

From a privacy standpoint, such a distinction between CGI and VGI is also very relevant. The opt-in approach of CGI makes sure that users have the choice whether or not to contribute data and that they are aware of it. Control over and awareness of data collection practices are two crucial aspects that affect the way people interact with online services [10, 117]. Usually, the higher is the offered control and transparency, the more comfortable are users with sharing information with the online platforms, especially because location data is one of the most sensitive types of personal data [10, 79]. In contrast to VGI, CGI is much less transparent when it comes to data collection, possible re-use and controls, because users may not be aware that such data is being collected at all [11]; a mobile device that is turned on and is connected to the Internet can continuously gather detailed data about the surroundings, such as radio identifiers (WiFi SSIDs, cellular antenna IDs, Bluetooth IDs), user identifiers (MAC adresses) and its position (GPS, WiFi trilateration). Based on results from such prior works in geography and privacy, Table 8.1 illustrates the differences between CGI, VGI and official geographic data curators and producers, with respect to different data and privacy properties. We define each of these properties as follows:

- *Quality*: It refers to the ability to ensure data-provenance [50]—attributes that allow one to assess the origin of the data as well as the processes used to collect and prepare it—as well as the trust in the contributors' accuracy when reporting geographic data. For example, geographic information produced by official entities is usually able to ensure both data-provenance and is assumed to be more trustworthy than data produced by an individual.
- *Coverage*: It refers to both the extent and detail contained in geographic information. For instance, the coverage provided by VGI contributors can be quite different depending on the region of the world that it pertains to. For example, regions in North America have a better coverage than those in southern Asia and Africa [84].

- *Freshness*: It refers to the update frequency of the geographic information. For instance, CGI data can be continuously collected and re-used, whereas official data relying on periodic surveys and census is usually more stale.
- *Legal liability*: It refers to the liability in case some geographic information offered by a service breaches contractual obligations or agreements, which could happen if, for example, a certain guarantee of accuracy was promised but not delivered [11].
- *Transparency*: It refers to the clear and open disclosure of data collection practices, processing and limits. For example, the presence of a privacy policy or informative content, describing the extent and use purposes of the data collection, contribute towards transparency.
- *Control*: It refers to the ability of users who engage with a service to be able to control the extent to which they are contributing information. It includes opt-in approaches, selective and granular information sharing and the ability to request information about oneself to be removed from the service. For example, opt-in approaches provide users with the choice of whether to contribute information to the service, whereas opt-out approaches usually require users to either accept all the conditions or not to use the service at all [50].
- *User benefits*: It refers to the presence of clear benefits for users, which derive from contributing geographic information to the service provider. For example, rescue operations after a natural disaster (such as the 2010 earthquake in Haiti [88]) have greatly benefited the affected population, as well as relatives, friends and organizations that were able to better monitor the evolution of the situation and to better prioritize the rescue efforts.

From Table 8.1 we notice that there is no single method that has the highest score in each of the aforementioned properties. With respect to privacy properties, the VGI method has clearly the highest aggregate score. However, it falls short in the data properties as data quality, coverage and legal liability, which are usually not satisfied. On the contrary, CGI has high score in data properties, thanks to the large number of samples that can be collected and their ubiquity. However, it falls short in the privacy properties, as the data collection methods, re-use practices and controls

Table 8.1 Properties of different methods for geographic content generation

Method	Data properties				Privacy properties		
	Quality	Coverage	Freshness	Legal liability	Transparency	Control	User benefits
VGI	−	−	+	−−	++	++	++
CGI	+	++	++	−−	−−	−−	+
Official	++	+	−−	++	+	−−	++

We assign scores on a 4-point scale from the lowest (−−) to the highest (++), reflecting the extent to which each method offers every listed property. For example, "Coverage" of VGI may be limited due to the lack of sufficient geographic data about certain regions but it may integrate environmental data collected from sensors which can enhance its value in specific cases (such as natural disasters or air quality monitoring) are less prominent or in some cases nonexistent. In one instance, according to a CNET news report, locations of laptops, cell-phones and WiFi devices have been released on the Internet without an adequate privacy protection and unbeknownst to the users who generated it [86].

Although the modern concept of "personal privacy" has been introduced in 1890 by Warren and Brandeis [134], it is not until the early 2010s that location privacy received a significant attention in the U.S. legislation [66]. The introduction of the bills in the U.S. Congress (such as the Location Privacy Protection Act, Geolocation Privacy and Surveillance Act, Electronic Communications Privacy Act Amendments, and Online Communications and Geolocation Protection Act [100]) have prohibited actions such as the unlawful acquisition and disclosure of geo-location information to government agencies and the unlawful acquisition and disclosure of geo-location information from electronic communication media without users' consent [66, 100].

8.2.2   Geo-Surveillance and Big Data

The availability of modern technologies and large amounts of data ("big data") have undoubtedly benefited both society and individual citizens, but it has also enabled a more detailed and granular insight into their social and personal lives. On the one hand, CGI and VGI have had a positive effect on society and helped save thousands human lives [111, 146], as they enabled organizations and governments to respond in a fast way to coordinate relief efforts in cases of natural disasters, thanks to the almost real-time updates to online maps by private citizens and organizations operating both in the affected areas and outside [88]. Similarly, the availability of detailed satellite imagery and street-level views on cities and neighborhoods have enabled a better distribution of limited resources for city planners and managers, improving the living conditions of their citizens [68, 69]. On the other hand, however, they have opened new surfaces for possible threats and attacks to citizens' privacy through surveillance [22] and inference [62, 71, 72].

8.2.2.1   Geo-Surveillance and Privacy

Surveillance has always been an important instrument to achieve security and safety for authorities and governments. Nowadays, the availability of inexpensive mobile devices equipped with miniaturized sensors (such as GPS, microphone, gyroscopes, accelerometers, etc.) has enabled the collection of vast amounts of detailed measurements about the physical and social environments. For example, GPS traces or cell-tower identifiers can be used to infer one's home/work locations [31, 44]; Bluetooth and WiFi interface identifiers can be recorded and processed to infer social circles of their owners, by only relying on co-presence [9]; such information can be complemented by mining conversations recorded by mobile devices [136];

accelerometer readings on smartphones and smartwatches can be used to infer passwords and PIN codes [82, 96], whereas data related to throughput can be used to determine the most likely trajectory that a user has traveled [113].

In addition to citizen-owned devices (such as smartphones and other mobile devices), people's behavior can be monitored through more conventional surveillance means such as closed-circuit television (CCTV), red-light and thermal cameras, as well as biometric systems and RFID tags. In 2015, it was estimated that there were 245 million active CCTV cameras worldwide, which are used for purposes including traffic monitoring, crime prevention, property and home surveillance [26]. For example, judicial authorities in the U.K. have tagged over 600 adults and about 6000 juveniles with RFID chips, in order to assess compliance with bail conditions [124]. Similarly, the U.S. Department of Homeland Security (DHS) is using RFID-based documents to facilitate the entry and exit from the U.S., which can be read from up to 30 ft away.[12] Uteck [124] argues that although there is no right not to be observed, surveillance assaults human dignity and can change behavioral patterns [38, 101]. In particular, as surveillance becomes "permanent in its effects, even if it is discontinuous in its action" [38], it "disturbs the victim's daily activities, alters her routines, destroys her solitude, and often makes her feel uncomfortable" [112].

Crampton [22], a geography scholar, explores the role of geospatial information systems (GIS) in geo-surveillance, which can be defined as the surveillance of geographic activities [23]. He studies how mapping and GIS are used in recent-day surveillance and security, by broadly applying Faucault's historical method on "governmentality", which describes how people have governed themselves and others [39]. Within that framework, Crampton argues how the rationales for geo-surveillance can be traced back to the nineteenth century, when they were directly concerned with "governing (counting, measuring, and establishing norms) individuals and populations in their distributions across territories". Crampton argues that when privacy is contrasted with security, the balance points in favor of the latter in times of threat, and sometimes in favor of the former in times of peace [22]. Moreover, he also argues that opposing surveillance by appealing to privacy (or civil rights) is problematic because the latter can be defined in different ways. For instance, [22, 116] report that after the attacks of September 11, 2001, Attorney General John Ashcroft stated on National Public Radio that "we're not sacrificing civil liberties. We're securing civil liberties". Crampton also makes an additional point in his essay, where he argues that civil liberties are increased for people who are "normal" in their behavior, but they are reduced for the others. Norms, in this sense, are determined by computing statistical averages and likelihoods of behavior, both at the individual as well as the group levels. Thematic maps, which we discuss in the next section, have emerged after such behavioral norms and statistics have been established.

---

[12]https://www.dhs.gov/radio-frequency-identification-rfid-what-it, last retrieved Dec. 4, 2016.

8.2.2.2 Thematic Maps, Big Data and Privacy

Thematic maps are usually designed to illustrate a specific type of data (such as socioeconomic, environmental or health data) related to a geographic area and for a single purpose.[13] In contrast, reference maps usually show a multitude of data types (such as political, geographical and geologic) together on the same map [119]. For example, Fig. 8.1 shows both kinds of maps: on the top, a thematic map illustrates the poverty rates of the total U.S. population in 2014, by County [123], whereas the map on the bottom depicts a reference map of the same geographic region. In the former, the county borders serve only as visual enhancements for the poverty information the map conveys, whereas in the latter, the data related to political boundaries, geological information and demographics serves its own purpose [119].

As shown in Fig. 8.1, thematic maps can be used to convey different types of geo-referenced data, with varying degrees of privacy sensitivity for the citizens. Information related to financial information, physical safety and health is usually considered to be more sensitive than data related to generic demographics such as age and gender [10, 79]. In the late 1990s and early 2000s, authorities in the U.S. and U.K, respectively, started releasing information related to crime statistics at a regional level through online crime maps [71]. For instance, Fig. 8.2 shows an online crime map for the region of Berkeley, California, for crimes reported by the Berkeley police between Oct. 18–24th, 2016. As it can be seen, the map shows that there were a total of 136 records during the time period under consideration in that region, and it is possible to select individual records to obtain the time at which it was reported and the place where it happened. Moreover, the interface allows the users to filter by type of crime, region, time period, and to visualize aggregate charts and reports.

Kounadi et al.[71] start the discussion on privacy issues related to crime maps by describing four main issues. When exact locations are attached to crime events, (1) the victims may fear that offenders would consider them as particularly easy targets, (2) they would not want to help the authorities with the investigation as a result, (3) they would be reluctant to report another similar offense to the police and finally (4) that their address and other information could be misused [135]. One on the first attempts to assess re-identification risks as well as to outline the implications of sharing sensitive crime-related information was published by the UK's public body "Information Commissioner's Office" (ICO) in 2012 [60]. The publications of crime-related data has started as a result of a transparency program of the U.K. police, which had three policy objectives [18]: (1) To improve the credibility of crime statistics for the citizens, (2) to provide a more community-focused police service and (3) to inform, engage and empower the public to participate in crime prevention efforts. In the official ICO report, the authors tie the release of crime information to the number of households and frequency of updates, in an effort to provide anonymity for the victims, obfuscate the precise locations of the reports

---

[13]http://guides.lib.uw.edu/c.php?g=341594&p=2304475, last retrieved Dec. 6, 2016.

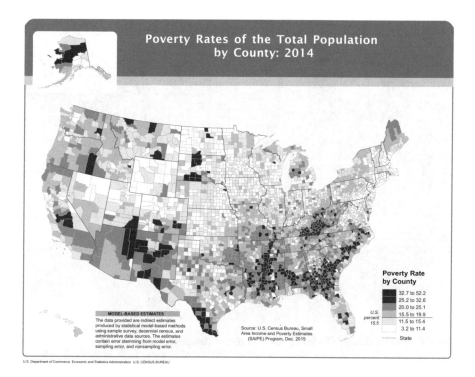

Fig. 8.1 A thematic map (top) and a reference map (bottom). The thematic maps shows the poverty rate by county in the U.S. in 2014 [123], whereas the reference map shows the U.S. territory by State, together with topographic, transportation and demographic information (images: (top) https://www.census.gov/did/www/saipe/data/statecounty/maps/iy2014/Tot_Pct_Poor2014.pdf, (bottom) https://upload.wikimedia.org/wikipedia/commons/7/7d/United_states_wall_2002_us.jpg, last retrieved Dec. 6, 2016)

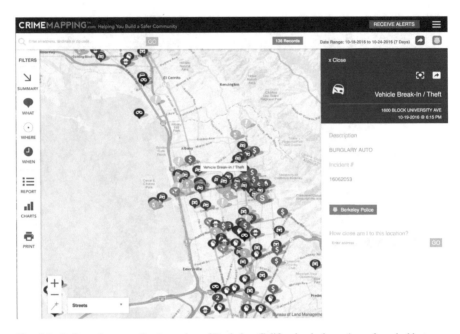

Fig. 8.2 Online crime map for the region of Berkeley, California, during a 1-week period between Oct. 18–24th, 2016. The map shows the different types of reports, such as thefts, burglaries, assaults, vandalisms, and the place where they were reported by the Berkeley Police (image shown with permission from crimemapping.com, http://www.crimemapping.com/, last retrieved Dec. 6, 2016)

and add statistical noise to them [71, 120]. Crime maps released by the U.K. police website[14] have to comply with such requirements. Although cime maps are being published, Kounadi argues that the policy objectives have not been fully achieved, in particular the one about citizen engagement and empowerment. Moreover, the participants to their study reported being more concerned with the risk implications of burglaries and violent crime statistics on maps than not, and they also expressed concerns that the released locations of burglaries could be used for commercial purposes by alarm and commercial companies (88%). However, when asked about the presence of privacy violations as a result of the release of exact burglary locations, one third of the participants that did not feel there were any violations. Such a number is certainly not insignificant, and it might indicate that for some people such information is indeed not sensitive, or it provides more benefits than risks, or that there is still misconception about the potential of geospatial tools and techniques [71].

Compared to other types of data that can be provided by any online user, the crime statistics are primarily collected by the police authorities in each country,

---

[14]https://www.police.uk/, last retrieved Dec 7, 2016.

and they are usually more trusted because it should be possible to verify their provenance, quality and truthfulness. Similarly, data related to socioeconomic factors such as income, education and occupation is usually collected by means of national or regional surveys by the respective governments. Often, the availability of such data is non-uniform across different countries and regions of the world, such as Africa or Asia [84]. One promising way to overcome the scarcity of official statistics about socioeconomic factors is to combine them with related data from other sources, such as satellite imagery. Jean et al.[62] have demonstrated how, by combining high-resolution satellite imagery with the survey data, they were able to explain up to 75% of the data related to economic factors such as average household expenditure or wealth. Their method relies on deep neural networks trained on both satellite images as well as existing survey data. One of the main properties that enabled authors to achieve such results is that the satellite images showed the shape and material of the rooftops, as well as the distance of the houses from the urban areas. Survey data showed that such features, which are visible in the daytime satellite images, varied roughly linearly with expenditure [62]. Moreover, the performance of the algorithm was degrading only modestly when data from one country was used to predict poverty in another country. As economic and financial data are considered to be some of the most sensitive data types, the privacy implications of fusing them with location (another sensitive data type) have only recently started to get attention by the research community. In particular, Bilogrevic and Ortlieb [10] have shown that, taken individually, location information was considered as the most personally identifying type of data, as compared to other types of data such as email address, web browsing and purchase history. However, when combined with other types of information, the combination that includes location was no longer considered as the most sensitive; a combination that included information related to online behavior, rather than offline, was considered as the most personally identifying, and thus sensitive.

Of particular concern to privacy in geospatial applications is data related to users' health conditions and their combination with location data, which can have negative effects on both the services users receive as well as on the value of their private properties [12, 14]. In many countries, medical data and records are regulated and their access and use is subject to strict access rules [4, 30]. For instance, in the U.S. the Privacy Rule in the Health Insurance Portability and Accountability Act (HIPAA) of 1996, which went into full effect in 2003 [12], applies to any individual's past, present and future data about both physical and mental health. It establishes limits to use of such data and defines which types of health data are considered "protected health information" (PHI).[15] For example, PHI includes patients' names, geographic identifiers that define a region smaller than a State (street, city, county, last three ZIP-code digits, etc.), dates (except years), telephone numbers, email addresses, vehicle identifiers, IP addresses, biometric

---

[15]https://www.hipaa.com/hipaa-protected-health-information-what-does-phi-include/, last retrieved Dec. 4, 2016.

data and images. In 2013, HIPAA was updated to cover additional entities, such as business associates, and it reinforced the need to disclose data breaches that previously would have been unreported [12]. Moreover, it increased penalties in case of PHI violations. Together with HIPAA, the recent legislation on location data has helped strengthen protections around two of the most sensitive data types [9], and to increase transparency in case of data breaches and leaks. In addition to HIPAA, independent institutional review boards (IRBs) are committees that have been formally created to approve, monitor, and review biomedical and behavioral research involving humans. In particular, such committees often perform a risk-benefit analysis to determine if a study should be conducted [97].

What is exempt from official IRB oversight are services that do not collect health data, and process non PHI data, such as search queries entered by online users on a search engine, in order to infer aggregate health-related trends. One such service is Google Trends,[16] which can be used to assess the popularity of different search terms over time and space. The precursor to Trends was Google Flu Trends, a service which provided flu prediction models based on patterns extracted from search queries, active between 2008 and 2015. Shortly after the launch of Google Flu Trends, the Electronic Privacy Information Center (EPIC) and the Patient Privacy Rights wrote a letter to the then Google CEO Eric Schmidt,[17] expressing concerns over the anonymity of the search queries and asking clarifications about the methods used to anonymize them. As of 2015, Google no longer publishes models directly, but it rather provides "Flu and Dengue signal data directly to partners", which include the Center for Disease Control and Prevention (CDC) [47].

So far, we have discussed how privacy concerns in geospatial applications have intensified and spread across multiple dimensions, fueled by the development of new mobile and Internet technologies, sensors, and interaction methods that allowed more and more data and people to contribute geographic information. In the next section, we focus on a more recent and very relevant subcategory of geospatial services that have received a large amount of attention and scrutiny by the privacy research community, i.e., Location-Based Social Networks (LBSNs).

8.3   Privacy in Location-Based Social Networks

Before online social networks became extremely popular over the first decade of the 2000s, Internet users relied on bulleting board systems (BBSs) instant messaging (IM) and forums in order to socialize online and exchange content [115]. Initially, online social networks such as Classmates.com [18] and Friendster[19] allowed users to

---

[16]https://www.google.com/trends/, last retrieved Dec 7, 2016.
[17]https://epic.org/privacy/flutrends/EPIC_ltr_FluTrends_11-08.pdf, last retrieved Dec 7, 2016.
[18]http://www.classmates.com/, last retrieved on Dec. 14, 2016.
[19]http://www.friendster.com/, last retrieved on Dec. 14, 2016. Friendster is no longer active as of Jun. 14, 2015.

search for other users they knew either by name or by affiliation to a group (such as school class or personal interests), but not much more. Later on, more recent social networks such as LinkedIn,[20] Myspace,[21] Facebook,[22] Gowalla[23] and Foursquare[24] started to integrate novel functionalities that would enable users to share more information with the service providers, and to search for and get recommendations about other people, places and activities. In particular, location APIs and location-sharing activities became more and more popular among users who were using their mobile devices to search for local content, places and people in their vicinity. By enabling users to share contextual and geographic information with the service providers, such social networks embraced the two concepts related to contributed and volunteered geographic information (CGI and VGI, respectively) discussed in the previous section: Users volunteer geographic information when they actively check-in to venues or share their locations with other users of the network, and they contribute information by simply connecting to the service from different places and devices.

There are several benefits that users enjoy if they share their location with OSNs. For instance, Foursquare users can receive location "badges" when they check-in very frequently to places and businesses. In turn, some of these businesses then provide incentives to users who have earned badges at their locations, in the form of coupons, discounts or prizes. Another popular example involves friend finder and online dating platforms. By sharing their locations, users can see other users in their proximity and engage with them, discover interesting events happening nearby and set location-based alerts that would inform them every time a given person is close to them. However, there are also downsides to location sharing. Exposing one's location renders the person more vulnerable to stalking, burglaries, physical harm and embarrassment [104]. For example, in 2010 three burglars relied on Facebook status updates to determine which houses to rob, and they managed to steal $ 200,000 worth of goods from 50 different locations [19]. A more comprehensive study conducted in 2011 showed that, based on the reports of 50 ex-burglars in England, 78% of them used Facebook, Twitter, Google Street View and Foursquare to prepare for the robberies [27]. The bridge between the online world and the physical one is clearly stated in the precise definition by Zheng of a location-based social network [144]:

> A location-based social network (LBSN) does not only mean adding a location to an existing social network so that people in the social structure can share location-embedded information, but also consists of the new social structure made up of individuals connected

---

[20] https://www.linkedin.com/.

[21] https://myspace.com/, last retrieved Dec. 14, 2016.

[22] https://www.facebook.com/, last retrieved Dec. 14, 2016.

[23] http://mashable.com/2012/03/11/gowalla-shuts-down/#sBOot7U3xSqf, last retrieved Dec. 14, 2016. Gowalla is no longer active as of 2012.

[24] https://foursquare.com/, last retrieved Dec 14, 2016.

by the interdependency derived from their locations in the physical world as well as their location-tagged media content, such as photos, video, and texts.

When interacting on LBSNs, users often face the question of how much location information to attach to the content they post, concerned with the possible privacy implications of their acts. While it is true that the platforms are usually designed to facilitate the sharing of geo-referenced content [114], users have very different attitudes and behavior towards sharing data online [3, 78, 118, 139]. For instance, it has been observed that, although users state they worry about the privacy of their data, they often reveal personal information on social networks [125]. The discrepancy between attitudes and behavior in the privacy domain was termed as "privacy paradox" by Barnes in 2006 [5], and is still relevant today [28]. On the one hand, some researchers argue that one way to re-conciliate attitudes with behaviors would be through the availability of better sharing controls and notices [6, 130]. On the other hand, however, some scholars believe that, although a necessary condition, better controls and notices have a limited effect on the information disclosure behavior on social networks [2].

Attitudes and behaviors aside, measuring privacy remains an open research topic. As opposed to network performance metrics such as throughput, latency, and error rate, metrics for privacy are highly dependent on the specific application and context being considered [7, 24, 54, 132]. Scholars from both the legal domain as well as engineering have attempted to classify and create taxonomies for the different ways in which privacy could be measured. For instance, Herrman [54] focused on the regulatory issues regarding compliance, operational resilience and returns on investments, whereas Wagner and Eckhoff [132] propose and categorize over 80 different privacy metrics for quantifying the privacy protection provided by privacy-enhancing technologies (PETs). In this section, we discuss privacy metrics that are directly related to the specific context of LBSNs and the privacy protection techniques that are used. More details about each of these metrics can be found in the respective paper, article or book.

In the remainder of this section, we first introduce the generic architecture of a LBSN. Next, we discuss privacy threats and protection mechanisms in five main categories: Location, absence, co-location, identity and demographics, and activity. We conclude the section with a discussion of open research challenges for privacy in LBSNs.

8.3.1 Architecture of Location-Based Social Networks

LBSNs inherit most of the standard architectural components from the traditional online social networks, which include entities (such as people and organizations) and resources (such as media or textual content), and relationships between them. Additionally, location-related information (such as location updates from users, check-ins and geotags) can be attached to both entities and resources [17, 129]. The service provider has a central role in enabling users to connect with each other and the other entities that have an account. In order to join the LBSN, users and other entities register with the service provider, which requires them to provide some personal information such as name and email address [17]. Once the registration is successful, usually after verifying the provided email address, users and other entities can start interacting with each other and post content on the platform. Social ties and group memberships are established by asking other users and groups to join their social circles. In addition to explicitly joining social circles of other entities, often users can also opt to simply stay up-to-date with other users' updates and public posts, by means of a follower-followee relationship model spearheaded by Twitter.

Figure 8.3 shows a generic architecture of a LBSN, with a particular emphasis on the location-related aspects. In the diagram, we can see that all registered entities (people and organizations) can provide location-related information to the LBSN. For instance, people can share their current location by means of a location update and by geotagging resources such as pictures, posts, status updates and other users. Similarly, organizations can geotag resources and other organizations (either directly or through a hashtag coupled to a geotagged post). Users can obtain their current location either locally, by relying on the GPS sensor on their devices, or

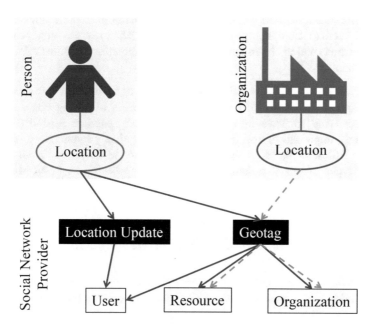

Fig. 8.3 System architecture of a generic Location-Based Social Network. The links represent possible ways by which location data can be attached to the content posted by either people or organizations (adapted from [129]). The solid and dashed lines correspond to the actions that people and organizations can perform, respectively

remotely by providing third-party services information about their current location context (such as the signal strengths and identifiers of nearby WiFi access points and cell-towers) [20, 141]).

In order to better classify the different types of LBSNs, Zheng defined three categories that capture the three main goals of a LBSN [104, 144]:

1. *Geotagged-media-based*: Service providers in this category allow users to attach location information to the content they share, such as text, pictures, videos and other types of media. For example, Twitter, Instagram, Facebook, Periscope all allow users to geo-tag content they post.
2. *Point-location-driven*: Services belonging to this category allow their users to share their current, real-time location, in order to enable a better convergence between physical and online presence, also by enabling users to discover the presence of friends (on the social network) that are in physical proximity. Moreover, such services allow users to share their experience about a certain place with other users by means of "tips" or reviews. Foursquare is an example of such a service.
3. *Trajectory-centric*: A trajectory-centric service enables users to not only share their punctual locations at different places, but also to share the route that connects them. Recently, such services have become increasingly popular, thanks to the availability of inexpensive activity tracking devices, such as Fitbit,[25] and to the increase in the type and number of sensors on mobile devices, which enable users to share their physical activity with other users, to engage in virtual competitions and to obtain virtual badges for completing activities with a certain performance [37].

Many of the popular LBSNs belong to either one or multiple of these categories. For instance, both Facebook and Twitter allow users to geotag content, check-in to places and to leave messages on a business' page or feed. Hence, it is increasingly important to understand the different privacy implications of sharing data on LBSNs. In the following section, we provide a systematic view of the different types of attackers and attacks on users' privacy in LBSNs, as well as mitigation strategies to help limit or prevent negative consequences of such attacks.

8.3.2 Privacy Threats and Protection Mechanisms

LBSN providers collect, process and store multiple types of users' data. By mining users' IP addresses, browser metadata, GPS coordinates, health data, photos, videos and audio recordings, service providers are in a unique position to capitalize on such wealth of information. Although there are techniques that allow online users to hide their true IP address—by connecting to proxies, VPNs and anonymization overlay

---

[25]https://www.fitbit.com/, last retrieved Dec 16, 2016.

networks such as Tor[26]—and protect their anonymity when browsing or searching for content [32], they lose effectiveness when used for LBSNs because users of these networks want to associate their account with the location information they provide.

For LBSN users, it is often challenging to assess the risks involved when sharing location and other type of information with LBSNs. One of the reasons is that people usually lack the awareness about the possible negative consequences due to the leak of sensitive data [104].

Most LBSNs offer users means to manage their data on the platform, usually through permission settings that allow users to specify the conditions under which certain types of data can be used and revealed to others [17]. While it is true that permissions are an important instrument for users to manage their privacy, their scope is often limited to other users of the LBSNs, with the assumption that the service provider is trusted to store and manage all of the data it collects—which is usually explained in a privacy policy. In addition to permission settings, Tsai et al. [122] identified different tools that can help users protect their privacy, which include blacklisting social contacts that should not access any of the users' information, using a restricted sharing approach where content would be visible to only a subset of a user's contacts, establishing "geo-fences", which are regions where location data should not be attached or shared with the LBSN [106], and by using time-based rules, as time is also highly-indicative of the type of visited place [87].

To reconciliate the different usability and privacy aspects in LBSNs, Carbunar et al. [17] have defined a set of five requirements that a LBSNs has to satisfy in order to preserve users' privacy. First, the LBSN should protect users' data from unauthorized access. Second, the privacy protection tools should not affect the functionality of the LBSN for the users. Third, they should enable providers and registered entities to be able to extract aggregate statistics and information that are relevant to their business. Fourth, privacy tools should minimize additional investments that need to be made to support them, both for the LBSN providers as well as for the registered entities. Fifth, such tools should minimize the additional effort that registered entities need to devote in order to use the LBSN, hence preserving its usability. The authors also note that some of these requirements can be contradicting, and that it can be challenging to satisfy all of them simultaneously.

To better understand possible attacks on users' privacy in LBSNs, in the following we characterize the adversary and the underlying assumptions of different mitigation strategies.

8.3.2.1 Adversaries, Threats and Solutions

We define as an *adversary* in a LBSN an entity (user, group, organization) that wishes to either (1) obtain access to data or derive information about a user or entity

---

[26]https://www.torproject.org/, last retrieved Dec. 16, 2016.

to which it does not have access, (2) modify that data, or (3) impersonate other users or entities. There are usually two types of adversaries in such systems, i.e., *internal* and *external*.

- *Internal adversary*: An internal adversary is an entity that has an existing relationship with the service provider. Examples of possible internal adversaries include other curious users of the LBSN who may want to infer additional information about victims by exploiting or misusing some functionality of the LBSN. It also includes curious or malicious employees of the LBSN who may try to access users' data without authorization, and the LBSN itself.
- *External adversary*: An external adversary does not have a direct relationship with the service provider. For instance, an external adversary could be a curious or malicious outsider or group who wants to steal personal data about users of the LBSN by either attacking the provider directly, or by aggregating data about potential victims from other sources that may be related to the victims. Examples of external adversaries include cyber-criminals, stalkers and groups who wish to steal users' data, or disrupt and block the service from functioning properly.

In addition to characterizing an adversary as either internal or external, privacy protection mechanisms in LBSNs are usually developed to counter a specific adversarial model, which can be either (a) *semi-honest* or (b) *malicious* [43, 58]. In the semi-honest model—also known as honest-but-curious—the adversary is assumed to follow the specified protocol but may try to learn information from the different operations it performs on the data. On the contrary, a malicious adversary can deviate from the specified protocol in any possible way, in order to maximize its success in reducing the users' privacy. In some scenarios, there could be an entity that is fully trusted by the user to execute the protocols correctly and not to reveal any personal information to adversaries. In such a scenario, the adversarial model can be either semi-honest or malicious, with the assumption that the trusted third-party (TTP) does not collude with the adversary and does not reveal any information to it.

An adversary might have several goals when conducting an attack. Vicente et al. [129] define four distinct categories of location privacy threats for LBSN users:

1. *Location privacy*: A location privacy loss occurs when the exact location of a LBSN user is revealed, and this create a significant concern for the users if it can be linked to her identity [48, 56], as it allows adversaries to infer other sensitive information such as the user's home and work locations, interests, political affiliations, and health issues [129]. We discuss location privacy threats and protection techniques in Sect. 8.3.2.2.
2. *Absence privacy*: Similarly to location privacy, absence privacy allows adversaries to learn that a user is not at a certain place during a given period of time. Although the possible consequences of this privacy threat are less obvious, there have been multiple cases where knowing that a user was not at a given place has enabled burglars to successfully rob several residences, multiple times (as described in Sect. 8.3) [19, 27]. We elaborate on absence privacy in Sect. 8.3.2.3.

3. *Co-location privacy*: A co-location privacy loss occurs when adversaries are able to infer the co-presence of multiple users at the same location at a given time. The type of privacy threat is of particular concern to users who either do not wish to reveal their presence at some location they consider to be sensitive, or who do not wish to reveal their proximity to other users. This type of threat is exacerbated due to the fact that, if the privacy settings are not properly configured, users can share their location and tag other users who might be unaware of this until it is too late. Co-location privacy can be extended to include the more generic aspect of interdependent privacy [8, 59, 95], in which the privacy of one user is threatened by the actions of other users on the LBSN. For brevity, we refer the reader to the individual articles for more information about interdependent privacy. We discuss co-location privacy threats in Sect. 8.3.2.4.

4. *Identity privacy*: An identity privacy loss occurs when it is possible for an adversary to link an account on a LBSN to a particular identity. Such threat is significant in many scenarios in which users wish to preserve their anonymity or pseudonymity with respect to other users and external parties. The loss of anonymity on such services can have devastating consequences: In the 2015, a data breach on a popular online dating site affected the account details of 35 million members, which resulted in hundreds of sentimental relationship being broken [81]. We discuss identity privacy in Sect. 8.3.2.5.

In addition to these four categories of threats, a fifth category started to become increasingly important thanks to the large growth of the number of devices capable of capturing fitness and activity data [61]. We therefore include the *activity privacy* category as well:

5. *Activity privacy*: An activity privacy loss occurs when an adversary is able to infer the type of activity that a user is doing at a given time. By using large-scale social media data, researchers have been able to accurately model the urban activities of individuals and to predict the sequence of activities only by relying on check-ins and geotagged posts on social media [51–53, 73, 77]. We discuss activity privacy threats in Sect. 8.3.2.6.

In order to tackle the privacy requirements and challenges in LBSNs, the research community has focused on several approaches based on different underlying techniques [63, 129]. One major category of privacy-preserving techniques are based on statistical methods that modify the reported location information in the space and time domains [99, 121]. In that category, we can find the following methods: (1) Query enlargement techniques [63], where instead of reporting the exact location of the user to the LBSN provider, the reported location is expanded to cover a larger geographic region, (2) fake or dummy location reports [121], where the users would report a set of fake locations together with the actual location in order to hide it among the dummy ones, and (3) progressive retrieval techniques that enable users to retrieve information by iteratively querying the provider [128]. In addition to statistical methods, another set of techniques rely on strong cryptography in order to design protocols to ensure that only the intended parties are able to obtain

Table 8.2 Categorization of different research works according to the adversary type (In: internal, E: external), adversarial model (M: malicious, S: semi-honest, T: trusted third-party), the goal of the adversary (L: location, Ab: Absence, C: co-location, Id: identity, Ac: activity) and the proposed or suggested privacy protection mechanism (spatial/temporal cloaking, elimination, fake data, cryptography)

Property	[74]	[49]	[64]	[143]	[105]	[40]	[75]	[70]	[57]	[102]	[103]	[76]
Adversary type	E	E	E	In/E	In	In/E	In/E	In	In	In/E	E	E
Adversary model	M	M	M	S	S	T	S	S	S	M	S	M
Goal	L	L	L	L	Ab	Ab	C	C	C	Id	Id	Ac
Privacy protection	Sp. cl.	Elim.	N/A	N/A	N/A	Sp&Tp. cl.	Crypt.	Crypt.	Crypt.	N/A	N/A	Sp. cl.

the information they require through secure computations, and nobody else [42]. For example, Private Information Retrieval techniques enable users to retrieve information without revealing what they are looking for to the provider [13].

In the following, we describe some examples of different privacy threats and proposed solutions for each of five different threat categories. Table 8.2 provides a summary of each of the works we present according to the different adversarial models, privacy threats and solution methodologies described (if any).

8.3.2.2 Location

Hereafter we discuss several techniques that threaten location privacy of LBSN users, which rely on one or several of the following data sources: users' location trajectories, textual content, location check-ins, social relationships and photo subjects.

With regards to location privacy, Li et al. [74] have recently conducted a study to measure the similarity between the real mobility pattern and the disclosed locations (through check-ins, for example) of LBSN users. Their results, based on a sample of 30 volunteers who have been providing their actual location samples as well as their disclosed locations on LBSNs, show that there is a substantial gap between the mobility pattern that can be extracted from the disclosed locations and the real mobility: Only in 16–33% of the cases, the authors were able to successfully derive the top-2 POIs (such as home and work). One possible reason is that it has been observed how users tend not to check-in at places considered to be "boring", such as home or work locations [142]. One of the implications of such a result is that even an adversary who has access to the location check-ins of a user on the LBSN—but not to the actual location trace—will have troubles in identifying the most visited POIs of a user with accuracy.

In contrast to studying location traces, Han et al. [49] focus on the textual analysis of Twitter posts in order to extract linguistic cues that can be linked to a specific location. In particular, the authors study the text and user profile information in order to predict the city where the user is located. The intuition is that, for example, users in London should be more likely to tweet about *piccadilly* and *tube* than users in New York or Bejing. Hence, the authors focus on identifying a small set of location indicative words (LIW) in order to increase the geolocalisation accuracy of their machine learning algorithm, both on the regional as well as global scales. The analysis, conducted using multi-lingual tweets, shows that it is possible to correctly predict the city for 49% of English users, with a median error distance of just 9 km. To preserve privacy, the authors suggest that users should reduce the usage of LIWs, particularly gazetted terms, and to delete location-sensitive data from their profiles (such as location and time-zone information).

In addition to the text of a post, shared media can also provide useful information for adversaries who want to infer the location of LBSN users. For instance, Zheng et al. [143] use the real scene captured in a photo in order to infer whether it represents a home or vacation location. Their algorithm, based on a convolutional neural network, examines both the scene of the pictures and their geotags in order to infer a user's home location within a cell of 100×100 m. The algorithm is able to correctly predict the home location of a user with an accuracy of 71%, within a 70.7 m error distance. With the shrinking cost of computational resources and the availability of machine learning models accessible on the cloud,[27] it is becoming increasingly affordable to process not just metadata but also the content of media in order to improve the predictive performance of location-inference algorithms.

LBSNs usually allow users to establish relationships with others, either in a symmetric (friendship) or asymmetric fashion (follower-followee). Such social networks can also be used in order to infer the location of the users, even if they do not reveal their location information. By studying the social relationships on Twitter, Jurgens [64] develops an algorithm based on spatial label propagation that is able to infer the locations of a group of Twitter users who generate 74% of all the daily message volumes. The inference algorithm starts from a small number of known locations from which it assigns the most likely location to users whose location is unknown. The algorithm is able to correctly predict the location of 50% of the users in a Twitter-based social network within 10 km. Moreover, the same technique is also able to infer the locations of 50% of users on a different social network (Foursquare), within a 25 km error.

8.3.2.3 Absence

As opposed to location inference, the goal of an absence inference attack is to infer whether a user was not physically present at a place during a given period of time. This attack can also have serious consequences in a scenario where the absence from a place is considered as sensitive information. For example, the absence of

---

[27]https://cloud.google.com/prediction/, \https://aws.amazon.com/machine-learning/, last retrieved Dec. 17, 2016.

an employee from her workplace during work hours could lead to disciplinary measures from the employer. In contrast to location privacy, where the privacy loss occurs if a user can be located at a given point in time, Saini and El Saddik [105] argue that for absence privacy, it is more appropriate to model the privacy loss during a period of time, because the absence from one place at a given point in time does not necessarily imply that the user was not there or in the vicinity at a different but very close time instant. In a first attempt to formalize absence privacy, Freni et al. [40] proposed a set of definitions and techniques to preserve absence privacy, which rely on spatio-temporal generalization of the reported location. Such techniques rely on a trusted third-party, which is responsible for enforcing users' privacy preferences through the notion of an absence privacy region, where an adversary cannot exclude any point as a possible location of the user. By means of temporal delays when publishing geotagged information, the authors show how the effect on the quality of service is relatively modest (16–26 min of delay), and that it largely depends on the amount of other users currently in a given area, as well as their posting frequency.

8.3.2.4   Co-location

One popular example of co-location privacy threat is represented by services that offer to notify users when they are in physical proximity to other users of a LBSN, usually referred to as "nearby friend alert" [75]. Solutions to such challenge are mostly based on cryptographic primitives [29, 57, 70, 85, 91, 93, 145], relying on secure multi-party computation, cryptographic hashing or either public- or symmetric-key encryption. For instance, Li et al. [78] propose a protocol for nearby friend alert that allows users to trade accuracy with communication overhead. Based on the grid-and-hashing approach [110]—which partitions the space is grids and compares the signatures of such grids between users in order to discover if they are in the same grid—the authors design a flexible algorithm that finds an optimal placement of such grids that reduces by more than 50% the number of required grids as compared to a random placement, hence saving communication and computation costs for the users and the service provider. Mascetti et al. [85] propose two cryptographic protocols (Hide&Crypt and Hide&Hash), based on set-inclusion, that rely on location obfuscation and encryption in order to provide secure proximity detection functionality that preserves the location privacy of the users with respect to other users and the service provider. Kotzanikolauo et al. [70] improve upon existing protocols based on private-equality testing by designing a lightweight solution that can be run on resource-constrained mobile devices. Similarly, Hu et al. [57] propose a novel scheme relying on homomorphic encryption—a set of cryptographic techniques allowing computations on encrypted data—and geo-hashing to enable users to determine whether they are in proximity without revealing their location to other users or the service provider, and to perform spatial cloaking over encrypted geographic coordinates.

8.3.2.5 Identity

In order to identify users on LBSNs, Rossi and Musolesi [102] developed a set of techniques based on the study of location check-in data. By using a Bayesian probabilistic model that relies on the sequence of check-ins, the frequency of their occurrence and the social ties of the users, the authors are able to correctly infer more than 90% of the identities of online users on different datasets of check-ins from existing LBSNs. Unsurprisingly, the authors show how the more unique a GPS position is, the more effective is their algorithm in identifying the user with a small number of check-ins. In a follow-up work, Rossi et al. [103] characterize the types of venues that an adversary should monitor in order to maximize its success. The results, based on a large dataset of more than 1 million check-ins from 17 urban regions of the U.S., show that unsurprisingly the type of venues in the category "Residence" have the highest re-identification potential. However, more surprisingly, the authors discover that users with a high location entropy— which means that they visit more distinct types of venues more frequently than other users with lower entropy—are not necessarily the hardest to re-identify. The authors claim that this result indicates how it is the collective behavior of many users that influences the complexity of re-identification, rather than the individual user's behavior.

In addition to re-identification, demographics inference can also pose a threat to users' privacy. In the work by Li et al. [74], the authors show that demographics inference is quite successful as it exploits similarities between check-in traces of different users, despite a relatively poor performance in predicting actual location traces from check-in data. Specifically, their algorithm is able to infer features such as age, occupation, living place, gender and education level with an accuracy of 69.2%, 53.8%, 54.5%, 73% and 76%, respectively, on a sample of over 22,000 volunteers.

8.3.2.6 Activity

An adversary might be able to infer the activity of a LBSN user from the type of place (such as "restaurant") that corresponds to the reported location, or from the sequence of location reports, which can happen at different time and space granularities. Lian and Xie [76] design and evaluate a method to infer the activity, i.e., the type of place a user is at, based on GPS readings, time, user identification and other contextual information. In such scenarios, one main challenge is the scarcity of sufficient samples that can be used for the inference. In order to overcome this, the authors propose to use data from other users' check-ins, provided that their check-in histories are similar. By leveraging clustering and matrix factorization techniques, the authors show that by training on all users' check-in data, instead of training only on the victim's check-in data, the prediction performance is reduced by only up to 10% (weighted F1-score).

In order to reduce the possible search space for the different types of places a user is likely to visit next, Ye et al. [138] build a technique which uses a mixed hidden Markov model for a 2-step prediction. First, the model predicts the category of the place a user is likely to visit next, and then it predicts a location given the category. Their approach reduces the number of possible location candidates by a factor of 5.45 and improves location prediction accuracy by over 13% on a dataset extracted from the LBSN Gowalla (which is no longer active [16]). More recently, Yang et al. [137] proposed a fusion model that combines two separate inference methods, one for spatial data and one for temporal data. Similarly to [76], the authors rely on affinities between different users' temporal activities and the specificity of one activity at any given location, and they show how their solution achieves consistently good performance on three different datasets from two LBSNs (Gowalla and Fourquare), improving upon various baseline methods.

8.3.3 Open Research Challenges

In this section, we have described the different privacy aspects that are relevant in a LBSN. From threat formalization, adversarial models, privacy requirements and protection techniques, the research community has studied a wide array of problems that have yet to find a unified framework and solution. The availability of large amounts of digital data that we leave by interacting with online services, known as "digital footprints" [80], coupled with the shrinking cost of computation and cloud-based machine-learning solutions, are already enabling powerful inferences about people's lives and affections. In Sect. 8.3.2.2 we have described how deep neural networks are able to enhance the performance of location inference by processing images collected from a LBSN, which is nowadays feasible for every adversary with a minimal cost. With more and more machine learning models available, it is important to assess the amount of private user information that is leaked from the model parameters themselves [108]. Hopefully, novel protection mechanisms are being developed to provide provable privacy guarantees against such adversaries, by combining data separation and adding statistical noise during the training process [1]. The utility implications of such methods have yet to be fully assessed, but it is clear that the more data about users' location-related activities are available, the greater is the risk of a potential misuse of such data.

A related open challenge remains the definitive measurement of privacy loss in LBSNs [107]. Currently, there are multiple ways of measuring privacy [7, 24, 54, 132], and researchers have yet to find a unified framework for measuring it. Progress has been achieved in the area of location privacy, where a unified framework based on accuracy, correctness and uncertainty has been proposed and validated [109]. However, more research is needed in other dimensions of privacy in LBSNs, such as co-location, absence, identity and activity.

No matter how effective privacy protection mechanisms can be, they would not achieve their fullest potential unless they are delivering a coherent, simple and functional user experience [21, 41, 67]. Managing privacy on LBSNs is nowadays challenging for many users, and the controls that are offered are often insufficient or too complex for most users to manipulate [41]. To help users feel more comfortable when sharing personal information on LBSN, better ways of presenting benefits and controls will have to be studied and developed [122], as well as clearer privacy policies that users can read and understand [10, 89].

8.4 Conclusion

Geospatial applications have witnessed a great revolution thanks to the development of modern collaborative technologies that enable users to both consume and contribute geographic information to the online community. In the first part of this chapter, we have introduced and discussed privacy issues that arise when location data is attached to different types of content shared with online services. We have introduced geo-spatial applications, such as interactive thematic maps, which can have a significant positive outcome for the people in scenarios including disaster relief efforts, transportation and urban resource management. However, we have also pointed out how geo-spatial information that is publicly accessible can also represent a source of privacy concern for citizens who might not want to have their locations associated with data that could be used in order to discriminate them or the places in which they live. In particular, we have shown how crime maps could be perceived both positively, when they increase transparency and awareness, as well as negatively, when they could influence the perception of property values in certain areas.

In the second part of this chapter, we have focused on a subcategory of geo-spatial applications, namely location-based social networks (LBSNs), as a recent phenomenon that has gained tremendous popularity among mobile users. To better understand the complex interaction patterns in such services, which comprise users, organizations and service providers, we have outlined a framework that enables researchers and practitioners to adopt a principled approach towards privacy threats and solutions. Such framework encompasses the network architecture, the threat categories as well as solution approaches. Although these categories cover several known attack goals, our analysis is not limited to the currently available solutions, as there are still important open questions that need to be addressed. We identified three research challenges that will benefit from a broader and systematic analysis in order to yield benefits for the users of LBSNs: big data processing with privacy guarantees, comparable and unified metrics across different privacy scenarios, and improved user experience through better and easier controls for managing privacy settings, as well as clearer notices related to the use and collection of users' data on LBSNs.

References

1. M. Abadi, A. Chu, I. Goodfellow, H. B. McMahan, I. Mironov, K. Talwar, and L. Zhang. Deep learning with differential privacy. In *Proceedings of the 2016 ACM SIGSAC Conference on Computer and Communications Security*, pages 308–318. ACM, 2016.
2. A. Acquisti, I. Adjerid, and L. Brandimarte. Gone in 15 seconds: The limits of privacy transparency and control. *IEEE Security & Privacy*, 4(11):72–74, 2013.
3. A. Acquisti and R. Gross. Imagined communities: Awareness, information sharing, and privacy on the facebook. In *International workshop on privacy enhancing technologies*, pages 36–58. Springer, 2006.
4. G. J. Annas. Hipaa regulations-a new era of medical-record privacy? *New England Journal of Medicine*, 348(15):1486–1490, 2003.
5. S. B. Barnes. A privacy paradox: Social networking in the united states. *First Monday*, 11(9), 2006.
6. V. Benson, G. Saridakis, and H. Tennakoon. Information disclosure of social media users: does control over personal information, user awareness and security notices matter? *Information Technology & People*, 28(3):426–441, 2015.
7. M. Bezzi. An information theoretic approach for privacy metrics. *Transactions on Data Privacy*, 3(3):199–215, 2010.
8. G. Biczók and P. H. Chia. Interdependent privacy: Let me share your data. In *International Conference on Financial Cryptography and Data Security*, pages 338–353. Springer, 2013.
9. I. Bilogrevic, M. Jadliwala, I. Lám, I. Aad, P. Ginzboorg, V. Niemi, L. Bindschaedler, and J.-P. Hubaux. Big brother knows your friends: on privacy of social communities in pervasive networks. In *International Conference on Pervasive Computing*, pages 370–387. Springer Berlin Heidelberg, 2012.
10. I. Bilogrevic and M. Ortlieb. If you put all the pieces together...: Attitudes towards data combination and sharing across services and companies. In *Proceedings of the 2016 CHI Conference on Human Factors in Computing Systems*, pages 5215–5227. ACM, 2016.
11. A. J. Blatt. The benefits and risks of volunteered geographic information. *Journal of Map & Geography Libraries*, 11(1):99–104, 2015.
12. A. J. Blatt. Data privacy and ethical uses of volunteered geographic information. In *Health, Science, and Place*, pages 49–59. Springer, 2015.
13. D. Boneh, E. Kushilevitz, R. Ostrovsky, and W. E. Skeith III. Public key encryption that allows pir queries. In *Annual International Cryptology Conference*, pages 50–67. Springer, 2007.
14. J. A. Bovenberg, B. M. Knoppers, A. Hansell, and K. de Hoogh. Exposing participants? population biobanks go geo. *European Journal of Human Genetics*, 2015.
15. T. Burghardt, E. Buchmann, J. Müller, and K. Böhm. Understanding user preferences and awareness: Privacy mechanisms in location-based services. In *OTM Confederated International Conferences" On the Move to Meaningful Internet Systems"*, pages 304–321. Springer, 2009.
16. J. Cabalona. Gowalla Is Officially Shut Down.
17. B. Carbunar, M. Rahman, and N. Pissinou. A survey of privacy vulnerabilities and defenses in geosocial networks. *IEEE Communications Magazine*, 51(11):114–119, 2013.
18. S. Chainey and L. Tompson. Engagement, empowerment and transparency: publishing crime statistics using online crime mapping. *Policing*, 6(3):228–239, 2012.
19. C. Chan. Robbers Checked Facebook Status Updates To See When People Weren't Home.
20. Y. Chen and H. Kobayashi. Signal strength based indoor geolocation. In *Communications, 2002. ICC 2002. IEEE International Conference on*, volume 1, pages 436–439. IEEE, 2002.
21. C. D. Cottrill et al. Location privacy preferences: A survey-based analysis of consumer awareness, trade-off and decision-making. *Transportation Research Part C: Emerging Technologies*, 56:132–148, 2015.

22. J. W. Crampton. Cartographic rationality and the politics of geosurveillance and security. *Cartography and Geographic Information Science*, 30(2):135–148, 2003.
23. J. W. Crampton. *Mapping: A critical introduction to cartography and GIS*, volume 11. John Wiley & Sons, 2011.
24. R. Dayarathna. Taxonomy for information privacy metrics. *J. Int'l Com. L. & Tech.*, 6:194, 2011.
25. C. Dempsey. Privacy in gis issues, 2008.
26. S. N. Desk. How many CCTV Cameras are there globally?, June 2015.
27. B. Dickinson. Infographic: 80% of robbers check Twitter, Facebook, Google Street View.
28. T. Dienlin and S. Trepte. Is the privacy paradox a relic of the past? an in-depth analysis of privacy attitudes and privacy behaviors. *European Journal of Social Psychology*, 45(3):285–297, 2015.
29. W. Dong, V. Dave, L. Qiu, and Y. Zhang. Secure friend discovery in mobile social networks. In *INFOCOM, 2011 Proceedings IEEE*, pages 1647–1655. IEEE, 2011.
30. S. O. Dyke, E. S. Dove, and B. M. Knoppers. Sharing health-related data: a privacy test? *NPJ Genomic Medicine*, 1:16024, 2016.
31. N. Eagle and A. S. Pentland. Reality mining: sensing complex social systems. *Personal and ubiquitous computing*, 10(4):255–268, 2006.
32. E. Erdin, C. Zachor, and M. H. Gunes. How to find hidden users: a survey of attacks on anonymity networks. *IEEE Communications Surveys & Tutorials*, 17(4):2296–2316, 2015.
33. Facebook. Help Center - Location Privacy.
34. Facebook. Privacy Settings and Tools.
35. Facebook. Facebook Reports Second Quarter 2016 Results, 2016.
36. Facebook. Stats, facebook newsroom, 2016.
37. Fitbit. What should i know about my fitbit badges?, 2016.
38. M. Foucault. *Discipline and punish: The birth of the prison*. Vintage, 1977.
39. M. Foucault, G. Burchell, C. Gordon, and P. Miller. *The Foucault effect: Studies in governmentality*. University of Chicago Press, 1991.
40. D. Freni, C. Ruiz Vicente, S. Mascetti, C. Bettini, and C. S. Jensen. Preserving location and absence privacy in geo-social networks. In *Proceedings of the 19th ACM international conference on Information and knowledge management*, pages 309–318. ACM, 2010.
41. M. Furini. Users behavior in location-aware services: Digital natives versus digital immigrants. *Advances in Human-Computer Interaction*, 2014, 2014.
42. G. Ghinita, P. Kalnis, A. Khoshgozaran, C. Shahabi, and K.-L. Tan. Private queries in location based services: anonymizers are not necessary. In *Proceedings of the 2008 ACM SIGMOD international conference on Management of data*, pages 121–132. ACM, 2008.
43. O. Goldreich. *Foundations of cryptography: volume 2, basic applications*. Cambridge university press, 2009.
44. P. Golle and K. Partridge. On the anonymity of home/work location pairs. In *International Conference on Pervasive Computing*, pages 390–397. Springer, 2009.
45. M. F. Goodchild. Citizens as sensors: the world of volunteered geography. *GeoJournal*, 69(4):211–221, 2007.
46. Google. Google My Account.
47. Google. The Next Chapter for Flu Trends.
48. M. Gruteser and B. Hoh. On the anonymity of periodic location samples. In *International Conference on Security in Pervasive Computing*, pages 179–192. Springer, 2005.
49. B. Han, P. Cook, and T. Baldwin. Text-based twitter user geolocation prediction. *Journal of Artificial Intelligence Research*, 49:451–500, 2014.
50. F. Harvey. To volunteer or to contribute locational information? towards truth in labeling for crowdsourced geographic information. In *Crowdsourcing Geographic Knowledge*, pages 31–42. Springer, 2013.
51. S. Hasan and S. V. Ukkusuri. Urban activity pattern classification using topic models from online geo-location data. *Transportation Research Part C: Emerging Technologies*, 44:363–381, 2014.

52. S. Hasan and S. V. Ukkusuri. Location contexts of user check-ins to model urban geo life-style patterns. *PloS one*, 10(5):e0124819, 2015.
53. S. Hasan, S. V. Ukkusuri, and X. Zhan. Understanding social influence in activity-location choice and life-style patterns using geo-location data from social media. *Frontiers in ICT*, 3:10, 2016.
54. D. S. Herrmann. *Complete guide to security and privacy metrics: measuring regulatory compliance, operational resilience, and ROI*. CRC Press, 2007.
55. H. Hodson. Baidu uses millions of users' location data to make predictions, 2016.
56. B. Hoh, M. Gruteser, H. Xiong, and A. Alrabady. Enhancing security and privacy in traffic-monitoring systems. *IEEE Pervasive Computing*, 5(4):38–46, 2006.
57. P. Hu, T. Mukherjee, A. Valliappan, and S. Radziszowski. Homomorphic proximity computation in geosocial networks. In *Computer Communications Workshops (INFOCOM WKSHPS), 2016 IEEE Conference on*, pages 616–621. IEEE, 2016.
58. Y. Huang, J. Katz, and D. Evans. Quid-pro-quo-tocols: Strengthening semi-honest protocols with dual execution. In *2012 IEEE Symposium on Security and Privacy*, pages 272–284. IEEE, 2012.
59. M. Humbert, E. Ayday, J.-P. Hubaux, and A. Telenti. On non-cooperative genomic privacy. In *International Conference on Financial Cryptography and Data Security*, pages 407–426. Springer, 2015.
60. ICO. Crime-mapping and geo-spatial crime data: Privacy and transparency principles., 2012.
61. IDC. Worldwide Wearables Market Increases 67.2% Amid Seasonal Retrenchment, According to IDC.
62. N. Jean, M. Burke, M. Xie, W. M. Davis, D. B. Lobell, and S. Ermon. Combining satellite imagery and machine learning to predict poverty. *Science*, 353(6301):790–794, Aug. 2016.
63. C. S. Jensen, H. Lu, and M. L. Yiu. Location privacy techniques in client-server architectures. In *Privacy in location-based applications*, pages 31–58. Springer, 2009.
64. D. Jurgens. That's what friends are for: Inferring location in online social media platforms based on social relationships. *ICWSM*, 13:273–282, 2013.
65. M. Kandias, L. Mitrou, V. Stavrou, and D. Gritzalis. Which side are you on? a new panopticon vs. privacy. In *Security and Cryptography (SECRYPT), 2013 International Conference on*, pages 1–13. IEEE, 2013.
66. B. Kar and R. Ghose. Is my information private? geo-privacy in the world of social media. In *GIO@ GIScience*, pages 28–31, 2014.
67. H.-S. Kim. What drives you to check in on facebook? motivations, privacy concerns, and mobile phone involvement for location-based information sharing. *Computers in Human Behavior*, 54:397–406, 2016.
68. R. Kitchin. The real-time city? big data and smart urbanism. *GeoJournal*, 79(1):1–14, 2014.
69. R. Kitchin. Data-driven, networked urbanism. In *Data and the City workshop*, 2015.
70. P. Kotzanikolaou, C. Patsakis, E. Magkos, and M. Korakakis. Lightweight private proximity testing for geospatial social networks. *Computer Communications*, 73:263–270, 2016.
71. O. Kounadi, K. Bowers, and M. Leitner. Crime Mapping On-line: Public Perception of Privacy Issues. *European Journal on Criminal Policy and Research*, 21(1):167–190, Mar. 2015.
72. D. Lazer, R. Kennedy, G. King, and A. Vespignani. The parable of google flu: traps in big data analysis. *Science*, 343(6176):1203–1205, 2014.
73. J. H. Lee, S. Gao, K. Janowicz, and K. G. Goulias. Can twitter data be used to validate travel demand models? In *IATBR 2015-WINDSOR*, 2015.
74. H. Li, H. Zhu, S. Du, X. Liang, and X. Shen. Privacy leakage of location sharing in mobile social networks: Attacks and defense. *IEEE Transactions on Dependable and Secure Computing*, (1):1–1, 2016.
75. H. P. Li, H. Hu, and J. Xu. Nearby friend alert: Location anonymity in mobile geosocial networks. *IEEE Pervasive Computing*, 12(4):62–70, 2013.

76. D. Lian and X. Xie. Collaborative activity recognition via check-in history. In *Proceedings of the 3rd ACM SIGSPATIAL International Workshop on Location-Based Social Networks*, pages 45–48. ACM, 2011.
77. F. Liu, D. Janssens, J. Cui, G. Wets, and M. Cools. Characterizing activity sequences using profile hidden markov models. *Expert Systems with Applications*, 42(13):5705–5722, 2015.
78. Y. Liu, K. P. Gummadi, B. Krishnamurthy, and A. Mislove. Analyzing facebook privacy settings: user expectations vs. reality. In *Proceedings of the 2011 ACM SIGCOMM conference on Internet measurement conference*, pages 61–70. ACM, 2011.
79. M. Madden. Americans Consider Certain Kinds of Data to be More Sensitive than Others, Nov. 2014.
80. M. Madden, S. Fox, A. Smith, and J. Vitak. *Digital Footprints: Online identity management and search in the age of transparency*. Pew Internet & American Life Project Washington, DC, 2007.
81. B. M. C. f. MailOnline. Ashley Madison members reveal the impact of last year's hack, Aug. 2016.
82. A. Maiti, M. Jadliwala, J. He, and I. Bilogrevic. (smart) watch your taps: side-channel keystroke inference attacks using smartwatches. In *Proceedings of the 2015 ACM International Symposium on Wearable Computers*, pages 27–30. ACM, 2015.
83. S. Mardenfeld, D. Boston, S. J. Pan, Q. Jones, A. Iamntichi, and C. Borcea. Gdc: Group discovery using co-location traces. In *Social computing (SocialCom), 2010 IEEE second international conference on*, pages 641–648. IEEE, 2010.
84. M. Maron. How complete is OpenStreetMap?
85. S. Mascetti, D. Freni, C. Bettini, X. S. Wang, and S. Jajodia. Privacy in geo-social networks: proximity notification with untrusted service providers and curious buddies. *The International Journal on Very Large Data Bases*, 20(4):541–566, 2011.
86. D. McCullagh. Microsoft's Web map exposes phone, PC locations, 2011.
87. G. McKenzie and K. Janowicz. Where is also about time: A location-distortion model to improve reverse geocoding using behavior-driven temporal semantic signatures. *Computers, Environment and Urban Systems*, 54:1–13, 2015.
88. P. Meier. Crisis mapping in action: How open source software and global volunteer networks are changing the world, one map at a time. *Journal of Map & Geography Libraries*, 8(2):89–100, 2012.
89. G. R. Milne and M. J. Culnan. Strategies for reducing online privacy risks: Why consumers read (or don't read) online privacy notices. *Journal of Interactive Marketing*, 18(3):15–29, 2004.
90. MIT. Real Time Rome.
91. A. Narayanan, N. Thiagarajan, M. Lakhani, M. Hamburg, and D. Boneh. Location privacy via private proximity testing. In *NDSS*, 2011.
92. B. News. Ice age star map discovered, 2000.
93. J. D. Nielsen, J. I. Pagter, and M. B. Stausholm. Location privacy via actively secure private proximity testing. In *Pervasive Computing and Communications Workshops (PERCOM Workshops), 2012 IEEE International Conference on*, pages 381–386. IEEE, 2012.
94. A. Noulas, M. Musolesi, M. Pontil, and C. Mascolo. Inferring interests from mobility and social interactions. In *NIPS Workshop on Analyzing Networks and Learning with Graphs*, pages 2–88, 2009.
95. A.-M. Olteanu, K. Huguenin, R. Shokri, M. Humbert, and J.-P. Hubaux. Quantifying interdependent privacy risks with location data. *IEEE Transactions on Mobile Computing*, 2016.
96. E. Owusu, J. Han, S. Das, A. Perrig, and J. Zhang. Accessory: password inference using accelerometers on smartphones. In *Proceedings of the Twelfth Workshop on Mobile Computing Systems & Applications*, page 9. ACM, 2012.
97. R. L. Penslar and J. Porter. Institutional review board guidebook. *Retrieved July*, 18:2003, 1993.

98. Pew Research Center. More Americans using smartphones for getting directions, streaming TV, Jan. 2016.
99. N. Polatidis, C. K. Georgiadis, E. Pimenidis, and E. Stiakakis. A method for privacy-preserving context-aware mobile recommendations. In *International Conference on e-Democracy*, pages 62–74. Springer, 2015.
100. K. Pomfret. Latitudes and attitudes: Zooming in on geospatial data, privacy and the law in the digital age. *Centre for Spatial Law and Policy*, 2013.
101. J. H. Reiman. Driving to the panopticon: A philosophical exploration of the risks to privacy posed by the highway technology of the future. *Santa Clara Computer & High Tech. LJ*, 11:27, 1995.
102. L. Rossi and M. Musolesi. It's the way you check-in: identifying users in location-based social networks. In *Proceedings of the second ACM conference on Online social networks*, pages 215–226. ACM, 2014.
103. L. Rossi, M. Williams, C. Stich, and M. Musolesi. Privacy and the city: User identification and location semantics in location-based social networks. In *Ninth International AAAI Conference on Web and Social Media*, 2015.
104. Z. Sahnoune, C. Y. Yep, and E. Aïmeur. Privacy issues in geosocial networks. In *International Conference on Risks and Security of Internet and Systems*, pages 67–82. Springer, 2014.
105. M. Saini and A. El Saddik. Absence privacy loss. *Computer*, 48(11):102–105, 2015.
106. D. Schoepe and A. Sabelfeld. Understanding and enforcing opacity. In *2015 IEEE 28th Computer Security Foundations Symposium*, pages 539–553. IEEE, 2015.
107. K. G. Shin, X. Ju, Z. Chen, and X. Hu. Privacy protection for users of location-based services. *IEEE Wireless Communications*, 19(1):30–39, 2012.
108. R. Shokri, M. Stronati, and V. Shmatikov. Membership inference attacks against machine learning models. *arXiv preprint arXiv:1610.05820*, 2016.
109. R. Shokri, G. Theodorakopoulos, J.-Y. Le Boudec, and J.-P. Hubaux. Quantifying location privacy. In *2011 IEEE Symposium on Security and Privacy*, pages 247–262. IEEE, 2011.
110. L. Šikšnys, J. R. Thomsen, S. Šaltenis, M. L. Yiu, and O. Andersen. A location privacy aware friend locator. In *International Symposium on Spatial and Temporal Databases*, pages 405–410. Springer, 2009.
111. R. Soden and L. Palen. From crowdsourced mapping to community mapping: the post-earthquake work of openstreetmap haiti. In *COOP 2014-Proceedings of the 11th International Conference on the Design of Cooperative Systems, 27–30 May 2014, Nice (France)*, pages 311–326. Springer, 2014.
112. D. J. Solove. A taxonomy of privacy. *University of Pennsylvania law review*, pages 477–564, 2006.
113. H. Soroush, K. Sung, E. Learned-Miller, B. N. Levine, and M. Liberatore. Turning off gps is not enough: Cellular location leaks over the internet. In *International Symposium on Privacy Enhancing Technologies Symposium*, pages 103–122. Springer, 2013.
114. A. Staff. Location Sharing Is Easier Than You Think, 2016.
115. D. T. Staff. The history of social networking, May 2016.
116. B. Steele. NPR's Gravitas, Sept. 2002.
117. F. Stutzman, R. Capra, and J. Thompson. Factors mediating disclosure in social network sites. *Computers in Human Behavior*, 27(1):590–598, 2011.
118. F. Stutzman, R. Gross, and A. Acquisti. Silent listeners: The evolution of privacy and disclosure on facebook. *Journal of privacy and confidentiality*, 4(2):2, 2013.
119. N. J. Thrower. *Maps and civilization: cartography in culture and society*. University of Chicago Press, 2008.
120. L. Tompson, S. Johnson, M. Ashby, C. Perkins, and P. Edwards. Uk open source crime data: accuracy and possibilities for research. *Cartography and Geographic Information Science*, 42(2):97–111, 2015.
121. M.-T. Tran, I. Echizen, and A.-D. Duong. Binomial-mix-based location anonymizer system with global dummy generation to preserve user location privacy in location-based services. In *Availability, Reliability, and Security, 2010. ARES'10 International Conference on*, pages 580–585. IEEE, 2010.

122. J. Y. Tsai, P. G. Kelley, L. F. Cranor, and N. Sadeh. Location-sharing technologies: Privacy risks and controls. *ISJLP*, 6:119, 2010.
123. U.S. Department of Commerce Economic and Statistics Administration, U.S. Census Bureau. Small area income and poverty estimates (saipe) program, 2014.
124. A. Uteck. Ubiquitous computing and spatial privacy. *Lessons from the Identity Trail: Anonymity, Privacy and Identity in a Networked Society*, pages 83–101, 2009.
125. S. Utz and N. Kramer. The privacy paradox on social network sites revisited: The role of individual characteristics and group norms. *Cyberpsychology: Journal of Psychosocial Research on Cyberspace*, 3(2):2, 2009.
126. VB Insight. Identity and marketing: Capturing, unifying, and using customer data to drive revenue growth | Insight | VentureBeat, 2015.
127. VCloudNews. Every Day Big Data Statistics – 2.5 Quintillion Bytes of Data Created Daily.
128. C. R. Vicente, I. Assent, and C. S. Jensen. Effective privacy-preserving online route planning. In *2011 IEEE 12th International Conference on Mobile Data Management*, volume 1, pages 119–128. IEEE, 2011.
129. C. R. Vicente, D. Freni, C. Bettini, and C. S. Jensen. Location-related privacy in geo-social networks. *IEEE Internet Computing*, 15(3):20–27, 2011.
130. B. Vladlena, G. Saridakis, H. Tennakoon, and J. N. Ezingeard. The role of security notices and online consumer behaviour: An empirical study of social networking users. *International Journal of Human-Computer Studies*, 80:36–44, 2015.
131. N. Vratonjic, K. Huguenin, V. Bindschaedler, and J.-P. Hubaux. How others compromise your location privacy: The case of shared public ips at hotspots. In *International Symposium on Privacy Enhancing Technologies Symposium*, pages 123–142. Springer, 2013.
132. I. Wagner and D. Eckhoff. Technical privacy metrics: A systematic survey. *arXiv preprint arXiv:1512.00327*, 2015.
133. M. Wall. Big Data: Are you ready for blast-off? *BBC News*, Mar. 2014.
134. S. D. Warren and L. D. Brandeis. The right to privacy. *Harvard law review*, pages 193–220, 1890.
135. J. Wartell and J. T. McEwen. Privacy in the information age: A guide for sharing crime maps and spatial data series: Research report. *Institute for Law and Justice, U.S. Department of Justice, Office of Justice Programs*, 2001.
136. D. Wyatt, T. Choudhury, J. Bilmes, and J. A. Kitts. Inferring colocation and conversation networks from privacy-sensitive audio with implications for computational social science. *ACM Transactions on Intelligent Systems and Technology (TIST)*, 2(1):7, 2011.
137. D. Yang, D. Zhang, V. W. Zheng, and Z. Yu. Modeling user activity preference by leveraging user spatial temporal characteristics in lbsns. *IEEE Transactions on Systems, Man, and Cybernetics: Systems*, 45(1):129–142, 2015.
138. J. Ye, Z. Zhu, and H. Cheng. What's your next move: User activity prediction in location-based social networks. In *Proceedings of the SIAM International Conference on Data Mining. SIAM*. SIAM, 2013.
139. A. L. Young and A. Quan-Haase. Privacy protection strategies on facebook: The internet privacy paradox revisited. *Information, Communication & Society*, 16(4):479–500, 2013.
140. M. Yutaka, O. Naoaki, I. Kiyoshi, et al. Inferring long-term user property based on users' location history. In *Proceedings of the IJCAI*. IJCAI, 2007.
141. P. A. Zandbergen. Accuracy of iphone locations: A comparison of assisted gps, wifi and cellular positioning. *Transactions in GIS*, 13(s1):5–25, 2009.
142. Z. Zhang, L. Zhou, X. Zhao, G. Wang, Y. Su, M. Metzger, H. Zheng, and B. Y. Zhao. On the validity of geosocial mobility traces. In *Proceedings of the Twelfth ACM Workshop on Hot Topics in Networks*, page 11. ACM, 2013.
143. D. Zheng, T. Hu, Q. You, H. Kautz, and J. Luo. Towards lifestyle understanding: Predicting home and vacation locations from user's online photo collections. In *Proceedings of the 9th International AAAI Conference on Web and Social Media*, pages 553–560, 2015.
144. Y. Zheng. Location-based social networks: Users. In *Computing with spatial trajectories*, pages 243–276. Springer, 2011.

145. G. Zhong, I. Goldberg, and U. Hengartner. Louis, lester and pierre: Three protocols for location privacy. In *International Workshop on Privacy Enhancing Technologies*, pages 62–76. Springer, 2007.
146. M. Zook, M. Graham, T. Shelton, and S. Gorman. Volunteered geographic information and crowdsourcing disaster relief: a case study of the haitian earthquake. *World Medical & Health Policy*, 2(2):7–33, 2010.

Chapter 9
Privacy of Connected Vehicles

Jonathan Petit, Stefan Dietzel, and Frank Kargl

Abstract By enabling vehicles to exchange information with infrastructure and other vehicles, connected vehicles enable new safety applications and services. Because this technology relies on vehicles to broadcast their location in clear text, it also raises location privacy concerns. In this chapter, we discuss the connected-car ecosystem and its underlying privacy threats. We further present the privacy protection approach of short-term identifiers, called pseudonyms, that is currently foreseen for emerging standards in car-to-X communication. To that end, we discuss the pseudonym lifecycle and analyze the trade-off between dependability and privacy requirements. We give examples of other privacy protection approaches for pay-as-you-drive insurance, sharing of trip data, and electric vehicle charging. We conclude the chapter by an outlook on open challenges.

9.1 Introduction

A lot of research on location privacy has focused on privacy of transportation systems and, particularly, of vehicles. As cars become more and more equipped with information and communication technology, they facilitate recording, storage, transmission, and processing of location data. Protecting driver privacy despite this information exchange is a particular challenge, because location information often has special semantics that can be leveraged by adversaries interested in tracking. For example, vehicles follow certain mobility patterns rather than moving randomly.

J. Petit (✉)
OnBoard Security, Wilmington, MA, USA
e-mail: jpetit@onboardsecurity.com

S. Dietzel
Department of Computer Science, Humboldt-Universität zu Berlin, Berlin, Germany
e-mail: stefan.dietzel@hu-berlin.de

F. Kargl
Institute of Distributed Systems, Ulm University, Ulm, Germany
e-mail: frank.kargl@uni-ulm.de

© Springer Nature Switzerland AG 2018
A. Gkoulalas-Divanis, C. Bettini (eds.), *Handbook of Mobile Data Privacy*,
https://doi.org/10.1007/978-3-319-98161-1_9

This behavior allows to predict future positions and, thereby, allows to link even otherwise perfectly anonymous position data. In addition, linked (and unlinked) location samples may be correlated with user-specific points of interest, such as their known home or work addresses. Finally, location samples may not be perfectly anonymous—but rather pseudonymous—in order to support information integrity or authenticity requirements.

Douriez et al. [4] provide an illustrative example for how knowledge of location traces can negatively affect privacy, even when data is seemingly anonymized. New York City's taxi and limousine commission (TLC) published purportedly anonymized historical data of yellow cab trips in New York City. The published information consists of pick-up and drop-off locations and times, together with other data, such as, the distance, duration, fare, and tip for millions of trips. Although the published information did not contain any direct identifiers, people quickly started to de-anonymize the published data and link trips to individuals using freely available information. For instance, public pictures of celebrities entering cabs were linked to trip information using the pictures' meta information.

Linking these taxi trips is just one example that used a—relatively speaking— small data set. Under the term "connected vehicle," car manufacturers, fleet operators, and public authorities are preparing to exploit the numerous benefits of always knowing where each vehicle is located at every point in time. Such data is often called floating car data (FCD) and basically consists of data records with timestamp, position, vehicle identifier or pseudonym, speed, heading, and potentially other data about a single vehicle or about large numbers of vehicles. Using FCD, logistics operators can track their fleet, rental cars can be prevented from leaving their allowed operation region, city-wide traffic can be analyzed and optimized, and vehicles on a colliding trajectory can warn their drivers to break—to name just a few of the many possible applications.

Car-to-car (C2C) communication—also called vehicle-to-vehicle (V2V) communication—characterizes a particular flavor of a connected car where cars use short-range radio communication or cellular networks to communicate FCD to other vehicles in their vicinity. In contrast to other FCD applications, C2C communication is particularly interesting from a privacy perspective, because the foreseen information exchange largely relies on broadcasts: all vehicles frequently make their current FCD information known to all vehicles within their vicinity openly, that is, without any encryption. The underlying message formats have been standardized in both the EU, where they are called cooperative awareness message (CAM) [8], and in the US, where they are called basic service message (BSM) [36].

As has been done for the taxi data set, a large body of work has repeatedly shown that even anonymized or pseudonymized position samples can often be linked [11, 15, 20, 32, 42]. Once linked, the information reveals complete vehicle trips. In many circumstances, it may also be attributed to specific vehicles or drivers using known information about their home or work places. These works point out how badly weak privacy protection in connected cars could influence drivers' privacy and, consequently, market acceptance of such systems.

In general, it depends a lot on the particular scenario and application whether FCD is only communicated to close-by vehicles or gathered in global databases. Likewise, application requirements dictate whether data is used and stored temporarily or retained more permanently. But no matter what the particular application at hand is, it is clear that the frequent exchange of location information by connected cars creates privacy issues that need to be investigated and solved before their widespread deployment. Therefore, research and standardization have early on worked on privacy solutions to better protect location privacy for connected cars in a multitude of scenarios.

In this chapter, we will provide an overview of solutions and challenges in many common applications of car-to-X (C2X) communication. We will mostly focus on technical solutions while being aware that complementing protection must be established at a regulatory and policy level to provide clear rules on when and how location data from connected cars may be used. Section 9.2 will introduce a system model for connected cars that provides the basis for our further discussion. Section 9.3 discusses attacker models for connected cars to show how location privacy may be infringed. Next, we discuss privacy protection mechanisms for vehicle-to-vehicle communication in Sect. 9.4, and we discuss solutions for other vehicular services—such as pay-as-you-drive (PAYD) insurance, traffic analysis, and electric vehicle charging—in Sect. 9.5. We conclude this chapter with an outlook on open challenges in Sect. 9.6.

9.2 System Model

Nowadays, vehicles increasingly connect with other vehicles, other road users, infrastructure, and Internet services. Interconnecting these systems has the potential to increase safety, efficiency, and comfort. But at the same time, making detailed information from a car's sensors available can uncover many details of the drivers' lives. In this section, we give an historic overview of the connected car ecosystem and discuss example applications. We also present representative information exchange paradigms, and we introduce dependability requirements, which render the trade-off between privacy and fitness for safety applications a particular challenge.

9.2.1 The Connected Car Ecosystem

The vision of "connected cars" today subsumes many different ideas, applications, and communication paradigms. The first application scenarios evolved around the idea to automate emergency calls in cases of accidents. Basically, cars were to be equipped with mobile communication units, positioning devices, and crash sensors. Once an accident was detected, all necessary and useful information would be automatically transmitted to emergency responders. Systems that implement such kinds of applications have since been proposed and built by numerous car manufacturers, and they have been mandated by the European Parliament under the name "eCall" to be implemented in all new cars starting in 2018 [33].

The EU's eCall initiative has met resistance by numerous privacy-conscious groups, which demonstrates the fundamental conflict of many connected car applications. If implemented properly, automatic emergency calls can, ultimately, help to save lives. To better help emergency responders, it is beneficial to acquire as much sensor information as possible about the accident's nature and the current state of involved passengers. On the other hand, many questions need to be addressed properly in order to avoid privacy issues. Some examples are:

- Who is allowed to access sensor information?
- Under what circumstances is sensor information transmitted?
- What measures need to be taken to avoid unauthorized access?
- How can tracking during normal driving be prevented?

These questions can—and should—be answered by legislation. But even when access is legally prohibited, collecting and transmitting sensor information remains possible. Therefore, it is important to discuss technical means to protect driver privacy and enforce data collection restrictions.

The emergency call application is just one example that demonstrates the trade-off between application utility and privacy requirements. In general, the connected car ecosystem can be coarsely subdivided into four categories:

1. safety applications,
2. driving efficiency and traffic management applications,
3. vehicular services, and
4. comfort and multimedia applications.

Safety applications aim to make driving safer and to reduce accidents or to provide better help in case of accidents. Some safety applications, such as the eCall discussed above, connect the vehicles to the service providers' backend infrastructure. Other applications depend on frequent exchange of sensor information directly between vehicles without involvement of additional infrastructure. Essentially, vehicles exchange broadcast messages to acquire a detailed view of their surroundings. This overview can be used to warn when drivers undertake dangerous driving maneuvers that may lead to crashes, and it can inform drivers about dangers that are not yet in the driver's field of view. Example applications are forward collision warning, intersection collision avoidance, and emergency electronic brake lights [18, Ch. 2]. Usually, safety applications use two types of information dissemination. Frequent sensor updates are pushed to all vehicles in the direct vicinity, which is typically assumed to be about 100 m in cities and up to 1000 m in highway scenarios. In addition, warnings about specific events may be transmitted to regions of affected vehicles with lower frequency, namely, only when the reported events occur. Besides connecting vehicles, safety applications can benefit from including other road users, such as pedestrians or cyclists, in the information exchange.

Efficiency applications provide support for navigation decisions and improve traffic flow. The simplest—and likely most privacy-preserving—example are traditional navigation systems, which use offline maps only. More advanced navigation systems may incorporate up-to-date traffic information from a centralized server to calculate better routes. At first, traffic information originated from manual observations using video surveillance or inductive loops that count vehicle flow. The acquired information was centrally managed and passively downloaded by individual vehicles. In recent years, navigation system providers have started to directly source information from each vehicle that uses their system. Indeed, navigation systems often come with a cellular data contract included, which is used to upload current location tracks to a centralized server, as well as to download current traffic predictions for requested routes. Current research aims to take live navigation one step further by calculating route recommendations that optimize the whole city's traffic flow rather than individual travel times [e.g., 2]. The more navigation systems use up-to-date sensor information, the higher their potential to infringe on user privacy. The potential danger to driver privacy is twofold: First, drivers that upload their current velocity, time, and location to improve travel time predictions may be subject to detailed surveillance of their whereabouts. Second, requests for current travel time information for specific routes may reveal the driver's destination to the navigation system provider. Sometimes, the way in which collected information affects drivers is surprising: in 2011, a manufacturer of navigation systems sold their gathered traffic information to the local police, which used it to optimize positioning of speed traps [25].

Under the term *vehicular services,* we subsume all kinds of applications that provide additional services based on location information. An increasingly popular example are pay-as-you-drive (PAYD) insurance models. In these models, drivers agree to base their insurance plan on real driving behavior rather than on surveys and statistical information. Some of these tariffs base their prices mainly on driven distance, but other influence factors, such as driving style or dangerous maneuvers can also be conceived to influence pricing. Besides insurance models, electric vehicle charging is another domain that introduces new information exchange patterns, which may influence driver privacy. For maximum convenience, drivers should be able to recharge their vehicles on arbitrary and widely available charging stations. One foreseen mode of operation is that the charging stations automatically detect the connected vehicle and, once the transaction is authorized, bill the consumed amount of energy to the driver's regular electricity plan.

Finally, *comfort and multimedia applications* generally aim to connect vehicles to the Internet. Usually relying to roadside infrastructure or cellular data connections, these applications aim to provide software downloads, updates, video streaming, or social applications to the drivers and passengers. Again, frequent requests for Internet content may enable infrastructure providers to track individual vehicles.

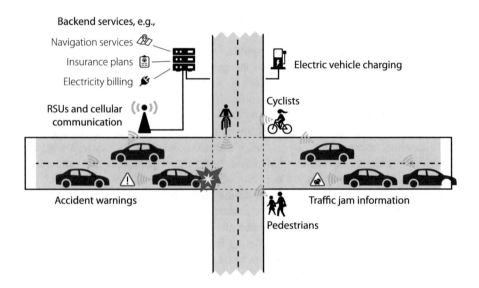

Fig. 9.1 The connected car ecosystem

Figure 9.1 shows an overview of the connected car ecosystem. To enable a wide range of applications, connected cars may exchange information with a wide range of potential communication partners, including

- infrastructure providers, such as, road-side units (RSUs) and traffic management centers (TMCs),
- providers of centralized services,
- other vehicles in their direct surrounding (using one-hop communication),
- other vehicles further away (using multi-hop or relayed communication),
- pedestrians, and cyclists (the latter two being especially vulnerable road users).

The transmitted information is often very detailed, especially when it targets safety applications. For example, very detailed timestamped location traces are required to calculate vehicle trajectories for crash avoidance warnings. Explicit identities of drivers, however, are typically not transmitted, as we will discuss in more detail in Sect. 9.4. Rather, short-term pseudonymous identifiers are used to link individual messages without directly referring to particular vehicles or drivers. Whether the transmitted information constitutes personal information, therefore, is subject to ongoing discussion in different legislations. In the USA, for instance, no federal law exists that governs such information, but existing recommendations and standards such as [6] would not regard most information transmitted by connect cars as personally identifiable. Contrarily in the European Union, the new general data protection regulation [27], which takes effect in 2018, specifically notes that even pseudonymous data falls under privacy regulations.

9.2.2 Information Exchange Paradigms

Researchers and practitioners have come up with a wide range of potential applications for connected cars, which require dissemination of different kinds of information with distinct granularity and frequency within certain regions or towards centralized servers. At the same time, the vehicular communication environment poses complex challenges for successful information transmission. Vehicles move at high speeds, which makes direct wireless communication between vehicles difficult. Also, cars roam within large areas, which makes complete wireless coverage by infrastructure, such as mobile network base stations, difficult. Finally, high vehicle density, such as found in traffic jams, poses scalability challenges for both vehicle-to-vehicle and vehicle-to-infrastructure communication.

As a result of these diverse requirements and challenges, a wide range of specialized information exchange protocols have been proposed [17, 38]. Here, we consider three representative categories of information exchange protocols, which imply different types of privacy properties and requirements:

- infrastructure-based networks,
- direct vehicle-to-vehicle communication, and
- information dissemination in geographic regions.

Infrastructure-based networks are often used for traffic optimization applications. Vehicles are either equipped with cellular network access, such as UMTS or LTE modems, or they use Wifi-style communication with dedicated road-side units. In these scenarios, information is typically collected by one or few centralized servers. As a result, information transmitted using infrastructure-based networks can be protected by encryption against overhearing by unauthorized entities, such as other cars or pedestrians. But since information is centrally collected, the operators of the infrastructure and servers can potentially access information from all vehicles that use the system.

Direct vehicle-to-vehicle communication is a core building block for many safety applications. Vehicles periodically broadcast information about their current location, time, velocity, heading, as well as a number of statistical parameters of their vehicle (e.g., length, width, type). These messages are transmitted up to ten times per second. They are received by all vehicles within wireless range, which is estimated to be between 100 and 1000 m, depending on the environment. Receiving vehicles do not forward these frequent updates, hence limiting their distribution to the direct wireless communication range. The received information is used to build a detailed, up-to-date virtual representation of the vehicles' direct surroundings. Based on this virtual representation, safety applications can calculate trajectories and issue warnings about potential collisions. While the message content does not contain direct identifiers, such as license plate numbers, all messages are signed using short-term cryptographic keys to fulfill dependability requirements. We will discuss signing strategies in more detail in Sect. 9.2.3. This direct type of information exchange potentially allows to build very detailed location tracks. But in contrast to infrastructure-based communication, messages—and consequently, location traces—can only be observed by close-by vehicles.

Geographic dissemination is used in all situations where information is useful for all vehicles within a certain geographic region. Warning all approaching vehicles about the end of a traffic jam or an accident is a typical example. These messages are forwarded using a number of proposed information dissemination and routing protocols (such as [9]) to eventually reach all vehicles within the pre-defined region. In contrast to direct communication, geographic messages are only triggered when specific events, such as accidents, occur. Therefore, geographically disseminated messages unlikely provide sufficient detail for longer location traces. They can, however, be observed by a larger number of vehicles, and they may contain personal information about sending vehicles.

9.2.3 Dependability Requirements

In this section, we focus specifically on dependability requirements for messages that are used for safety applications, as the tension between dependability requirements and drivers' privacy requirements renders protocol design for vehicle-to-vehicle communications particularly challenging. We regard dependability as an overarching design goal that encompasses security features, such as information integrity and accountability; safety requirements, such as required message frequency and real time constraints; and legal issues, such as liability requirements. Schaub et al. [37] presented a number of requirements for privacy-preserving protocol design. Here, we survey those requirements that are, at first sight, particularly contradictory to privacy requirements. We will discuss in Sect. 9.4 how thoughtful protocol design approaches can jointly support these seemingly contradictory requirements.

Real Time Constraints Many applications, including safety applications, require information transmission with low latency and high frequency. Due to their high relative speeds, vehicles may only have a short window to transmit information before they move out of their mutual communication range. In addition, safety applications may need to react quickly in order to prevent accidents, so information should be transmitted with low delay. Finally, safety applications may need to process a number of messages to detect vehicle trajectories or otherwise correlate information. Therefore, information should be transmitted with high frequency to provide sufficient information for trajectory detection.

Linkability In order to process several messages and determine trajectories that may lead to collisions, safety applications may need to link several messages from the same vehicle. If all messages appear to stem from different vehicles, applications may not be able to deduct sufficient information about dangerous situations, as shown by Lefèvre et al. [24].

Authentication Many connected car applications require authentication of participants to prevent unauthorized use. Some applications may be subject to membership fees and may want to avoid that freeloaders use their services without paying. For safety and other vehicle-to-vehicle applications, it may be desirable to exclude adversaries from the network that try to inject or modify messages in the network. Besides authenticating vehicle identities as message senders, authentication may also pertain to specific properties, such as to identify police cars or other vehicle attributes like length, permissible load weight, and so forth.

Accountability Closely related to the authentication requirement is accountability. In certain situations, such as when malicious messages lead to accidents, attacks on connected car applications may constitute crimes. Under those circumstances, it may be desirable to be able to identify and hold liable the individuals that committed those crimes.

Restricted Credential Use When vehicles use credentials, such as asymmetric cryptographic key pairs, it may be desirable to restrict their usage in time and to avoid parallel use. Credentials may be issued for a certain validity period only, and that period should be confirmable by message recipients. Preventing parallel use is a paramount feature to avoid so-called Sybil attacks [3] where a single vehicle could otherwise simultaneously transmit messages under multiple identities. Such Sybil attacks may otherwise lead to false warnings or manipulated navigation decisions.

Revocation In cases where credentials were used to conduct crimes or otherwise interfere with correct system operation, it may be desirable to revoke credentials before their originally intended usage period is over.

Obviously, these dependability requirements influence driver privacy. Many requirements call for identification of the driver's identity under certain circumstances. Or they may necessitate to transmit certain attributes or pseudonymous identities that reduce the potential search space for adversaries that aim to link information to identities.

9.3 Attacker Model

The previous sections illustrated the privacy risks of location tracking and re-identification. In order to understand who can perform such attacks on privacy, we have to define the privacy attacker model. We distinguish three types of attackers [31, 32]:

Local observer: An attacker that is in the vicinity of the target vehicle and can collect its broadcast messages or simply stalk it.

Mid-sized observer: An attacker that does not have a full coverage of the area but rather has sniffing stations located at deemed-strategic spots. This type of attacker collects floating car data (FCD) and may employ algorithms to try and fill their gaps to obtain a real-time location tracking.

Global observer: An all-seeing attacker that has universal coverage and can collect every broadcast messages.

Protection against a local observer is counter-intuitive, as the main benefit of FCD is to create *local awareness*. Therefore, neighboring entities (e.g., vehicles, pedestrians, and cyclists) must be able to track a vehicle in order to avoid collision or create a platoon for example.

The mid-sized attacker is likely to be interested in a zone-level tracking, i.e., knowing in which region a target is instead of an exact street. More sophisticated mid-sized attackers may use forms of re-identification to fill gaps in recorded FCD data. To that extend, the attacker may use computer vision or fingerprinting techniques that are able to distinguish communication devices based on their radio properties [13].

To defend against mid-sized observers, an objective is to harden inferences (e.g. medical condition, relationships, religion). Indeed, the goal of the attacker is to guess movements within gaps in coverage in order to reconstruct tracks (and gain similar knowledge as a global observer). Therefore, by using so-called pseudonyms as short-term identifiers and enforcing change of pseudonyms (see Sect. 9.4), we create more uncertainty, making it harder to perform location tracking. The global observer is even more challenging to prevent, because it can be seen as a constant local observer. So the goal is to create gaps in tracks to shift her toward a mid-sized observer.

Attacks on FCD have already been demonstrated. Petit et al. [32] presented an attack that can be mounted by a mid-sized observer who installs sniffing stations in order to track a target vehicle at a road-level and at a zone-level. This work demonstrates why pseudonyms are mandatory to preserve privacy, and it gives a cost model for frequent pseudonym change strategy.

Wiedersheim et al. [42] analyzed how effectively a global observer can create location profiles. That is, it determined the maximum length of tracks for the same vehicle. Utilizing an approach based on multi-target tracking, the authors found that linking samples under different pseudonyms for the same vehicle can be surprisingly successful under various system setups. Bissmeyer et al. [1] also demonstrated that by solely using the content of cooperative awareness message (CAM) [8] messages they were able to accurately recreate individual vehicles' paths.

Thus, even if pseudonyms are mandatory, one can see that the key question is how to change them so that linking pseudonyms consumes prohibitively time-consuming, requires massive amounts of data, or is computationally infeasible. In the following section, we will delve further into the details of how pseudonyms are used and how their lifecycle can influence or prevent different types of attacks.

9.4   Privacy Protection Using Pseudonyms

Pseudonyms are a wide-spread strategy to combine authentication and account-ability requirements with suitable privacy protection. In connected car systems, pseudonyms are the predominant solution to combine dependability and privacy requirements of safety and efficiency applications in car-to-car communication. Pseudonyms' main feature is to prevent trivial linking of all messages from an individual car. In contrast to completely anonymous transmissions, pseudonyms can help to provide authentication and accountability under well-defined circumstances, despite maintaining driver privacy. They allow to link messages that have been sent with the same pseudonym and only break linkability when a vehicle changes its pseudonym. This is important to allow local tracking of vehicles, e.g., to calculate their trajectories. Given a suitable pseudonym scheme, a car's transmitted messages could, for instance, remain completely anonymous and unlinkable until proof for mischievous behavior is brought forward, whereafter the originator of all messages could be identified.

Combining the seemingly contradictory requirements of anonymity (or pseudonymity), linkability for dependability, and accountability, often requires the combination of complex cryptographic primitives, and many such proposals have been discussed in literature [30]. In addition, standardization for vehicle-to-vehicle communication in both the USA [21] and in Europe [7] include pseudonym architectures. Basically, they all follow a similar lifecycle: Each vehicle is first assigned cryptographic credentials—e.g., an elliptic-curve digital signature algorithm (ECDSA) key pair—that are bound to its long term identity, such as a license plate or the car holder's identity. The key pairs can be generated locally and, together with their identifying attributes, are signed by a trusted authority. To prevent trivial privacy leakage, the long term identity is, however, not used to sign outgoing messages. Rather, the vehicle periodically uses its long term identity to obtain one or more certificates for short term credentials, again this can be ECDSA key pairs. These certificates are issued by another trusted authority, and they attest the holder's authenticity but not their identity. Vehicles then use their short term key pairs to sign outgoing messages. Receivers can verify the signature and attached certificate without learning the sender's identity. In cases of misuse, the short term keys' certificates can be used in cooperation with authorized authorities to prevent issuing fresh certificates.

While similar from a bird's eye perspective, many different proposals with distinct features and restrictions exist for each step of the pseudonym lifecycle. Petit et al. [30] provide a comprehensive survey including details on individual pseudonym schemes. Here, we provide an overview of the canonical pseudonym lifecycle.

9.4.1 Canonical Pseudonym Lifecycle

Today, many different proposals for pseudonym schemes exist, and their implementations vary greatly in the used cryptographic primitives. Petit et al. [30] identifies the following generic steps of a pseudonym's lifecycle, as shown in Fig. 9.2.

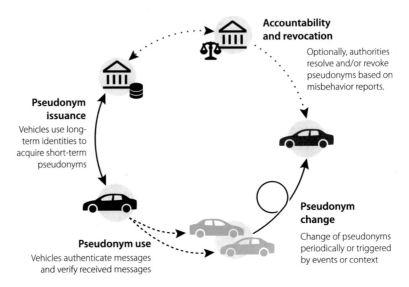

Fig. 9.2 Abstract pseudonym lifecycle

Pseudonym Issuance First, vehicles contact a centralized authority to obtain one or more pseudonyms. Vehicles typically use their long term identifier to authenticate towards the pseudonym provider. The long term identity can be a form of an electronic license plate that is issued by the same authorities that also manage vehicle registrations. To prevent privacy issues at the Public Key Infrastructure level, another authority acts as pseudonym provider. Usually, more than one pseudonym is requested at once. Pseudonyms are then stored locally in the cars; as part of their certification attributes, they may contain a maximum permitted usage period in order to avoid parallel pseudonym use and pseudonym reuse.

Pseudonym Use Vehicles then use pseudonyms to sign all outgoing messages. Likewise, vehicles use the attached signatures and pseudonymous certificates to verify the authenticity of received messages. The major challenge during pseudonym use is scalability: Vehicles may be required to perform hundreds, perhaps thousands, of signature verifications per second, and they may need to generate ten or more signatures per second. As some pseudonym schemes require computationally expensive cryptographic primitives, it is a challenge to implement sufficiently fast cryptographic processors [35].

Pseudonym Change To prevent the creation of detailed location traces, vehicles should change pseudonyms frequently. One challenge of pseudonym change is scalability: The more often vehicles change their pseudonyms, the more pseudonyms need to be acquired and stored for future use. In terms of privacy, a major challenge is when to perform pseudonym change such that adversaries cannot match the old with the new pseudonym. If, for instance, a sole vehicle changes its pseudonym on an empty road, it is trivial for adversaries to match both pseudonyms.

Accountability and Revocation Finally, it may be desirable to hold drivers account-able for their messages in certain well-defined scenarios. For instance, vehicle-to-vehicle messages may prove involvement in accidents, or it may prove injection of manipulated messages. Depending on legislative requirements, pseudonym schemes should support mechanisms that reveal drivers' identities under these circumstances. Ideally, revealing identities should be *technically* restricted to the scenarios required by law. That is, underlying cryptography—rather than laws and regulations alone—should prevent unlawful pseudonym resolution.

From a privacy perspective, strategies for pseudonym change and mechanisms for accountability and revocation are the most challenging aspects of the pseudonym lifecycle. In the following, we discuss these aspects in more detail.

9.4.2 Pseudonym Change Strategies

When pseudonyms are used to sign messages, messages signed with the same pseudonym can be linked to each other. To achieve privacy, it is therefore necessary to frequently change pseudonyms. It is a difficult challenge, however, to decide in what context and how frequently to change pseudonyms. This difficulty arises, because pseudonym change affects both dependability and privacy. Certain safety applications may require to determine short vehicle trajectories in order to work correctly [24]. For instance, consider an application that warns about potential collisions. To determine whether two vehicles would collide if the drivers do not alter their routes, trajectories are an important source of information. If pseudonyms are changed during intersection crossing, the application will necessarily regard messages signed with the new pseudonyms as originating from a different vehicle. As a result, both false warnings and omitted warnings may occur, which signifi-cantly reduces the application's dependability.

In addition, frequent pseudonym change may affect scalability. When pseudonyms are changed frequently—perhaps even after each message—, significant communication capacity, as well as storage and computational capacity is required in order to manage and certify each car's fresh pseudonym pool. From a scalability standpoint, it is, therefore, beneficial to change pseudonyms with lower frequency.

Interestingly, it is also desirable from a privacy standpoint to curate pseudonym changes rather than performing them frequently at will [14]. If only a single vehicle is present on a road segment, it is very likely that an adversary can link its messages despite frequent pseudonym changes. Therefore, it is desirable that vehicles change their pseudonyms in situations where sufficient other vehicles are present, which increase the size of the anonymity set. Also, vehicles should change their pseudonyms within an agreed time period that is preceded by a silence period. That way, attackers can only observe a larger set of pseudonyms before and after the change period, which makes it harder to correlate pseudonyms of individual vehicles. This concept of synchronizing pseudonym change in time and location is known as mix zones [12]. Mix zone placement is complicated by the contradicting requirements of safety applications. For example, intersections are good candidates for ideal mix zone locations, because vehicle density tends to be high in the vicinity of intersections and vehicles change directions there. But as discussed above, intersections are also points with high accident potential where safety applications can benefit from analyzing trajectories. It is a topic of ongoing research where (geographically) and how to implement mix zones for vehicular communication.

9.4.3 Accountability and Revocation

In some situations, it may be desirable to resolve pseudonyms. For example, pseudonyms may help to determine whether drivers were involved in an accident or in a crime scene. Moreover, resolving identities can help to identify people that misuse vehicular communication for their own benefit or to disturb normal system operation. When such misuse is detected, resolved identities can be used to revoke other active pseudonyms of the same user or to invalidate their long term identity. Whether and to what extent such pseudonym resolution and revocation functionality should be implemented is a topic of active debate, and it is a question that cannot be answered technically. Here, we give an overview of technical solutions that can be implemented to support a pseudonym resolution and revocation mechanism that prevents misuse by network operators and authorities.

The simplest solution for pseudonym resolution is to keep a mapping from all issued pseudonyms to their corresponding long term identity at a centralized entity. This implementation, however, would allow operators of the centralized service to reveal identities at will. More advanced resolution mechanisms, as proposed in the US by the Crash Avoidance Metrics Partnership (CAMP), are based on the idea to distribute pseudonym resolution authority over several entities to avoid misuse by individuals. For example, pseudonym distribution can be distributed over regional pseudonym authorities, so that these distributed authorities need to be contacted for pseudonym resolution. Similarly, secret sharing techniques can be used to encrypt pseudonym-identity links such that at least k out of n authorities need to cooperate before a pseudonym's corresponding identity can be decrypted.

Once pseudonyms are linked, it depends on the pseudonym lifetime how their revocation should be implemented. If pseudonyms are restricted to short lifetimes anyways, the central authority can simply revoke the vehicle's long term identity in order to prevent further misuse. Otherwise, so-called certificate revocation lists can be used to revoke individual pseudonyms before they expire. These lists contain— usually in an efficiently encoded form—the identifiers of all pseudonyms that are to be revoked. It is, however, challenging to implement timely and scalable dissemination of such certificate revocation lists.

9.5   Privacy Protection for Vehicular Services

In the previous section, we discussed how pseudonyms can improve privacy in safety-oriented vehicle-to-vehicle communication. In other scenarios, where FCD is collected, stored, and processed in backend systems, different solutions are required for privacy protection. Here, application-specific privacy protection designs are required, which are engineered individually on a case-by-case basis following a privacy-by-design approach.

Exemplarily, we will discuss solutions for three increasingly common example applications: pay-as-you-drive (PAYD) insurance, collection of trip data for traffic analysis, and automated charging for electric vehicles.

9.5.1   Pay-as-You-Drive Insurance

Troncoso et al. [40, 41] discussed the concept of so-called pay-as-you-drive (PAYD) insurance and its implications for drivers' privacy. The basic idea of PAYD systems is that you can earn an additional discount on your car insurance fee by adhering to certain rules laid out in your insurance policy. You may, for example, only drive a certain maximum distance per year or not speed more often than twice a year. Rather than relying on statistical information, your insurance company verifies that you comply with these rules before granting you the discount.

First, the authors surveyed a number of PAYD insurance providers and conclude that a common approach is to install a tracking device in the car that monitors driving behavior and reports this data via cellular network to a central database where it is evaluated for compliance. This architecture is shown in Fig. 9.3 (left).

In order to check eligibility for the discount, the insurance company evaluates the data sent by the vehicles. Some of the surveyed companies also evaluate the data for secondary purposes or provide access to third parties, typically in anonymized or aggregated form. Obviously, this approach requires substantial trust of users in insurance companies to handle the data correctly and keep it secure from malicious access.

The authors therefore propose an alternative scheme called PriPAYD, which is illustrated in Fig. 9.3 (right). It basically relies on a trustworthy black box being installed in the car, which will—locally and offline—determine the appropriate insurance fee and report it to the insurance company. The company then uses the aggregated data for billing, and therefore, no position information leaves the car. Both the insurance company and the user, however, have to trust the black box to correctly calculate the fee. Both would have a rational interest in cheating with the box, the insurance company to raise the fee, the driver to lower it.

Therefore, PriPAYD foresees an audit mechanism, which enable both parties to verify that the correct fee was calculated. The black box inside the car records all data necessary for calculating the insurance fee, such as, distance driven, speeds,

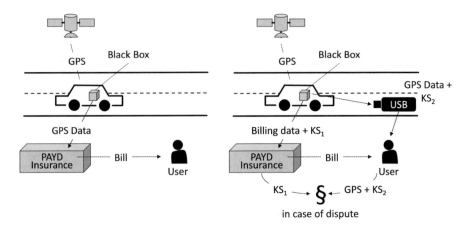

Fig. 9.3 Pay-as-You-Drive insurance models, once with the classical, privacy-invasive model (left) and once with the PriPAYD model that ensures no user data is inadvertently leaked (right)

GPS positions, and so forth, in encrypted form on a removable storage device, such as, a USB stick. The encryption key is split into two key shares ks_1 and ks_2; one is given to the driver, and the other one is sent to the insurance company. In case of a dispute, both parties have to combine the key shares and to decrypt the raw data in order to verify whether the correct fee was calculated.

Kargl et al. [22, 23] report on an architecture for enforcement of privacy policies relying on trusted computing mechanisms that provide a different approach to build generic systems that can enforce privacy policies of arbitrary kinds which is also suitable for PAYD insurance scenarios.

9.5.2 Privacy-Preserving Sharing of Trip Data

Next, we want to focus on another common privacy problem in transportation systems. Municipalities and other organizations are often interested in trip data. Especially, they want to find out which vehicles went from which origin to which destination. Knowing how many vehicles travel from one part of the city to another at certain times of the day is very useful for, amongst others, road capacity planing. So far, inductive loops or manual traffic counts are often used to get this kind of data. But using connected vehicles, the vehicles themselves could report where they are traveling to provide more detailed and accurate information.

Most people will happily contribute to such a data collection in anonymized form, but may feel uneasy with the idea that every trip they do is recorded and may potentially be deanonymized based on their specific origin-destination pairs. One potential solution to the problem is to report trip origins and destinations in coarser detail. For instance, a vehicle may only report to have driven from on district of the town to another. Ideally, the granularity would be determined such that your own trip becomes *k-anonymous* with *k* other trips with same origin-destination pairs. If many vehicles do similar trips, you can report more precise data, but if you are the only one going from place A to place B, you will reduce the level of detail accordingly.

Mechanisms like this and similar ideas have been proposed based on a central proxy that collects all the data and then adjusts the spatial and temporal granularity of data accordingly [16]. Eliminating the need for a central trusted entity, Förster et al. [11] propose a distributed scheme that achieves the same goal. The distributed scheme consists of three phases:

1. Participants establish location- and time-specific keys, both at the start and destination of their trips. They do this by exchanging key shares with other nearby vehicles, eventually converging towards the same keys for certain spatial and temporal granularity levels. The scheme assumes a global spatial and temporal granularity hierarchy to be a pre-defined system parameter. The authors show via simulations that the success rate of this decentralized key agreement scheme is reasonably close to the theoretically achievable maximum.
2. Participants upload copies of their trip reports with different accuracy levels, encrypted with the appropriate keys from step 1, to the trip database. The system defines a decentralized, non-interactive secret sharing scheme by which each vehicle additionally uploads one share of each key to a central database.
3. Traffic authorities query the trip database. If, for a certain temporal and spatial granularity level, enough key shares have been uploaded, they will be able to reconstruct this key and can decrypt the corresponding reports. This is true only if at least *k* vehicles have uploaded trip reports for the same origin-destination pair, and thus, provided key shares to the corresponding location- and time-specific key. Therefore, the scheme naturally ensures *k*-anonymity. Obviously, the chance of collecting sufficient key shares for decryption increases with coarser spatial and temporal resolution.

The interesting aspect of this scheme is that, while the application requires central collection of mobility data, the scheme itself does not require trust in any central entity to ensure privacy. Establishing keys locally among neighboring cars using direct car-to-car communication together with a secret-sharing-scheme is sufficient to provide a fully de-centralized privacy protection mechanism that only reveals data if *k*-anonymity can be maintained.

9.5.3 Privacy-Preserving Charging of Electric Vehicles

Another domain of connected vehicles is communication of electric vehicles with charging infrastructure. The ISO/IEC norm 15118 [34] defines standards for smart charging where vehicles communicate with the road-side charging units and backend systems in order to authenticate the vehicle, control the charging process, and digitally sign the charging bill in order to automate payment.

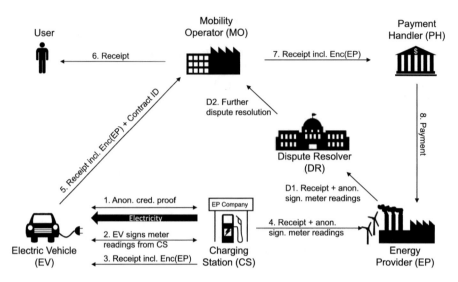

Fig. 9.4 POPCORN protocol for privacy-preserving charging of electric vehicles

Figure 9.4 illustrates a privacy-preserving variant of the ISO's protocol called POPCORN. The norm itself foresees the electric vehicle (EV), charging station (CS), the mobility operator (MO), and the electricity provider (EP) as acting entities. POPCORN in its full version adds a payment handler (PH) and dispute resolver (DR). EV uses a communication link to CS to authenticate via MO with whom it has a contractual relationship to settle the bills incurred during charging. During charging, V will periodically digitally sign partial bills that CS will forward to MO and EP after charging completed. Finally, MO will use the billing information to pay EP.

While this high degree of automation is very convenient from a usability point of view—the driver just plugs the car in and out of the charging station and everything else happens automatically—, the norm lacks a proper treatment of privacy concerns. Particularly, MO and EP will both learn about every charging process, including the location of the CS and thus the EV. Hence, they are in the position to generate fine-grained mobility traces, particularly if one assumes widely spread charging stations and frequent vehicle charging. Current discussions on inductive charging, where an enhanced version of ISO 15118 will be used, will foster this trend.

Höfer et al. [19] have taken these privacy concerns as a motivation to first conduct a privacy impact assessment (a PIA) to clearly identify the privacy shortcomings of the standard. They furthermore propose a step-wise reengineering of the protocols. The result is a privacy-enhanced version of ISO 15118, which they call POPCORN.

As illustrated in Fig. 9.4, POPCORN applies group signatures (GS) and anonymous credentials (AC) to build a protocol that is functionally identical with the original ISO protocol in that drivers can plug in and out from the charging station and everything else happens automatically in the backend systems. At the same time, vehicles will remain anonymous to the CS and EP, and the MO will not learn where its customers have been charging. So location privacy is fully provided except in the case of payment irregularities where a trusted dispute resolver will be provided with encrypted evidence that can be used to link vehicles to a charging process.

In an initialization phase, EVs get equipped with credentials for the GS and AC schemes. When an EV connects to a CS (1.), it creates an AC proof that it is eligible to charge and provisioning of electricity starts. The EV will periodically receive meter readings from the CS which it signs with its GS credentials (2.). When charging ends, the CS provides a receipt to the EV (3.) and the EP (4.). Here, it is important to note that receipt in step 3 contains the recipient of the payment in encrypted form that only the PH can decrypt and that the receipt in step 4 does not reveal the identity of the EV or details on the MO (this would only be deducible through the GS by the DR which is the group manager for the GS). EV forwards the receipt augmented with its contract ID to the MO (5.) which will send the receipt to the user for information purposes (6.) and will trigger the payment via the PH (7.). Here, MO does not learn the recipient of the payment, this is only revealed to the PH when it decrypts $Enc(EP)$. On the other hand, PH will not receive details about the EV involved in this payment and therefore the scheme achieves unlinkability of payments. Payment is then forwarded to the EP (8.) and the process ends. If EP detects unpaid bills, it can trigger dispute resolution (D1.) by sending the (group-) signed meter readings, receipt and other proof data to DH. As key distribution center of the group signature scheme, DH (and only DH) is able to reveal the identify from the signed meter readings and can inform MO about the missing payment (MO) to further investigate and resolve the issue.

Fazouane et al. [10] verified the protocol using a model-checking approach to formally verify the privacy properties of POPCORN, thereby identifying one collusion attack that the original paper missed to notice and proposed a fix to the protocol to resolve the issue.

POPCORN and its enhanced version illustrate how existing protocols that have deficiencies in location privacy can be re-engineered to come up with functionally equivalent solutions that provide strong privacy guarantees.

9.6 Open Challenges and Conclusion

In this chapter, the connected-car ecosystem and its underlying privacy threats were discussed. We presented the privacy protection approach of short-term identifiers, called pseudonyms, and discussed its lifecycle. Then, we analyzed the trade-off between dependability and privacy requirements before presenting examples of other privacy protection approaches for pay-as-you-drive insurance, sharing of trip data, and electric vehicle charging.

Despite the large body of work on location privacy protection for FCD in vehicular systems, researchers have not yet found the optimal solution to jointly maximize privacy, dependability, and utility. In this section, we highlight a couple of open challenges and give directions to address them.

Pseudonym Change Strategy As discussed earlier in this chapter, pseudonym changes have to be carefully orchestrated to be efficient against location tracking by mid-sized and global attackers. Privacy is context-dependent, and so should be the pseudonym change strategy. Depending on the activity performed or the area passed by, the pseudonym change strategy can be more or less effective. For example, grid-style road network patterns offer a higher intrinsic level of privacy than other road networks because of its high density of intersections [28]. Therefore, researchers should define a context-adaptive pseudonym change system. For example, Emara, Woerndl, and Schlichter [5] proposed a scheme that adapts the strategy according to the density of neighboring vehicles and the user privacy preferences. This work could be extended by also considering the type of road network.

Impact of Privacy Protection Lefèvre et al. [24] were the first to investigate the impact of privacy protection techniques on safety applications. Their impact, however, extends beyond safety applications. For example, pseudonym changes and silent periods can affect the computation of estimated travel time (which is an important metric for traffic planning) [43]. Thus, one should take a holistic approach on privacy and perform a comprehensive analysis of its impact, individually for each application and also on the whole communication stack, as noted by Schoch et al. [39]. The impact of pseudonymity on safety raises the question of its impact on FCD utility as a whole. Analyzing how the use of pseudonyms could affect data analyses using collected FCD is an open research question.

Cross-Reference and Re-identification of FCD FCD are being shared between stakeholders (e.g., original equipment manufacturers, service providers, data aggregators). It is a challenge, however, to prevent cross referencing of FCD datasets with each other and with external information that would lead to re-identification of drivers or inference of sensitive information [26, 29]. A survey of location data stakeholders in automotive systems should be performed in order to identify threats and to design corresponding privacy controls.

Privacy of Automated Vehicles automated vehicles (AVs) require a rich data set in order to fully exploit their potential. For example, AVs will form a platoon and, thus, should share their final destinations to ensure stable groups. So, by sharing rich data sets, the privacy concerns increase. Also, because AV cannot rely on a human operator anymore, it is important to maximize predictability, which may render pseudonyms less effective. Knowing how an AV reacts makes profiling (and prediction) easier and more accurate. Therefore, sharing of AV data exhibits more stringent privacy requirements than connected vehicle data. One challenge is to design a privacy-preserving AV data sharing protocol while ensuring a high level of dependability.

Research and industry are well aware of these and other issues involved in making connected cars a success, and there are strong efforts to protect privacy and, particularly, to ensure driver acceptance of such new technologies. At the same time, there are a hard challenges that need to be solved, and a constant privacy engineering effort is required to make sure that future connected vehicles will not become a "big brother" on wheels. With this chapter, we have provided a broad overview on the various aspects of location privacy for connected vehicles, and we have shown where contradictions between dependability and privacy requirements can be solved using clever protocol designs and where further work is required.

References

1. Norbert Bissmeyer et al. "Assessment of Node Trustworthiness in VANETs Using Data Plausibility Checks with Particle Filters". In: Nov 2012. https://doi.org/10.1109/VNC.2012. 6407448
2. D. Cagara, B. Scheuermann, and A. L. C. Bazzan. "Traffic Optimization on Islands". In: *2015 IEEE Vehicular Networking Conference (VNC)* Dec. 2015, pp. 175–182. https://doi.org/10. 1109/VNC.2015.7385574
3. John Douceur. "The Sybil Attack". In: *Iptps '01: First International Workshop on Peer-to-Peer Systems* Springer, 2002, pp. 251–260.
4. Marie Douriez et al. "Anonymizing NYC Taxi Data: Does It Matter?" In: *Proc. of IEEE Intl. Conf. on Data Science and Advanced Analytics (DSAA '16)* Montreal, Canada, Oct. 2016.
5. Karim Emara, Wolfgang Woerndl, and Johann Schlichter. "CAPS: Context-Aware Privacy Scheme for VANET Safety Applications". In: *Proceedings of the 8th ACM Conference on Security & Privacy in Wireless and Mobile Networks* WiSec '15. New York, NY USA: ACM, 2015, 21:1–21:12. ISBN: 978-1-4503-3623-9. https://doi.org/10.1145/2766498.2766500
6. Erika McCallister Tim Grance, and Karen Scarfone. *Guide to Protecting the Confidentiality of Personally Identifiable Information (PII)* Special Publication SP 800-122. NIST, 2010. URL: https://doi.org/10.6028/NIST.SP.800-122
7. ETSI. *Intelligent Transport Systems (ITS); Security; ITS Communications Se- curity Architecture and Security Management* TS 102 940. 2012.
8. ETSI. *Intelligent Transport Systems (ITS); Vehicular Communications; Basic Set of Applications; Part 2: Specification of Cooperative Awareness Basic Service* EN 302 637–2. 2013.
9. ETSI. *Intelligent Transport Systems (ITS); Vehicular Communications; Basic Set of Applications; Part 3: Specifications of Decentralized Environmental Notification Basic Service* EN 302 637–3. 2013.
10. Marouane Fazouane et al. "Formal Verification of Privacy Properties in Electric Vehicle Charging". In: *Engineering Secure Software and Systems* Springer Cham, Mar 4, 2015, pp. 17–33. https://doi.org/10.1007/9783319156187_2
11. David Förster, Frank Kargl, and Hans Löhr. "A Framework for Evaluating Pseudonym Strategies in Vehicular Ad-Hoc Networks". In: *Proceedings of the 8th ACM Conference on Security & Privacy in Wireless and Mobile Net- works* WiSec '15. New York, NY USA: ACM, 2015, 19:1–19:6. ISBN: 978-1-4503-3623-9. https://doi.org/10.1145/2766498.2766520
12. J. Freudiger et al. "Mix-Zones for Location Privacy in Vehicular Networks". In: Vehicular Networks (VNs) seek to provide, among other applications, safer driving conditions. To do so, vehicles need to periodically broadcast safety messages providing preciseposition information ...2007.
13. Ryan M. Gerdes et al. "Device Identification via Analog Signal Fingerprint- ing: A Matched Filter Approach." In: *NDSS* 2006.
14. M. Gerlach and F. Guttler "Privacy in VANETs Using Changing Pseudonyms Ideal and Real". In: *Vehicular Technology Conference 2007. VTC2007- Spring. IEEE 65th* Apr 2007, pp. 2521–2525. https://doi.org/10.1109/VETECS2007.519

15. Philippe Golle and Kurt Partridge. "On the Anonymity of Home/Work Loca- tion Pairs". In: *Pervasive Computing* Springer Berlin, Heidelberg, May 11, 2009, pp. 390–397. https://doi.org/10.1007/9783642015168_26

16. Marco Gruteser and Dirk Grunwald. "Anonymous Usage of Location-Based Services Through Spatial and Temporal Cloaking". In: *Proceedings of the 1st International Conference on Mobile Systems, Applications and Services* MobiSys '03. New York, NY USA: ACM, 2003, pp. 31–42. https://doi.org/10.1145/1066116.1189037

17. H. Hartenstein and L. P. Laberteaux. "A Tutorial Survey on Vehicular Ad Hoc Networks". In: *IEEE Communications Magazine* 46.6 (June 2008), pp. 164–171. ISSN: 0163-6804. https://doi.org/10.1109/MCOM.2008.4539481

18. Hannes Hartenstein and Kenneth Laberteaux, eds. *VANET Vehicular Applica- tions and Inter-Networking Technologies* 1 edition. Chichester U.K: Wiley Feb 15, 2010. 466 pp. ISBN: 978-0-470-74056-9.

19. Christina Höfer et al. "POPCORN: Privacy-Preserving Charging for Emobility". In: *Proceedings of the 2013 ACM Workshop on Security Privacy & Dependability for Cyber Vehicles* CyCAR '13. New York, NY USA: ACM, 2013, pp. 37–48. ISBN: 978-1-4503-2487-8. https://doi.org/101145/25179682517971

20. Baik Hoh et al. "Enhancing Security and Privacy in Traffic-Monitoring Systems". In: *IEEE Pervasive Computing* 5.4 (Oct. 2006), pp. 38–46. ISSN: 1536-1268. https://doi.org/10.1109/MPRV.2006.69

21. "IEEE Standard for Wireless Access in Vehicular Environments Security Ser vices for Applications and Management Messages". In: *IEEE Std 1609.2-2016* (2016), pp. 1–289. https://doi.org/10.1109/IEEESTD.2016.7426684

22. Frank Kargl, Florian Schaub, and Stefan Dietzel. "Mandatory Enforcement of Privacy Poli- cies Using Trusted Computing Principles". In: *Intelligent Information Privacy Management Symposium (Privacy 2010)* Stanford University USA: AAAI, Mar 2010.

23. Frank Kargl et al. "Enforcing Privacy Policies in Cooperative Intelligent Transportation Sys- tems". In: *ACM 15th Annual International Conference on Mobile Computing and Networking (ACM Mobicom 2009) Poster Session* Beijing, China, Sept. 2009.

24. S. Lefevre et al. "Impact of V2X Privacy Strategies on Intersection Collision Avoidance Systems". In: *2013 IEEE Vehicular Networking Conference* Dec. 2013, pp. 71–78. https://doi.org/10.1109/VNC.2013.6737592

25. Maija Palmer. *TomTom Sorry for Selling Driver Data to Police* Financial Times. 2011. URL: https://wwwft.com/content/3f80e432719911e09b7a00144feabdc0 (visited on 01/09/2017).

26. Min Mun et al. "PDVLoc: A Personal Data Vault for Controlled Location Data Sharing". In: *ACM Transactions on Sensor Networks* 10.4 (2014).

27. *On the Protection of Natural Persons with Regard to the Processing of Per sonal Data and on the Free Movement of Such Data, and Repealing Directive 95/46/EC (General Data Protection Regulation)* 2016.

28. Balaji Palanisamy and Liu Ling. "Attack-Resilient Mix-Zones over Road Networks: Architec- ture and Algorithms". In: *IEEE Transactions on Mobile Computing* 14.3 (2015), pp. 495–508.

29. Fayola Peters et al. "Balancing Privacy and Utility in Cross-Company De- fect Prediction". In: *IEEE Transactions on Software Engineering* 39.8 (2013), pp. 1054–1068.

30. J. Petit et al. "Pseudonym Schemes in Vehicular Networks: A Survey". In: *IEEE Communi- cations Surveys Tutorials* PP.99 (2014), pp. 1–1. ISSN: 1553-877X. https://doi.org/10.1109/COMST.2014.2345420

31. Jonathan Petit, Michael Feiri, and Frank Kargl. "Revisiting Attacker Model for Smart Vehicles". In: Sept. 2014. https://doi.org/10.1109/WIVEC.2014.6953258

32. Jonathan Petit et al. "Connected Vehicles: Surveillance Threat and Mitigation". In: *Black Hat Europe* Nov 2015.

33. *Regulation (EU) 2015/758 of the European Parliament and of the Council of 29 April 2015 Concerning Type-Approval Requirements for the Deployment of the eCall in-Vehicle System Based on the 112 Service and Amending Directive 2007/46/EC* 2015.

34. *Road Vehicles – Vehicle to Grid Communication Interface* ISO 15118. ISO/IEC.

35. Carsten Rolfes et al. *PRESERVE Deliverable 3.2: FOT Trial 2 Results* July 31, 2015. URL: https://www.preserve-project.eu/deliverables
36. SAE. *Dedicated Short Range Communications (DSRC) Message Set Dictionary* Standard J2735. SAE, 2016.
37. F Schaub, Zhendong Ma, and F Kargl. "Privacy Requirements in Vehicular Communication Systems". In: *International Conference on Computational Science and Engineering, 2009. CSE '09* Vol. 3. Aug. 2009, pp. 139–145. https://doi.org/10.1109/CSE.2009.135
38. E. Schoch, F. Kargl, and M. Weber. "Communication Patterns in VANETs". In: *IEEE Communications Magazine* 46.11 (Nov 2008), pp. 119–125. ISSN: 0163-6804. https://doi.org/10.1109/MCOM.2008.4689254
39. Elmar Schoch et al. "Impact of Pseudonym Changes on Geographic Routing in VANETs". In: *Security and Privacy in Ad-Hoc and Sensor Networks* Springer Berlin, Heidelberg, Sept. 20, 2006, pp. 43–57. https://doi.org/101007/11964254_6
40. C. Troncoso et al. "PriPAYD: Privacy-Friendly Pay-As-You-Drive Insurance". In: *IEEE Transactions on Dependable and Secure Computing* 8.5 (Sept. 2011), pp. 742–755. ISSN: 1545-5971. https://doi.org/10.1109/TDSC.2010.71
41. Carmela Troncoso et al. "Pripayd: Privacy Friendly Pay-as-You-Drive Insur ance". In: *Proceedings of the 2007 ACM Workshop on Privacy in Electronic Society* WPES '07. New York, NY USA: ACM, 2007, pp. 99–107. ISBN: 978-1-59593-883-1. https://doi.org/10.1145/1314333.1314353
42. Björn Wiedersheim et al. "Privacy in InterVehicular Networks: Why Sim- ple Pseudonym Change Is Not Enough". In: *Wireless On-Demand Network Systems and Services (WONS), 2010 Seventh International Conference on* IEEE, 2010, pp. 176–183.
43. Fangfang Zheng and Henk Van Zuylen. "Urban Link Travel Time Estimation Based on Sparse Probe Vehicle Data". In: *Transportation Research Part C: Emerging Technologies* 31 (June 2013), pp. 145–157.

Chapter 10
Privacy by Design for Mobility Data Analytics

Francesca Pratesi, Anna Monreale, and Dino Pedreschi

Abstract Privacy is an ever-growing concern in our society and is becoming a fundamental aspect to take into account when one wants to use, publish and analyze data involving human personal sensitive information, like data referring to individual mobility. Unfortunately, it is increasingly hard to transform the data in a way that it protects sensitive information: we live in the era of big data characterized by unprecedented opportunities to sense, store and analyze social data describing human activities in great detail and resolution. This is especially true when we work on mobility data, that are characterized by the fact that there is no longer a clear distinction between quasi-identifiers and sensitive attributes. Therefore, protecting privacy in this context is a significant challenge. As a result, privacy preservation simply cannot be accomplished by de-identification alone. In this chapter, we propose the Privacy by Design paradigm to develop technological frameworks for countering the threats of undesirable, unlawful effects of privacy violation, without obstructing the knowledge discovery opportunities of social mining and big data analytical technologies. Our main idea is to inscribe privacy protection into the knowledge discovery technology by design, so that the analysis incorporates the relevant privacy requirements from the start. We show three applications of the Privacy by Design principle on mobility data analytics. First we present a framework based on a data-driven spatial generalization, which is suitable for the privacy-aware publication of movement data in order to enable clustering analysis. Second, we present a method for sanitizing semantic trajectories, using a generalization of visited places based on a taxonomy of locations. The private data then may be used for extracting frequent sequential patterns. Lastly, we show how to apply the idea of Privacy by Design in a distributed setting in which movement data from

F. Pratesi (✉)
ISTI-CNR of Pisa, Pisa, Italy
Computer Science Department of University of Pisa, Pisa, Italy
e-mail: francesca.pratesi@isti.cnr.it

A. Monreale · D. Pedreschi
Computer Science Department of University of Pisa, Pisa, Italy
e-mail: annam@di.unipi.it; pedre@di.unipi.it

© Springer Nature Switzerland AG 2018
A. Gkoulalas-Divanis, C. Bettini (eds.), *Handbook of Mobile Data Privacy*,
https://doi.org/10.1007/978-3-319-98161-1_10

253

individual vehicles is made private through differential privacy manipulations and then is collected, aggregated and analyzed by a centralized station.

10.1 Introduction

The big data originating from the digital breadcrumbs of human activities, generated by ICT systems that we use every day, record the multiple facets of social life: automated payment systems record the tracks of our purchases; search engines record the logs of our queries for finding information on the web; social networks record our connections and communications with friends and colleagues; mobile devices record the traces of our movements. These big data are at the heart of the vision of a "knowledge society", where the understanding of social phenomena is sustained by the knowledge extracted from data describing human activities across the various social dimensions by using social mining technologies. Thus, the analysis of our digital traces can create new opportunities to understand complex aspects, such as mobility behaviors, economic and financial crises, the spread of epidemics, the diffusion of opinions and so on.

The worrying side of the story is that big data contain personal sensitive information, so that the occasions of discovering knowledge increase with the risks of privacy violation. Indeed, when personal sensitive data are published and/or analyzed, it must be checked if this may violate the right of individuals to have full control of their personal sphere. It is clear that maintaining control of personal data is increasingly difficult and it cannot simply be achieved by de-identification (i.e., by removing the direct or explicit identifiers contained in the data, such as name, address and phone number [33]).[1] In the scientific literature and in the media, many examples of re-identification from supposedly anonymous data have been reported, from health records to querylogs to GPS trajectories. In the past years, several techniques have been developed for countering privacy violations, without losing the benefits of big data analytics technology [4, 12, 22, 28, 34]. Despite these results, no general method exists that is able of handling both general personal data and preserving general analytical results. Anonymity in a global sense is believed to be a chimera, and the concern about infringement of the private sphere by means of big data is now in news headlines of major media. Nevertheless, big data analytics and privacy are not necessarily enemies: the goal of this chapter is exactly to show that practical and effective services based on big data analytics can be proposed in such a way that the quality of results can coexist with high protection of personal data. The magic word is Privacy by Design. Here, we review a methodology for purpose-driven privacy protection, where the purpose is a target knowledge service to be deployed on top of data analysis. The key observation is that providing a

---

[1]This definition of de-identified data is compliant with the General Data Protection Regulation (GDPR) [18], especially referring to Recital 26. Indeed, with the de-identification process we are going to transform identified persons in identifiable persons.

reasonable trade-off between a measurable protection of individual privacy together with a measurable quality of service is unfeasible in general, but it becomes feasible in context, i.e., if we have a previous knowledge of the desired analytical goal and the expected level of privacy.

In this chapter, we elaborate on the above ideas the Privacy by Design paradigm, introduced by Anne Cavoukian, in the 1990s, to deploy big data analytical services. Firstly, we discuss the Privacy by Design principle highlighting and how it has been embraced by the United States and Europe.

Secondly, we introduce the idea of Privacy by Design in mobility data analytics domain and show how inscribing privacy "by design" in three different specific scenarios assuring a good balance between privacy protection and quality of data analysis. As first example, we present a framework based on a data-driven spatial generalization, which is suitable for the privacy-aware publication of movement data in order to enable clustering analysis [23]. Then, we present a method for sanitizing semantic trajectories [25], using a generalization of visited places based on a taxonomy of locations. The private data then may be used for extracting frequent sequential patterns.

Lastly, we show how to apply the idea of Privacy by Design in a distributed setting in which movement data from individual vehicles is made private through differential privacy manipulations and then is collected, aggregated and analyzed by a centralized station [26].

The remaining of the chapter is organized as follows. In Sect. 10.2 we discuss the Privacy by Design paradigm and its articulation in data analytics. Sections 10.3 and 10.4 discuss the application of the Privacy by Design principle in the case of publication of personal mobility trajectories, regarding clustering analyses and semantic trajectories respectively, while in Sect. 10.5 we show a possible distributed scenario for privacy preserving mobility analytics. Lastly, Sect. 10.6 concludes the chapter.

10.2 Privacy by Design

Privacy by Design is a paradigm developed by Dr. Ann Cavoukian, the former Ontario's Information and Privacy Commissioner, in the 1990s, to address the emerging and growing threats to online privacy. The key idea of this model is to inscribe the privacy protection into the design of information technologies from the very start. It represents a significant innovation w.r.t. traditional privacy protection approaches since it requires a significant shift from a reactive model to a proactive one. In other words, the idea is preventing privacy issues, instead of remedying to them.

Given the ever growing availability and diffusion of big data and also the impact of big data analytics on both human privacy risks and the possibility of comprehending relevant phenomena, many companies are understanding the necessity to consider privacy at every stage of their business and, thus, to integrate privacy requirements "by design" into their business model. Unfortunately, in many contexts, it is not completely clear which are the methodologies for incorporating Privacy by Design.

10.2.1 Privacy by Design in Law

The Privacy by Design paradigm has been recognized in legislation, and in the last years, privacy officials in Europe and the United States are embracing this attitude.

In 2010, at the annual conference of "Data Protection and Privacy Commissioners" the International Privacy Commissioners and Data Protection Authorities approved a resolution recognizing Privacy by Design as an *essential component of fundamental privacy protection* [1] and advocates the adoption of this principle as part of an organization's default mode of operation. In 2009, the EU Article 29 Data Protection Working Party and the Working Party on Police and Justice released a joint Opinion, encouraging the incorporation of Privacy by Design principles into a new EU privacy framework [2]. In March 2010, the European Data Protection Supervisor advocated to "include unequivocally and explicitly the principle of Privacy by Design into the existing data protection regulatory framework" [17]. This recommendation was taken into consideration in the reform of Data Protection Rules, entered into force on 5 May 2016. Indeed, in this new European Directive [3], in particular in Article 20, there is an explicit reference to data protection "by design" and "by default".

Privacy by Design has been embraced with the same enthusiasm in the United States. Indeed, the U.S. Federal Trade Commission hosted a series of public discussions on privacy issues in the digital age and in a recent staff report [19] it describes a proposed framework with three main recommendations: *privacy by design, simplified consumer choice*, and *increased transparency of data practices*. Moreover, in April 2011, Senators John Kerry (D-MA) and John McCain (R-AZ) proposed their legislation entitled "Commercial Privacy Bill of Rights Act of 2011" that would require companies that collect, use, store or transfer consumer information to implement a version of Privacy by Design when developing products.

10.2.2 Privacy by Design in Big Data Analytics and Social Mining

Unfortunately, it is not always clear what means applying the Privacy by Design principle and which is the best way to apply it for obtaining the desired result. In this section, we discuss the articulation of the general "by design" principle in the big data analytics domain.

Our key idea is to consider privacy protection into any analytical process by design, so that the analysis incorporates the relevant privacy requirements from the very start, evoking the concept of Privacy by Design discussed above.

The application of the general "by design" principle in the big data analytics domain is based on a key concept: higher protection and quality can be better achieved in a goal-oriented approach. Indeed, the data analytical process is designed with assumptions about:

(a) the sensitive personal data subject of the analysis;
(b) the attack model, i.e., the knowledge and purpose of a malicious party that has an interest in discovering the sensitive data of certain individuals;
(c) the category of analytical queries that are to be answered with the data.

These assumptions are essential for the design of a privacy-aware technology. First of all, privacy preservation techniques strongly depend on the nature of the data to be protected, e.g., an algorithm suitable for social networking data could not be appropriate for trajectory data. Second, a valid framework has to define the attack model based on a specific adversary's background knowledge and correspondent countermeasure: different assumptions on the background knowledge require different defense strategies. For example, an attacker could possess an approximated information about the mobility behavior of an individual and exploit it to infer all his movements. It is worth noting that a defense strategy designed for counter attacks with approximate knowledge could be too weak in case of more detailed knowledge. Finally, a privacy-aware solution should find an acceptable trade-off between data privacy and data utility. Thus, it is fundamental to consider the kind of analytical queries to be answered for understanding which data properties must be preserved. As an example, a defense strategy for spatio-temporal data should consider that these data might be useful for collective mobility analyses in an urban area.

Under the above hypotheses, we claim that it is possible to design a privacy-aware analytical process able to:

1. transform the data into an anonymous version with a quantifiable privacy guarantee, i.e., measuring the probability that the malicious attack fails;
2. guarantee that a category of analytical queries can be answered correctly, within a quantifiable approximation that specifies the data utility, using the transformed data instead of the original ones.

We want to point out that different legal frameworks could imply different techniques that are considered to be sufficient for data protection. To define an adequate anonymization level, we mainly rely on the GDPR [18]. Indeed, Privacy by Design is compliant with the GDPR also regarding the principle of reasonableness stated in GDPR (Article 26), where it is stated that "to determine whether a natural person is identifiable, account should be taken of all the means reasonably likely to be used", where the reasonableness should consider some objective factors, such as the costs and the amount of time required for identification, taking into consideration the available technology and technological developments.

In the following, we show three existing ways to apply the Privacy by Design for the design of the same amount of analytical frameworks: one for clustering analysis, one for the publication of trajectory data and one for computing aggregation of movement data in a distributed setting. In the three scenarios we first analyze the privacy issues related to this kind of data; second, we identify the attack model; and third, we provide a method for assuring data privacy taking into consideration the data analysis to be maintained valid. However, these are not the unique privacy-preserving frameworks adopting the Privacy by Design principle, many approaches proposed in the literature can be seen as instances of this promising paradigm (see [4, 12, 27, 28, 34]).

10.3   Privacy by Design in Mobility Data Publishing

In this section, we present a framework that offers an instance of the privacy by design paradigm concerning personal mobility trajectories, obtained from GPS devices or cell phones [23]. It is convenient for the privacy-aware publication of movement data, and its focus is on clustering analysis useful for the comprehension of human mobility behavior in specific urban areas. The released trajectories are made anonymous by a suitable process that realizes a generalized version of the original trajectories.

In the following, we consider a mobility dataset as a collection of trajectories $D = \{T_1, T_2, \ldots, T_m\}$ where each T_i is a trajectory represented by a sequence of spatio-temporal points.

Definition 10.1 (Trajectory) A Trajectory or spatio-temporal sequence is a sequence of triples $T = \langle x_1, y_1, t_1 \rangle, \ldots, \langle x_n, y_n, t_n \rangle$, where t_i $(i = 1 \ldots n)$ denotes a timestamp such that $\forall_{1 < i < n}\ t_i < t_{i+1}$ and (x_i, y_i) are points in $\mathbf{R}^2$.

Intuitively, each triple $\langle x_i, y_i, t_i \rangle$ indicates that the object is in the position (x_i, y_i) at time t_i.

Definition 10.2 (Sub-Trajectory) Let $T = \langle x_1, y_1, t_1 \rangle, \ldots, \langle x_n, y_n, t_n \rangle$ be a trajectory. A trajectory $S = \langle x'_1, y'_1, t'_1 \rangle, \ldots, \langle x'_m, y'_m, t'_m \rangle$ is a *sub-trajectory* of T or *is contained* in T $(S \preceq T)$ if there exist integers $1 < i_1 < \ldots < i_m <= n$ such that $\forall 1 \leq j \leq m\ \langle x'_j, y'_j, t'_j \rangle = \langle x_{i_j}, y_{i_j}, t_{i_j} \rangle$.

We use g to denote the function that applies the spatial generalization to a trajectory. Given a trajectory $T \in D$, the generalized version of T is generated by a function g that applies the spatial generalization to the trajectory. It is represented by the centroid sequence of areas crossed by T. More formally,

Definition 10.3 (Generalized Trajectory) Let $T = \langle x_1, y_1, t_1 \rangle, \ldots, \langle x_n, y_n, t_n \rangle$ a trajectory. A generalized version of T is a sequence of pairs $T_g = \langle x_{c_1}, y_{c_1} \rangle, \ldots, \langle x_{c_m}, y_{c_m} \rangle$ with $m <= n$ where each x_{c_i}, y_{c_i} is the centroid of an area crossed by T.

The privacy by design framework presented in the following is based on a data-driven spatial generalization of the dataset of trajectories and the obtained results put in evidence how trajectories can be anonymized to a high level of protection against re-identification attacks, preserving, at the same time, the possibility of mining clusters of trajectories, which enables powerful analytic services for info-mobility or location based services.

10.3.1 Attack and Privacy Model

Here, it is evaluated the *linkage attack model*, i.e., the ability to link the published data to external information, which enables some respondents associated with the data to be re-identified. In relational data, linking is made possible by *quasi-identifiers*, i.e., attributes that, in combination, can uniquely identify individuals, such as birth date and gender [30]. The remaining attributes represent the private respondent's information (PI), sometimes called sensitive attributes (SA), that may be violated by the linkage attack. In privacy-preserving data publishing techniques, such as *k*-anonymity, the goal is to find countermeasures to this particular attack and to release person-specific data in such a way that the ability to link to other information using the quasi-identifier(s) is limited. However, in the case of mobility data, where each record is a temporal sequence of locations visited by a specific person, the above dichotomy of attributes into quasi-identifiers (QI) and private information (PI) does not hold anymore: here, a (sub)trajectory can play both the role of QI and the role of PI. To understand this point, assume the attacker may know a sequence of places visited by some specific person P: e.g., by shadowing P for some time, the attacker may learn that P was in the shopping mall, then in the gym, and then at the pub. The adversary could exploit such knowledge to retrieve the complete trajectory of P in the released dataset: this attempt would succeed, provided that the attacker knows that P's trajectory is actually present in the dataset and the known sub-trajectory is compatible with (i.e., is a sub-trajectory of) just one trajectory in the dataset. In this example of a linkage attack in the movement data domain, the sub-trajectory known by the attacker serves as QI, while the entire trajectory is the PI that is disclosed after the re-identification of the respondent. Clearly, as the example suggests, it is rather difficult to distinguish QI and PI: in principle, any specific location can be the theater of a shadowing action by a spy, and therefore any possible sequence of locations can be used as a QI, i.e., as a means for re-identification. As a consequence of this remark, it is reasonable to contemplate the radical assumption that any (sub)trajectory that can be linked to a small number of individuals is a potentially dangerous QI and a potentially sensitive PI. Therefore, in the *trajectory linkage attack*, the adversary M knows a sub-trajectory of a respondent R (e.g., a sequence of locations where R has been seen by M) and M would try to discover the whole trajectory belonging to R in the data, i.e., learn all places visited by R. In particular, we assume the following adversary knowledge.

Definition 10.4 (Adversary Knowledge) The attacker has access to the anonymized dataset D^* and knows: (a) the details of the schema used to anonymize the data, (b) the fact that a given user R is in the mobility dataset D and (c) a sub-trajectory S relative to R.

This background knowledge is used in the following attack.

Definition 10.5 (Attack Model) Given the anonymized dataset D^* and a sub-trajectory S relative to a user R, the attacker: *(i)* generates the partition of the territory starting from the trajectories in D^*; *(ii)* computes $g(S)$ generating the sequence of centroids of the areas containing the points of S; *(iii)* constructs a set of candidate trajectories in D^* containing the generalized sub-trajectory $g(S)$ and tries to identify the whole trajectory relative to R. The probability of identifying the whole trajectory by a sub-trajectory S is denoted by $prob(S)$.

10.3.2 Privacy-Preserving Technique

How is it possible to guarantee that the probability of success of the above attack is very low while preserving the utility of the data for meaningful analyses? Consider the source trajectories represented in Fig. 10.1a, obtained from a massive dataset

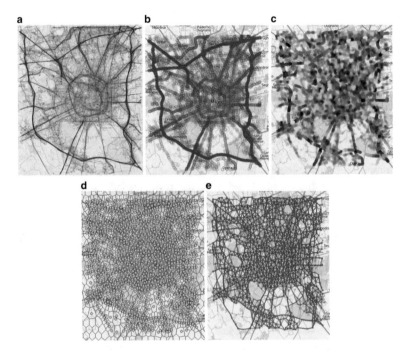

Fig. 10.1 (**a**) Milan GPS Trajectories, (**b**) characteristic points, (**c**) spatial clusters, (**d**) tessellation of the territory, and (**e**) generalized trajectories

of GPS traces (17,000 private vehicles tracked in the city of Milan, Italy, during a week).

Each trajectory is a de-identified sequence of time-stamped locations, visited by one of the tracked individuals or vehicles. Although de-identified, each trajectory is essentially unique—two different trajectories seldom are exactly the same, due to the extremely fine spatio-temporal resolution involved. Therefore, the chances of success for this attack are not low. If the attacker M has access to a sufficiently long sub-sequence S of locations visited by the individual R, it is possible that only a few trajectories in the dataset match with S, possibly just one. Indeed, publishing raw trajectory data such as those depicted in Fig. 10.1a is an unsafe practice, which leads to a high risk of infringement on the private sphere of the tracked drivers (e.g., guessing the home place and the work place of most respondents is very easy). Now, assume that one wants to discover the trajectory clusters emerging from the data through data mining, i.e., the groups of trajectories sharing common mobility behavior, such as the commuters following similar routes in their home-work and work-home trips. A privacy transformation of the trajectories consists of the following steps:

1. characteristic points are extracted from the original trajectories: starting points, ending points, points of significant turn, points of significant stop (Fig. 10.1b);
2. characteristic points are clustered into small groups by spatial proximity (Fig. 10.1c);
3. the central points of the groups are used to partition the space by means of Voronoi tessellation (Fig. 10.1d);
4. each original trajectory is transformed into the sequence of Voronoi cells that it crosses (Fig. 10.1e).

As a consequence of this data-driven transformation, where trajectories are generalized from sequences of points to sequences of cells, the re-identification probability already drops significantly. Further transformation can be applied to lower this probability even more, obtaining a safe theoretical upper bound for the worst case (i.e., the maximal probability that the linkage attack succeeds), and an extremely low average probability. A possible technique is to ensure that for any sub-trajectory used by the attacker, the re-identification probability is always controlled below a given threshold $\frac{1}{k}$; in other words, ensuring the k-anonymity property in the released dataset. Here, the notion of k-anonymity is based on the definition of k-*harmful trajectory*, i.e., a trajectory occurring in the database with a frequency less than k. Thus, a trajectory database D^* is considered a k-anonymous version of a database D if: each k-harmful trajectory in D appears at least k times in D^* or if it does not appear in D^* anymore. To obtain this k-anonymous database, the generalized trajectories, produced after the data-driven transformation, are transformed in such a way that all the k-harmful sub-trajectories in D are not k-harmful in D^*. In the example shown in Fig. 10.1a, the probability of success is theoretically bounded by $\frac{1}{20}$ (i.e., 20-anonymity is achieved), but the real upper bound for 95% of attacks is below 10^{-3}.

10.3.3 Analytics Quality

The above results highlight that the transformed trajectories are orders of magnitude safer than the original ones in a measurable sense: *but are they still useful to achieve the desired result, i.e., discovering trajectory clusters?*

Figure 10.2(top) and (down) listed the most relevant clusters found by mining the original trajectories and the anonymized trajectories, respectively.

A direct consequence of the anonymization process is an increase in the concentration of trajectories, i.e., many original trajectories are bundled on the

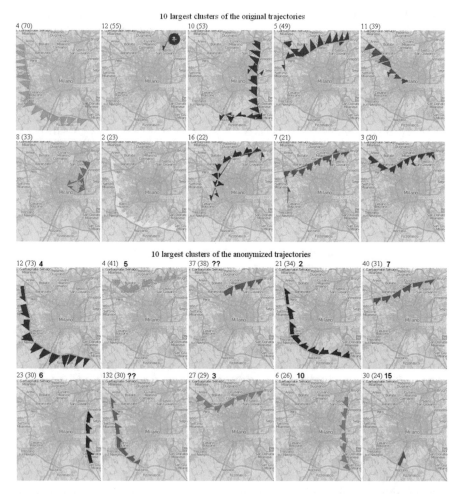

Fig. 10.2 10 largest clusters of the original trajectories (top) and of the anonymized trajectories (down)

same route; the clustering method will be influenced by the variation in the density distribution. This change is mainly caused by the reduction of noisy data. In fact, the anonymization procedure tends to render each trajectory similar to the neighboring ones. This means that the original trajectories, initially classified as noise, can now be "promoted" as members of a cluster. This phenomenon may produce an enlarged version of the original clusters. F-measure is adopted to evaluate quantitively the clustering preservation. This measure is usually used to express the combined values of precision and recall and is defined as the harmonic mean of the two measures. Here, the recall measures how the cohesion of a cluster is preserved: if the whole original cluster is mapped into a single anonymized cluster its value is 1; otherwise, the value tends to zero if the original elements are scattered among several anonymized clusters. The precision indicates how the singularity of a cluster is mapped into the anonymized version: it is 1 if the anonymized cluster contains only elements corresponding to the original cluster, it tends to zero if there are other elements corresponding to other clusters. The contamination of an anonymized cluster may depend on two factors: (1) there are elements corresponding to other original clusters or (2) there are elements that were formerly noise and have been promoted to members of an anonymized cluster.

The immediate visual perception that the resulting clusters are very similar in the two cases in Fig. 10.2(top) and (down) is also confirmed by various cluster comparisons by F-measure, re-defined for clustering comparison (Fig. 10.3).

The conclusion is that in the illustrated process the desired quality of the analytical results can be achieved in a privacy-preserving setting with concrete formal guarantees and the protection w.r.t. the linkage attack can be quantified.

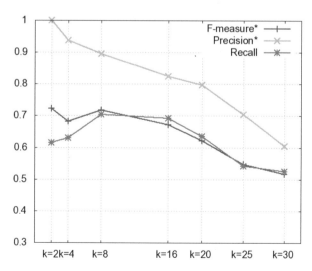

Fig. 10.3 F-measure for comparison of the clusterings of the anonymized dataset versus the clustering of the original trajectories

10.4 Privacy by Design in Semantic Trajectories Anonymization

In this section, we present a framework that offers an instance of the Privacy by Design paradigm concerning mobility trajectories enriched with semantic information, i.e., *semantic trajectories* introduced in [31] for reasoning over trajectories from a semantic point of view.

In detail, a semantic trajectory is a sequence of stops and moves of an individual during her movement. Stops are the *important parts* of a trajectory where the moving object has stayed for a minimal amount of time. Moves are the sub-trajectories describing the movements between two consecutive stops. Each location of the stop can be attached to some contextual information such as the visited place or the purpose—either by explicit sensing or by inference. An example of semantic trajectory is the sequence of places visited by a moving individual such as *Supermarket, Restaurant, Gym, Hospital, Museum.*

Important parts of a trajectory, i.e., stops, correspond to the set of x, y, t points of a trajectory that are important from an application point of view. A set of important places characterizes a semantic trajectory.

Definition 10.6 (Semantic Trajectory) Given a set of important places $\mathscr{I}$, a *semantic trajectory* $T = p_1, p_2, \ldots, p_n$ with $p_i \in \mathscr{I}$ is a temporally ordered sequence of important places, that the moving object has visited.

The Privacy by Design framework presented in this section (introduced in [25]) enables sophisticated reasoning on the scope of people's movements by maintaining under control the individual privacy. In particular, the released semantic trajectories are made *safe* concerning the inference of sensitive information derived from the knowledge of the reason of the individual's movement and from the knowledge of the place that the individual visited. The framework is based on a data transformation that generalizes places driven by a place taxonomy, thus providing a way to preserve the semantics of the generalized trajectories.

The results obtained with the application of this framework show how it possible to preserve the semantics of trajectories making them useful for extracting valid mobility semantic patterns while guaranteeing the limitation of sensitive information inferences from the individual visits.

10.4.1 Attack and Privacy Model

The use of a domain taxonomy for generalizing places enables the identification of *sensitive* and *non-sensitive* places. A place is considered sensitive when it allows inferring personal information about the individual who has stopped there. For example, a stop at an oncology clinic may indicate that the user has some health problem. Other places (such as parks, restaurants, cinemas, etc.) are considered as

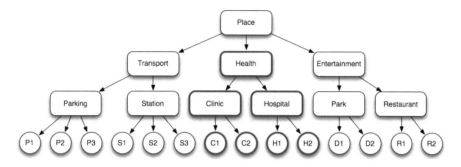

Fig. 10.4 The places taxonomy

quasi-identifiers. The labeled taxonomy is given by the domain expert who tags each concept with the corresponding "sensitivity" label.

In this context, the attack model considers an attacker with the following adversary knowledge:

Definition 10.7 (Adversary Knowledge) The attacker has access to the generalized dataset D^* and knows: (a) the algorithm used to anonymize the data, (b) the privacy place taxonomy $PTax$, (c) that a given user is in the dataset and (d) a quasi-identifier place sequence S_Q visited by the given user R.

In this model, the idea is to keep private all the sensitive places visited by a given user. As a consequence, the attack model considers the ability to link the released data to other external information enabling the inference of visited sensitive places.

In practice, given the quasi-identifier sequence S_Q, the attacker constructs a set of candidate semantic trajectories in D^* containing S_Q and tries to infer the sensitive leaf places related to R. $Prob(S_Q, S)$ denotes the probability that, given a quasi-identifier place sequence S_Q related to a user R, the attacker infers his/her set of sensitive places S which are the leaves of the taxonomy $PTax$. An example of labelled taxonomy is depicted in Fig. 10.4.

10.4.2   Privacy-Preserving Technique

How to guarantee that the probability of success of the above attack is very low while preserving the utility of the data for meaningful analyses? From a data protection perspective it is necessary to control the probability $Prob(S_Q, S)$ and a solution is to release a c-safe dataset, i.e., a dataset where for every quasi-identifier place sequence S_Q, we have that for each set of sensitive places S the $Prob(S_Q, S) \leq c$ with $c \in [0, 1]$. On the contrary, for a data utility point of view, a data analyst might use the semantic trajectories to extract common and frequent human behaviors by sequential pattern mining analyses, having in this way the possibility to reason on the semantic of the human movements. Therefore, we need a privacy transformation that tries to minimize the cost of a trajectory generalization. A privacy transformation of semantic trajectories consists of the following steps:

1. suppressing from the original semantic trajectories each *sensitive place* when, for that given user, that place is a *quasi-identifier*;
2. grouping semantic trajectories in groups of a predefined size, m;
3. building a generalized version of each semantic trajectory in the group generalizing the quasi-identifier places. In each group, the quasi-identifiers of the generalized trajectories should be identical. Sensitive places are generalized when the quasi-identifiers generalization is not enough to get a c-safe dataset. The generalization is performed with the support of the taxonomy $PTax$.

This method generates a c-safe version of a dataset of semantic trajectories keeping under control both the probability to infer sensitive places and the generalization level (thus the information loss) introduced in the data. In other words, the obtained dataset guarantees the c-safety and maintains the information useful for the data mining tasks, as much as possible. The taxonomy defined by the domain expert is crucial in this process. In fact, having more levels of abstraction allows the method in finding a better generalization in terms of information loss. In order to consider the generalization cost it is possible to use distance functions that measure the cost to transform an original semantic trajectory into a generalized one, based on the taxonomy. A measure might be the distance in steps from two places in the taxonomy tree, the so called *Hops-based distance*.

If we consider the dataset in Fig. 10.1a, after the privacy transformation where the probability of success is theoretically bounded by 0.3 we have an empirical upper bound of 0.07 in average on 10,000 attacks using as background knowledge 5 places (see Fig. 10.5).

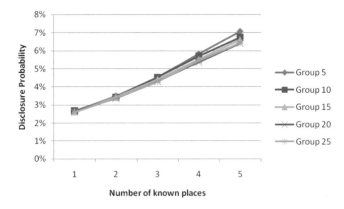

Fig. 10.5 The empirical disclosure probability on Milano dataset

10.4.3   Analytics Quality

Now, the point is to understand if the guaranteed privacy protection also allows the possibility to perform some analysis based on the sequential pattern mining extractable from the c-safe data. To evaluate this, it is necessary to measure the quality of the sequential patterns. Figure 10.6a shows the effect of the privacy transformation on the number of patterns extractable from the dataset after the sanitization. The figure highlights the fact that the generalization has a double effect on the patterns: (1) the frequency of generalized places increases, (2) the frequency of leaf places of the taxonomy decreases. Therefore, with a high support threshold, the difference between the patterns created and removed by the generalization phase is positive, and this increases the size of the resulting patterns set. Figure 10.6b depicts instead the trend of the *coverage coefficient*. This index measures how many

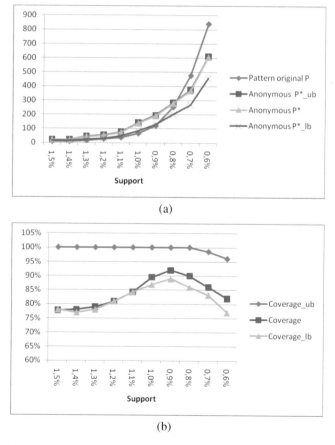

(a)

(b)

Fig. 10.6 (a) Number of patterns extracted from Milan data and (b) coverage of the patterns varying the support threshold

patterns extracted from the original dataset are covered at least by the patterns extracted from the anonymized dataset with a certain level of generalization. It is important to notice that the coverage does not measure *how much* the patterns are generalized, but only if they are covered by a pattern obtained from the anonymized dataset or not. The results highlight that the coverage guaranteed by the patterns after the privacy transformation is not 100% but the levels are high enough to enable analyses; in fact, by changing the support (i.e., the minimum frequency used for the pattern extraction) the coverage is always greater than 75%.

10.5 Privacy by Design in Distributed Systems

The previous scenarios (Sects. 10.3 and 10.4) are based on centralized environments, where the privacy preservation step is performed by a central entity; in fact, we showed two variants of k-anonymity which can be used only by a trusted aggregation center. However, Privacy by Design paradigm can also be applied with success to distributed systems. In this section, we discuss an instance of this case [26]; in particular, we analyze the handling of personal mobility trajectories, generated by several vehicles distributed in a territory and collected by a central entity, called *coordinator*. Streams of data updates arrive continuously at remote sites (i.e., vehicles), while the coordinator is responsible for computing the aggregation of movement data on a territory by combining the information received by each node.

We show how privacy can be obtained before data leave users, ensuring the utility of some data analysis performed at the collective level, also after the transformation. This example brings evidence to the fact that the Privacy by Design model has the potential of delivering high data protection combined with high quality even in massively distributed techno-social systems.

10.5.1 Attack and Privacy Model

As in the case analyzed in Sect. 10.3, any data from which the typical mobility behavior of a user may be inferred is assumed as sensitive information. This information is considered sensitive for two main reasons: (1) typical movements can be used to identify drivers even when a simple de-identification of the individual in the system is applied; and (2) the places visited could identify distinguishing sensitive areas such as clinics, hospitals and routine locations such as the user's home and workplace.

The assumption is that each node in the system is honest; in other words, attacks at the node level are not considered. Instead, potential attacks are from any adversary between the node and the coordinator (i.e., attacks during the communications), and from any adversary at coordinator site, so this privacy preserving technique has to guarantee privacy even against a malicious behavior of the coordinator. For example, the coordinator may be able to obtain real mobility statistic information from other sources, such as from public datasets on the web, or through personal knowledge about a specific individual, like in the previously (and diffusely) discussed linking attack.

The solution proposed in [26] is based on *Differential Privacy*, a recent paradigm of randomization presented in [14] by Dwork. The general idea of this model is that the privacy risks should not increase for a respondent as a result of occurring in a statistical database; differential privacy ensures, in fact, that the ability of an adversary to inflict harm should be essentially the same, independently of whether any individual opts in to, or opts out of, the dataset. This privacy model is called ϵ-differential privacy, due to the level of privacy guaranteed ϵ, also called *privacy budget*. Note that when ϵ grows very little perturbation is introduced and this yields a low privacy protection; on the contrary, better privacy guarantees are obtained when ϵ tends to zero. Differential privacy guarantees a record owner that any privacy breach will not be a result of participating in the database since nothing, or almost nothing, that can be learned from the database with his record can be learned from the database without his data. Moreover, in [14] is formally proved that ϵ-differential privacy can provide a guarantee against adversaries with arbitrary background knowledge, thus, in this case, we do not need to define any explicit background knowledge for attackers.

In a nutshell, the differential privacy mechanism works by adding appropriately chosen random noise (from a specific distribution) to the true answer, then returning the perturbed answer. The formal definition of differential privacy [14] is the following. Here the parameter, ϵ, specifies the level of guaranteed privacy.

Definition 10.8 (ϵ-Differential Privacy) [14] A privacy mechanism A gives ϵ-differential privacy if for any dataset D_1 and D_2 differing on at most one record, and for any possible output D' of A we have $Pr[A(D_1) = D'] \leq e^\epsilon \times Pr[A(D_2) = D']$ where the probability is taken over the randomness of A.

A basic notion used by differential privacy mechanisms is the *sensitivity* of a query, which provides a way to set the noise distribution in order to calibrate the noise magnitude on the basis of the type of query.

Definition 10.9 (Global Sensitivity) [13] For any function $f : D \rightarrow R^d$, its sensitivity is $\Delta f = max_{D_1,D_2} ||f(D_1) - f(D_2)||_1$ for all D_1, D_2 differing in at most one record.

Intuitively, the sensitivity measures the maximum distance between the same query executed on two close datasets, i.e., datasets differing on one single element (either a user or an event). As an example, consider a count query on a medical dataset, which returns the number of patients having a particular disease. The result of the query performed on two close datasets, i.e., differing exactly on one patient, can change at most by 1; thus, in this case (or, more generally, in count query cases), the sensitivity is 1.

A little variant of this model is the (ϵ, δ)-differential privacy [16], where the noise is bounded at the cost of introducing a privacy loss, δ. (ϵ, δ)-differential privacy allows a small amount of privacy loss due to a variation in the output distribution for the privacy mechanism A.

Definition 10.10 ((ϵ, δ)-Differential Privacy) [16] A privacy mechanism A gives (ϵ, δ)-differential privacy if for any dataset D_1 and D_2 differing on at most one record, and for any possible output D' of A we have $Pr[A(D_1) = D'] \le e^\epsilon \times Pr[A(D_2) = D'] + \delta$ where the probability is taken over the randomness of A.

The questions are: *How can we hide the event that the user moved from a location* a *to a location* b *in a time interval* τ? *And how can we hide the real count of moves in that time window?* In other words, *How can we enable collective movement data aggregation for mobility analysis while guaranteeing individual privacy protection?* The solution that we report is based on (ϵ,δ)-differential privacy, and provides a good balance between privacy and data utility.

10.5.2 Privacy-Preserving Technique

First of all, each participant must share a common partitioning of the considered area; for this purpose, it is possible to use an existing division of the territory (e.g., census sectors, road segments, etc.) or to determine a data-driven partition as the Voronoi tessellation introduced in Sect. 10.3.2. Once this is accomplished, each trajectory is generalized as a sequence of crossed areas (i.e., a sequence of movements). For the sake of convenience, this information is mapped onto a *frequency vector*, linked to the partition.

In order to perform this mapping task, we firstly need a function *Move Frequency* (MF) to compute how many times the move appears in a generalized trajectory T_g within a given time interval.

Definition 10.11 (Move Frequency) Let T_g be a generalized trajectory and let (l_{c_i}, l_{c_j}) be a move. Given the temporal interval τ the move frequency function is defined as:

$$MF(T_g, (l_{c_i}, l_{c_j}), \tau) = |\{(l_{c_i}, l_{c_j}, t_i, t_j) \in T_g : t_i \in \tau \wedge t_j \in \tau\}|.$$

This function can be easily extended to take into consideration a set of generalized trajectories $\mathscr{T}^\mathscr{G}$. In this case, computed information represents the total number of movements from the cell c_i to the cell c_j in a time interval in the set of trajectories.

Definition 10.12 (Global Move Frequency) Let $\mathscr{T}^\mathscr{G}$ be a set of generalized trajectories and let (l_{c_i}, l_{c_j}) be a move. Let τ be a time interval. The global move frequency function is defined as:

$$GMF(\mathscr{T}^\mathscr{G}, (l_{c_i}, l_{c_j}), \tau) = \sum_{\forall T_g \in \mathscr{T}^\mathscr{G}} MF(T_g, (l_{c_i}, l_{c_j}), \tau).$$

The number of movements between two cells computed by either the function MF or GMF describes the amount of traffic flow between the two cells in a specific time interval τ. This information can be represented by a frequency vector. To define the frequency vector, we first define *vector of moves*.

Definition 10.13 (Vector of Moves) Let $C = \{c_1, c_2, \ldots, c_p\}$ be the set of the cells composing the territory partition. The vector of moves M is a vector of size $s = |\{(c_i, c_j)|c_i \text{ is adjacent to } c_j\}|$, in which each element $M[k] = (l_{c_i}, l_{c_j})$, where $1 \leq k \leq s$, is the move from the cell c_i to the adjacent cell c_j.

At this point, we can define the frequency vector.

Definition 10.14 (Frequency Vector) Let $C = \{c_1, c_2, \ldots, c_p\}$ be the cells that compose the territory partition and let M be its vector of moves. Given a set of generalized trajectories $\mathscr{T}^{\mathscr{G}}$ in a time interval τ, its *frequency vector f* is a vector of size $s = |\{(c_i, c_j)|c_i \text{ is adjacent to } c_j\}|$, in which each element $f[k] = GMF(\mathscr{T}^{\mathscr{G}}, M[k], \tau)$.

Unfortunately, releasing frequency of moves instead of raw trajectory data to the coordinator is still not privacy-preserving, as the intruder may still infer the sensitive typical movement information of the driver. As an example, the attacker could discover the driver's most frequent move; this information can be very sensitive because it usually corresponds to a user's transportation between home and workplace. Thus, the proposed solution is based on the differential privacy model, relying on a Laplace distribution [15]. At the end of a preset time interval τ, each node, before sending the frequency vector to the coordinator and for each element in the vector, extracts the noise from the Laplace distribution and adds it to the original value in that position of the vector. At the end of this operation, the node V_j converted its frequency vector f_{V_j} into its private version $\tilde{f}_{V_j}$. This ensures the respect of the ϵ-differential privacy. This simple general strategy has some inconveniences: first, it could lead to a large amount of noise that, although with small probability, can be arbitrarily large; second, adding noise drawn from the Laplace distribution could produce negative frequency counts of moves, which does not make sense in mobility scenarios. In order to fix these two problems, it is possible to bound the noise drawn from the Laplace distribution, reducing to an (ϵ, δ) differential privacy schema. In particular, for each value x of the vector f_{V_j}, it is possible to draw the noise bounding it in the interval $[-x, x]$. In other words, for any original frequency $f_{V_j}[i] = x$, its perturbed version after adding noise falls in the interval $[0, 2x]$. This approach satisfies (ϵ, δ)-differential privacy, where δ measures the privacy loss. Note that, since in a distributed environment communications need to be quite limited, it is possible to reduce the amount of transmitted information, i.e., the size of frequency vectors. A possible solution to this problem is reported in [26], but this discussion is omitted here because is beyond the purpose of our review.

10.5.3 Analytical Quality

So far we reported the formal guarantees to individual privacy preservation, now we have to show how individually transformed values are still useful once they are collected and aggregated by the coordinator, i.e., they are still suitable at a collective level for analysis. In the proposed framework, the coordinator gathers the perturbed frequency vectors from all the vehicles in the time interval τ and sums them movement by movement. This achieves to obtain the resulting global frequency vector, which indicates the flow values for each possible link of the spatial tessellation. Since the privacy transformation operates on the entries of the frequency vectors, and hence on the flows, we report the comparison (before and after the transformation) of two measures: (1) the *Flow per Link*, i.e., the directed volume of traffic between two adjacent zones; (2) the *Flow per Zone*, i.e., the sum of the incoming and outgoing flows in a zone. The following results refer to the application of this technique on a large dataset of GPS vehicles traces, collected in a period from 1st May to 31st May 2011, in the geographical areas around Pisa, in central Italy. It counts for around 4200 vehicles, generating around 15,700 trips in total. The τ interval is 1 day, so the global frequency vector represents the sum all the trajectories crossing any link, at the end of each day. The reported results are relative to 25th May 2011, but they are similar to ones obtained in the other working days.

Figure 10.7 shows the resulting Complementary Cumulative Distribution Functions (CCDFs) of different privacy transformation varying ϵ from 0.9 to 0.01. Figure 10.7-Left shows the global (approximated) *Flow per Link* distribution: fixed a value of flow (x) is counted the number of links (y) that have that flow. Figure 10.7-Right shows the distribution of sum of flows passing for each zone, i.e., *Flow per Zone*: given a flow value (x) it shows how many zones (y) present that total flow. From the distributions, we can observe that the privacy transformation preserves

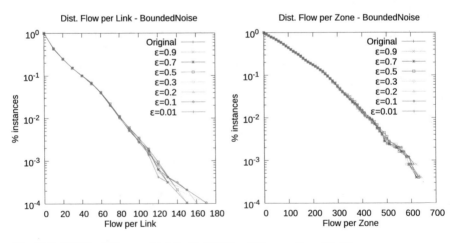

Fig. 10.7 CCDFs of *Flow per Link* (Left); CCDFs of *Flow per Zone* (Right)

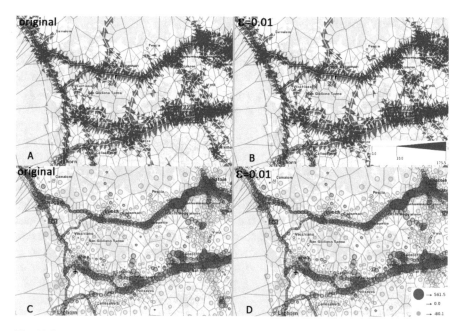

Fig. 10.8 Visualization of *Flow per Link* (A-B) and *Flow per Zone* (C-D)

very well the distribution of the original flows, even for more restrictive values of the parameter ϵ. Also considering several flows together, like those ones that are incident to a given zone (Fig. 10.7-Right), the distributions are well preserved for all the privacy transformations. These results reveal how a method which *locally* perturbs values, at a *collective* level permits to obtain a very good utility.

Qualitatively, Fig. 10.8 shows a visual comparison of results of the privacy transformation with the original ones. This is an example of analyses that can be carried out with mobility data. Since the global complementary cumulative distribution functions are comparable, it is possible to choose a very low epsilon ($\epsilon = 0.01$) with the aim to emphasize the very good quality of mobility analysis that an analyst can obtain even if the data are transformed by using this low ϵ value, i.e. obtaining a better privacy protection. In Fig. 10.8a, b each flow is drawn with arrows with the thickness proportional to the volume of trajectories observed on a link. From the figure it is evident how the relevant flows are maintained in the transformed global frequency vector, highlighting the major highways and urban centers. The *Flow per Zone* is also preserved, as it is shown in Fig. 10.8c, d, where the flow per each cell is rendered with a circle of radius proportional to the difference from the median value of each global frequency vector. The maps allow us to recognize the dense areas (red circles, above the median) separated by sparse areas (blue circles, below the median). The high density traffic zones follow the highways and the major city centers along their routes. These two comparisons confirm the intuition that, while the transformations protect individual sensitive information, the utility of data is preserved.

10.6   Conclusion

The potential impact of the big data analytics and social mining is high because it could generate enormous value to society. Unfortunately, often big data describe sensitive human activities and the privacy of people is always more at risk. The danger is also increasing thanks to the emerging capability to integrate diversified data. In this chapter, we have introduced the articulation of the Privacy by Design in big data analytics and social mining for enabling the design of analytical processes that minimize, or even prevent, the privacy harm. We have discussed how applying the Privacy by Design principle to three different scenarios showing that under suitable conditions is feasible to reach a good trade-off between data privacy and good quality of the data. We believe with the Privacy by Design principle social mining has the potential to provide a privacy-respectful social microscope, or socioscope, needed to observe the hidden mechanisms of socio-economic complexity.

10.7   Bibliographic Notes

In the following, we provide a quick overview of some techniques and solutions adopted in privacy-preserving data mining for mobility data. The Privacy by Design model was applied in data mining in several contexts [24, 27], with special treatment to mobility data, due to their complex nature, their sensitivity and their importance for understanding human behaviors. Privacy issue in mobility data mining and sharing have been intensively studied in literature [8, 20, 22], and the existing methods of privacy-aware releasing and sharing of (trajectory) data can be classified into two main classes: (1) generalization/suppression based data perturbation, and (2) randomization/differential privacy perturbation.

The most widely used privacy model for generalization and suppression perturbation is adapted from what so called k-anonymity [30, 32], which requires that an individual should not be identifiable from a group of size smaller than k based on their quasi-identifies (QIDs), i.e., a set of attributes that can be used to identify uniquely the individuals. Unfortunately, in trajectory data, it often impossible to distinguish clearly between quasi-identifiers and sensitive attribute. In [36], Yarovoy et al. deeply analyze the problem of quasi-identifiers in mobility data: they show that the anonymization groups may not be disjoint. Thus there may exist objects that can be identified explicitly by combining different anonymization groups. They suggest that QIs may be provided directly by personal settings or found by means of statistical data analysis. In [4], Abul et al. propose the notion of (k,δ)-anonymity for moving object databases, where δ represents the possible location imprecision. This is an innovative concept of k-anonymity based on co-localization, which takes advantage of the inherent uncertainty of the whereabouts of the moving objects. The authors also proposed an approach, called Never Walk Alone, based on trajectory clustering and spatial translation, and they present its improvement, Wait for Me, in [5]. This method is very similar to the previous one, but it is based on EDR distance (instead of Euclidean distance), which is time-tolerant, so Wait for Me can recognize similar trajectories even if they are (slightly) shifted in time. Finally, Domingo Ferrer and Trujillo-Rasua [12] show a solution based on perturbation and micro-aggregation: this method k-anonymizes each location independently, using the whole set of trajectories. Particularly, the algorithm creates clusters of locations (close in time and space) in such a way that the locations in each group belong to k different trajectories. The result of this transformation is that the probability that a location of a true trajectory appears in its anonymized version is at most $\frac{1}{k}$ while guaranteeing that the anonymized trajectories are suitable for range query for every value of k.

Regarding the application of Differential Privacy mechanisms to mobility data, many works have been proposed in last years. In [35] authors provide an algorithm, based on Markov Chain and Differential Privacy, which aims to protect the continual location sharing of perturbed locations in the context of Location Based Services. In particular, they select a set of locations that are highly probable for a user, guaranteeing that the probability of these locations is similar to the other, and chooses one of these locations to be released outside. In this case, the event protected by Differential Privacy is a specific request to a service, instead of a specific move. However, they do not provide guarantees if the attacker has a stronger external knowledge w.r.t. the history of the released locations. This additional constraint is analyzed in [7], where Andrés et al. show a technique for Location Based Services independent from the side information of users. They use an extension of Laplace distribution for the continuous plane and promise a privacy level which is distance-dependent, i.e., guarantees are stronger if you get closer to the real location of the user. A very promising research line about Differential Privacy on spatio-temporal data is the one related to space partitioning. Ho and Ruan [21] apply Differential Privacy to interesting locations to perform location pattern discovery, granting protection at the user-level also when a user contributes to more than one record. They partition the space of the data into smaller ones, in order to limit the total number of events and, consequently, the events connected with each individual in each dataset, in order to overcome the problem of the presence of a clear upper-bound to the events related to a single user. In [10], Cormode et al. describe a solution to publish differentially private spatial index (e.g., quadtrees and kd-trees) to provide a private description of the data distribution [10]. Its main utility concern is the accuracy of multi-dimensional range queries (e.g., how many individuals fall within a given region). Therefore, the spatial index only stores the counts of a specific spatial decomposition, even their solution does not store the movement information (e.g., how many individuals move from location i to location j). In [9], authors rely on a prefix tree of trajectories with injected Laplace noise; the prefix tree is data-dependent, i.e., it should have a different structure when the underlying database changes. Qardaji et al. [29] provide an adaptive uniform partition method, considering different density-regions, i.e., depending on the total number of points in the dataset. In Acs et al. [6], authors apply Geometrical mechanism to a partition of a territory, taking advantage of a Voronoi tessellation to keep track of the presence of individuals and use clustering and sampling with Fourier-based perturbation Finally, Cormode et al. [11] propose to publish a contingency table of trajectory data, that can be indexed by specific locations so that each cell in the table contains the number of people who commute from the given source to the given destination. The purpose of this work is to address the sparsity issue of the contingency table and presents a method of releasing a compact summary of the contingency table with Laplace noise.

References

1. *Privacy by Design Resolution*, International Conference of Data Protection and Privacy Commissioners (Jerusalem, Israel, October 27–29, 2010)
2. *Article 29 Data Protection Working Party and Working Party on Police and Justice, The Future of Privacy: Joint Contribution to the Consultation of the European Commission on the Legal Framework for the Fundamental Right to Protection of Personal Data*, 02356/09/EN, WP 168 (Dec. 1, 2009)
3. *Directive (EU) 2016/680 of the European Parliament and of the Council of 27 April 2016*, Official Journal of the European Union (2016)
4. O. Abul and F. Bonchi and M. Nanni, in *Never Walk Alone: Uncertainty for Anonymity in Moving Objects Databases*, ICDE 2008, pp. 376–385
5. O. Abul and F. Bonchi and M. Nanni, in *Anonymization of Moving Objects Databases by Clustering and Perturbation*, Inf. Syst., vol 35, num 8, pp. 884–910 (Elsevier Science Ltd., December, 2010), doi: 10.1016/j.is.2010.05.003
6. G. Ács and C. Castelluccia, *A case study: privacy preserving release of spatio-temporal density in Paris*, KDD 2014, pp. 1679–1688
7. M. E. Andrés and N.E. Bordenabe and K. Chatzikokolakis and C. Palamidessi, *Geo-indistinguishability: differential privacy for location-based systems*, ACM Conference on Computer and Communications Security, p. 901–914, 2013
8. F. Bonchi and L.V.S. Lakshmanan and H. W. Wang, in *Trajectory anonymity in publishing personal mobility data*, SIGKDD Explor. Newsl. 2011, pp. 30–42, https://doi.org/10.1145/2031331.2031336
9. R. Chen and B.C.M. Fung and B.C. Desai and N.M. Sossou, *Differentially private transit data publication: a case study on the Montreal transportation system*, KDD 2012, pp. 213–221
10. G. Cormode and C. Procopiuc and D. Srivastava and E. Shen and T. Yu, *Differentially private spatial decompositions*, ICDE 2012, pp. 20–31
11. G. Cormode and C. Procopiuc and D. Srivastava and T. Tran, *Differentially private summaries for sparse data*, ICDT 2012, pp. 299–311
12. J. Domingo-Ferrer and R. Trujillo-Rasua, in *Microaggregation- and permutation-based anonymization of movement data*, Inf. Sci. 2012, volume 208 pp. 55–80
13. C. Dwork, in *Differential privacy: A survey of results*, International Conference on Theory and Applications of Models of Computation, pages 1–19. Springer, 2008
14. C. Dwork, in *Differential Privacy*, ICALP 2006, Lecture Notes in Computer Science, vol 4052, https://doi.org/10.1007/11787006_1
15. C. Dwork and F. Mcsherry and K. Nissim and A. Smith, in *Calibrating noise to sensitivity in private data analysis*, Proceedings of the 3rd Theory of Cryptography Conference (Springer, 2006), pp. 265–284

16. C. Dwork, K. Kenthapadi, F. McSherry, I. Mironov, M. Naor, in *Our Data, Ourselves: Privacy Via Distributed Noise Generation*, Advances in Cryptology-EUROCRYPT 2006, pp. 486–503. Springer Berlin Heidelberg, 2006
17. European Data Protection Supervisor in *Opinion of the European Data Protection Supervisor on Promoting Trust in the Information Society by Fostering Data Protection and Privacy* (Mar. 18, 2010)
18. European Parliament & Council. General data protection regulation, 2016. L119, 4/5/2016
19. Federal Trade Commission (Bureau of Consumer Protection, in *Preliminary Staff Report, Protecting Consumer Privacy in an Era of Rapid Change: A Proposed Framework for Business and Policy Makers* (Dec. 2010)
20. F. Giannotti and D. Pedreschi, in *Mobility, Data Mining and Privacy: A Vision of Convergence*, Mobility, Data Mining and Privacy - Geographic Knowledge Discovery (2008), https://doi.org/10.1007/978-3-540-75177-9_1
21. S. Ho and S. Ruan, *Preserving Privacy for Interesting Location Pattern Mining from Trajectory Data*, Transactions Data Privacy (2013) 6(1): 87–106, 2013
22. A. Monreale and D. Pedreschi and R. G. Pensa, in *Anonymity Technologies for Privacy-Preserving Data Publishing and Mining*, Privacy-Aware Knowledge Discovery: Novel Applications and New Techniques (2010), pp. 3–33
23. A. Monreale and G. L. Andrienko and N. V. Andrienko and F. Giannotti and D. Pedreschi and S. Rinzivillo and S. Wrobel, in *Movement Data Anonymity through Generalization*, Transactions on Data Privacy (2010), volume 3 number 2, pp. 91–121
24. A. Monreale, in *Privacy by Design in Data Mining*, PhD Thesis, Dept. of Computer Science (University of Pisa, 2011)
25. A. Monreale and R. Trasarti and D. Pedreschi and C. Renso and V. Bogorny in *C-safety: a framework for the anonymization of semantic trajectories*, Transactions on Data Privacy (2011), volume 4 number 2 pp. 73–101
26. A. Monreale and W. Hui Wang and F. Pratesi and·S. Rinzivillo and D. Pedreschi and G. Andrienko and N. Andrienko, in *Privacy-preserving Distributed Movement Data Aggregation*, AGILE (Springer 2013), https://doi.org/10.1007/978-3-319-00615-4_13
27. A. Monreale and S. Rinzivillo and F. Pratesi and F. Giannotti and D. Pedreschi, in *Privacy-by-design in big data analytics and social mining*, EPJ Data Science (2014)
28. R. G. Pensa and A. Monreale and F. Pinelli and D. Pedreschi, in *Pattern-Preserving k-Anonymization of Sequences and its Application to Mobility Data Mining* PiLBA 2008
29. W. H. Qardaji and W. Yang and N. Li, in *Differentially private grids for geospatial data*, ICDE 2013, pp 757–768
30. P. Samarati and L. Sweeney, in *Protecting privacy when disclosing information: k-anonymity and its enforcement through generalization and suppression* (SRI International, 1998)
31. S. Spaccapietra, C. Parent M.L. Damiani, J. Macedo, F. Porto, C. Vangenot. *A conceptual view on trajectories*. DKE Journal 65(1): 126–146 (2008).
32. L. Sweeney, in *Computational disclosure control: a primer*, Ph.D. thesis, Dept. of Electrical Eng. and Computer Science (MIT, 2001)
33. L. Sweeney, in *Simple Demographics Often Identify People Uniquely*, Carnegie Mellon University, Data Privacy Working Paper 3. Pittsburgh 2000
34. W.K. Wong and D.W. Cheung and E. Hung and B. Kao and Nikos Mamoulis, in *Security in Outsourcing of Association Rule Mining*, VLDB 2007, pp. 111–122
35. Y. Xiao and L. Xiong, in *Protecting Locations with Differential Privacy under Temporal Correlations*, ACM Conference on Computer and Communications Security, p.298–1309, 2015
36. R. Yarovoy and F. Bonchi and L.V. S. Lakshmanan and W. H. Wang, in *Anonymizing moving objects: how to hide a mob in a crowd?*, International Conference on *Extending DataBase Technology* (2009), pp. 72–83

Part III
Usability, Systems and Applications

Chapter 11
Systems for Privacy-Preserving Mobility Data Management

Despina Kopanaki, Nikos Pelekis, and Yannis Theodoridis

Abstract The increasing availability of data due to the explosion of mobile devices and positioning technologies has led to the development of efficient management and mining techniques for mobility data. However, the analysis of such data may result in significant risks regarding individuals' privacy. A typical approach for privacy-aware mobility data sharing aims at publishing an anonymized version of the mobility dataset, operating under the assumption that most of the information in the original dataset can be disclosed without causing any privacy violation. On the other hand, an alternative strategy considers that data stays in-house to the hosting organization and privacy-preserving mobility data management systems are in charge of privacy-aware sharing of the mobility data. In this chapter, we present the state-of-the-art of the latter approach, including systems such as HipStream, Hermes++, and Private-Hermes.

11.1 Introduction

Recent advances in mobile devices, positioning technologies and spatiotemporal database research, have made possible the tracking of mobile devices at a high accuracy, while supporting the efficient storage of mobility data in databases. From this perspective, we have nowadays the means to collect, store and process mobility data of an unprecedented quantity, quality and timeliness. As ubiquitous computing pervades our society, user mobility data represents a very useful but also sensitive source of information. On the one hand, the movement traces of the users can aid traffic engineers, city managers and environmentalists towards decision making in a wide spectrum of applications, such as urban planning, traffic engineering

D. Kopanaki (✉) · Y. Theodoridis
Department of Informatics, University of Piraeus, Piraeus, Greece
e-mail: dkopanak@unipi.gr; ytheod@unipi.gr

N. Pelekis
Department of Statistics & Insurance Science, University of Piraeus, Piraeus, Greece
e-mail: npelekis@unipi.gr

© Springer Nature Switzerland AG 2018
A. Gkoulalas-Divanis, C. Bettini (eds.), *Handbook of Mobile Data Privacy*,
https://doi.org/10.1007/978-3-319-98161-1_11

and environmental pollution. On the other hand, the disclosure of mobility data to untrusted parties may jeopardize the privacy of the users whose movement is recorded, leading the way to abuse scenarios such as user tailing and profiling. As it becomes evident, the sharing of user mobility data for analysis purposes has to be done only after the data has been protected against potential privacy breaches.

In this chapter, we consider the following data sharing scenario: a data holder (telecom operator, governmental agency, etc.) collects movement information about a community of people. The raw movement data, capturing the location of each individual in the course of time, is processed to generate user trajectories that are subsequently stored in a database. Apart from the analysis that this data undergoes within the premises of the hosting organization, we assume that at least part of the data has to be made available to external, possibly untrusted, parties for querying and analysis purposes. As is evident, direct publishing of this information, even if the data is first deprived from any explicit identifiers, would severely compromise the privacy of the individuals whose movement is recorded in the database. This is due to the fact that malevolent end-users could potentially link the published trajectories to sensitive locations of the individuals (such as their houses), thus identify the users. To ensure privacy-aware sharing of in-house mobility data, a mechanism is necessary to control the information that is made available to external parties when they query the database, so that only nonsensitive information leaves the premises of the hosting organization.

Recently, several methodologies have been proposed to enable privacy-preserving mobility data sharing. Existing approaches, such as [1, 2, 9, 10, 16, 30], aim at publishing an anonymous counterpart of the original dataset in which adversaries can no longer match the recorded movement of each user to the real identity of the user. A common assumption that is implicitly made in these approaches is that most of the information stored in the original dataset can be disclosed without causing any privacy violations. However, this assumption can be proven unrealistic in certain data sharing scenarios. In order to avoid privacy breaches a more conservative approach can be employed by assuming that the majority of the information that is captured in the mobility dataset should remain private and that the data should stay in-house to the hosting organization. This assumption is primarily based on the following arguments:

- The data owner may be reluctant to publish the entire mobility dataset, or conformance to certain business regulations may require that the dataset resides in-house to the hosting organization.
- Mobility datasets typically support many types of data analysis. In order for the anonymous dataset to be useful in practical applications, it is necessary that the anonymization approach can offer specific utility guarantees and this, in turn, requires knowledge of the intended workload. When data resides in-house, the privacy preservation algorithms can support many types of data analysis (which may be unknown apriori) by guaranteeing at the same time the privacy of the users, whose information is recorded in the dataset.

- Data sharing policies may change from time to time and new types of privacy attacks to mobility data may be identified, yielding previously released data unprotected. In such events, it is crucial for the data owner to have knowledge of the sensitive information that was leaked, as well as be capable of safeguarding the data based on the new evidence. When data resides in-house, the privacy-aware query engine can be updated to conform to the new policies and block new types of attack. Additionally, the auditing of queries allows the data owner to have knowledge about the extent of the data leakage by examining the history of user queries to the database and keeping track of the returned answers.

In this chapter, we present the state-of-the-art systems which are based on the assumption that data should stay in-house to the hosting organization in order to ensure that no privacy violation may occur during analysis processes. First, Gkoulalas-Divanis and Verykios [8] proposed a query engine that offers k-anonymous answers to user queries. The engine generates fake records to guarantee about what can be found by untrusted third parties. Based on the same notion, Hermes++ which was proposed by Pelekis et al. [22], is a novel query engine for sensitive trajectory data that allows subscribed end-users to gain restricted access to the database to accomplish various analysis tasks. Hermes++ can shield the trajectory database from potential attacks to user privacy, while supporting popular queries for mobility data analysis, such as range queries, distance queries and nearest neighbor queries. Hermes++ operates by retrieving real user trajectories from the database and generating carefully crafted fake trajectories in order to reduce the confidence of attackers regarding the information of the real trajectories in the query result. Hermes++ achieves to (a) audit end-user queries and block an extended set of attacks to user privacy, securing the database against user identification, sensitive location tracking, and sequential tracking attacks, (b) generate smooth and more realistic fake trajectories that preserve the trend of the original data, and (c) ensure that no sensitive locations that would lead to user identification are reported as part of the returned trajectories. The latter goal is achieved by modifying parts of the trajectories that are close to sensitive locations, such as the houses of the users.

Moreover, we present Private-Hermes [23], a benchmark framework for privacy-preserving mobility data querying and mining methods. The first dimension of this benchmark with respect to privacy issues involves in-house stored data and privacy-aware query answering. Private-Hermes incorporates Hermes [21], a query engine based on a powerful query language for trajectory databases, which enables the support of aggregative queries. Hermes supports a variety of well-known queries such as range, nearest neighbor, topological, directional queries, etc. On top of this functionality, Hermes++ audits queries for trajectory data to block potential attacks to user privacy, supports the most popular spatiotemporal queries (range, distance, $k - NN$) and preserves user privacy by generating carefully crafted, realistic fake trajectories. The second dimension with respect to privacy that is supported by this benchmark involves privacy-preserving MOD publishing. Two state-of-the-art algorithms, namely NWA [1] and W4M [2], have been integrated in Private-Hermes to help anonymize trajectories. The objective is to support the evaluation of such anonymization techniques and to study their effect in the utility of the sanitized data, when compared with queries into the original MOD.

Finally, HipStream [31] is a privacy-preserving system for managing mobility data streams. The system enforces three fundamental Hippocratic principles introduced by Agrawal et al. [4] of limited use, limited disclosure and limited collection of data during data stream management. Hippocratic databases extend the functionalities of traditional databases with privacy-preserving capabilities. Service providers have limited access to the data w.r.t. the privacy requirements that the data owner has enforced. Queries are modified if needed from the system and data are partially anonymized if necessary before being processed.

The rest of this chapter is organized as follows. Section 11.2 provides a description of the background of privacy-preserving mobility data management, highlighting the design principles of a privacy-aware trajectory query engine and the types of attacks to user privacy that such an engine should be able to block. Section 11.3 discusses Hermes++ putting emphasis on the auditing and the fake trajectory generation algorithms that are implemented as part of the query engine to support its functionality. Section 11.4 demonstrates the Private-Hermes benchmark framework. In Sect. 11.5, HipStream privacy-preserving system is presented. Section 11.6 summarizes this chapter.

11.2 Background

Research in the domain of privacy-preserving data publishing has progressed along two main directions: providing off-site publication of sanitized data and providing on-site, restricted access to in-house data.

The first direction in privacy-preserving data publishing collects methodologies that provide off-site publication of sanitized data. Several methodologies have been proposed to support different data types and analysis tasks [1, 2, 13, 16, 27, 29, 30].

Hoh and Gruteser [9] present a data perturbation algorithm that is based on path crossing. The approach identifies when two nonintersecting trajectories that belong to different users are "sufficiently" close to each other in the original dataset and generates a fake crossing of these trajectories in the sanitized counterpart to prevent adversaries from tracking a complete user's trajectory. Terrovitis and Mamoulis [30] consider datasets that depict user movement in the form of sequences of places that each user has visited, set out in the order of visit. They propose an anonymization approach that suppresses selected places from user trajectories to protect users from adversaries who hold projections of the data on specific sets of places. Nergiz et al. [16] also rely on the sequential nature of mobility data and propose a coarsening strategy to generate a sanitized dataset that consists of k-anonymous [27, 29] sequences. The algorithm consolidates the trajectories of the original dataset into clusters of k and then anonymizes the trajectories in each cluster. Abul et al. [1] propose a k-anonymity approach that relies on the inherent uncertainty that exists with respect to the whereabouts of the users in historical datasets representing user mobility. The anonymity algorithm identifies trajectories that lie close to each other in time, employs space translation and generates clusters of at least k trajectories. Each cluster of k trajectories forms an anonymity region and the co-clustered trajectories can be released. In order to achieve space-time translation, the authors proposed W4M [2], which uses a different distance measure that allows time-warping.

In the second category, methodologies have been proposed for disclosure control in statistical databases [3]. These approaches protect sensitive information in a database while allowing statistical queries such as count and/or sum queries but no other information can be made available to the inquirer. According to the authors, addressing privacy violation problems can be classified into four main categories: (1) conceptual, (2) query restriction, (3) data perturbation, and (4) output perturbation. In the conceptual approach, two different data models are included. The conceptual model explores the privacy problem at the conceptual level while the lattice model comprises a framework for data represented in tabular form. Query restriction approach provides answer either by restricting the size of the set of the query or by controlling the overlap between successive queries. In the third category, attacks can be handled through data perturbation. Queries are answered according to a perturbed database. Essentially, a set of alteration / modification methodologies is used aiming for the best possible result w.r.t. privacy-preservation and data utility. Contrary, in the output perturbation approach the answer of the query is computed and then noise is added to the answer.

A privacy-aware query engine, as a protection mechanism, was first envisioned by Gkoulalas-Divanis and Verykios [8] (Fig. 11.1). The design principles of a query engine that protects users' privacy by generating fake trajectories are described. The idea behind that work is that malevolent users who query the trajectory database should not be able to discover (with high confidence) any real trajectories that are returned as part of the answer set of their queries, while they can use the returned

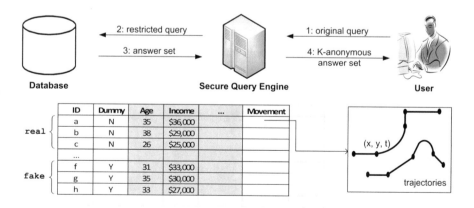

Fig. 11.1 A big picture of the system architecture [8]

data to support their analytic tasks. The engine allows subscribed end-users to gain restricted access to the trajectory data in order to perform various analysis tasks while preserving users' privacy from several types of attacks. It supports range, distance, k-nearest neighbor, landmark, route and queries for aggregative statistics for both trajectory (movement) and non-trajectory (relational) data. When a user poses a query, the engine retrieves the real trajectories r that belong to the answer set and combine them with $k - r$ fake trajectories in order to maintain k-anonymity principle by ensuring that the malevolent is not able to distinguish the real trajectories with high confidence. The necessary fake trajectories are generated based on an interpolation technique applied on pairs of real trajectories without though taking into consideration the time dimension.

Regarding the attacks that malevolent users may try to pursue in the original database, they are classified in three types:

- *User identification attack*: the identity of the user can be exposed by ad-hoc queries involving overlapping spatiotemporal regions.
- *Sensitive location tracking attack*: the malevolent user tries to map match one or more locations in a user trajectory to known locations that can effectively expose the identity of the user (e.g., the address of a house or a betting office). Such locations are called sensitive for the user as they should not be disclosed to the attackers.
- *Sequential tracking attack*: the user is tracked down through his trajectory by a set of focused queries on regions that are near to each other, in terms of space and time. The attacker can "follow" the user and learn the places that she has visited.

In the section that follows, we present Hermes++, a privacy-aware query engine, and we pay particular attention to the specific procedures it performs in order to block these types of attacks.

11.3 Hermes++ Query Engine

In this section, we present the architecture of Hermes++ query engine proposed by Pelekis et al. [22] and the algorithms that deliver its functionality. In particular, Sect. 11.3.1 provides details about Hermes++ architecture, Sect. 11.3.2 describes the algorithm that generates realistic fake trajectories, and Sect. 11.3.3 presents the auditing technique that is used to audit user queries and preserve the privacy in the answers to the queries.

11.3.1 Hermes++ Architecture

Hermes++ exploits on the trajectory storage functionality and the spatiotemporal query processing capabilities of Hermes for providing privacy-aware queries to

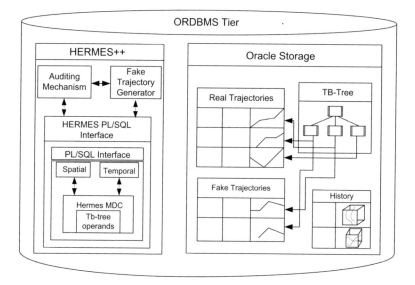

Fig. 11.2 The architecture of Hermes++ [22]

end-users. More specifically, Hermes defines a trajectory data type and a collection of operations as an Oracle data cartridge, which is further enhanced by the TB-tree access method [26] for efficient querying on trajectory data. Hermes++ directly utilizes this functionality at the ORDBMS level to store fake trajectories, as well as any historical information of all the users' queries (and the corresponding responses), in order to avoid different types of tracking attacks (e.g., sequential tracking). It succeeds so by the embedded auditing module, which invokes the Hermes queries and the fake trajectory generator algorithm. Since the entire framework is built at the ORDBMS level, end-users are also able to pose their queries through PL/SQL (i.e. not only via the GUI). As such, from an architectural point of view, Hermes++ acts as a wrapper over the Hermes query engine and not as a secure middleware. Figure 11.2 illustrates the Hermes++ architectural framework.

As observed in this architecture, the two key components of Hermes++ functionality are the fake trajectory generator and the auditing mechanism (see the top left part of the architecture). These components are crucial for Hermes++ performance and will be described in detail in the sections that follow.

11.3.2   Fake Trajectory Generation

The Fake Trajectory Generation algorithm, originally presented in [22], aims to produce trajectories that follow the trend of the input set of real trajectories, thus minimize the potential of privacy breaches when query results are released to the end-users. This algorithm plays a central role in the privacy-aware query mechanism. When a user poses a query to the database, the engine provides the answer only if at least L real user trajectories exist in the area. Lower bounding the number of users is a simple way to prevent answering queries whose original result set is very small (e.g., a range query in a region with very few trajectories), as in this case the generated fake trajectories may fail to capture the trend of the real trajectories. Prior to releasing any real trajectory, an approach is employed (see Sect. 11.3.3) to protect any sensitive locations in the trajectory that could be used by malevolent end-users to identify the corresponding user. To produce the answer set for the query, the engine generates N fake trajectories, where N is an owner-specified threshold. The algorithm has the ability to produce fake trajectories for different types of queries, such as range, nearest neighbor and distance queries, while it is used by the auditing mechanism (to be presented in Sect. 11.3.3) to handle different types of attacks from malevolent users.

The fake trajectory generation algorithm is based on the idea of the *Representative Trajectory Generation* (*RTG* for short) algorithm, introduced by Lee et al. [12]. The main idea of this algorithm is that the resulting representative trajectory describes the overall movement of a set of directed segments, produced after the partitioning of a set of trajectories. The partitioned trajectories (i.e., directed segments) are clustered according to a distance function taking into account the parallel, perpendicular and angle distance of the segments. The outcome of the *RTG* algorithm, applied on each cluster, produces a smooth (more or less) linear trajectory that best describes the corresponding cluster. However, the original *RTG* algorithm fails to consider the temporal dimension of the generated trajectory. Therefore, fake trajectory generation algorithm transforms the *RTG* output by appropriately integrating the time dimension into the fake trajectory generation process.

Algorithm 11.1 provides the details of the fake trajectory generation approach. The algorithm takes as an input a set of line segments S_i resulting from a set of trajectories which form the answer to a user query. In the first step (line 1), the representative trajectory is produced based on this set of line segments of trajectories. For simplicity, in Fig. 11.3 segments are depicted as consecutive parts

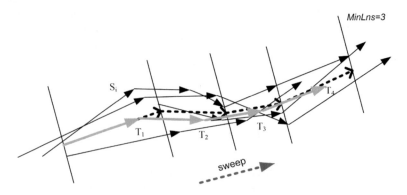

Fig. 11.3 Generating a fake trajectory over a set of line segments [22]

of trajectories; however, in the general case, they could be disconnected and independent segments that are filtered in a way that all move towards (more or less) the same direction. This is because *RTG* assumes that all segments follow the same directional pattern. Then, *RTG* sweeps a rotated vertical line according to the average direction vector towards the major axis, counting the number of line segments that are either the starting or the ending point of a line segment.

If the resulted number is equal to or greater than a threshold *MinLns*, the algorithm calculates the average coordinate of those points and assigns the average into the set of representative trajectory; otherwise, it proceeds to the next point. To avoid segments that are too close to each other, a smoothing parameter γ is utilized. The final outcome of this step is the trajectory with the dotted line shown in Fig. 11.3.

After calculating the representative trajectory, the algorithm inserts the time dimension to each line segment and performs additional computations to adjust it and make it more plausible. In detail, a realistic length and speed for the 3D segments of the fake trajectory are examined and required. In Fig. 11.3, the grey solid line depicts the final fake trajectory after assigning the time dimension to the segments and adjusting them to be more realistic. In order to achieve this, the algorithm takes as parameter the spatiotemporal *Minimum Bounding Box* (MBB), which is set by the auditing mechanism and may be either the MBB of the user's query parameter (in the case of range queries), or the MBB that is formed by the whole trajectories whose parts belong to the results of user's query. An additional set of input parameters that is provided by the auditing mechanism corresponds to statistical computations regarding d_{min}, d_{max}, l_{min}, l_{max}, which are the minimum and maximum trajectories' duration and segments' length, respectively, and $avgU_{min}$, $avgU_{max}$, l_{avg}, which are the average minimum and maximum speed, as well as, the average length of the segments, respectively. The *Timestep* parameter is the duration of a line segment and is considered to be constant indicating that the moving object transmits its location update at regular temporal intervals. The outcome of the algorithm is a set of line segments forming a trajectory, which are stored in the array *fake_trajectory*.

Having calculated the set of line segments, the algorithm computes the initial timestamp t_0 that the fake trajectory will start at (line 2). The initial timestamp is defined as: $t_0 = t_{MBBmin} + random(0, SP)$, where $SP = (t_{max} - t_{min}) - random(d_{min}, d_{max})$ corresponds to a value used to ensure that time t_0 of the first point of the fake trajectory will not be placed near t_{MBBmax}. Moreover, the maximum timestamp of the fake trajectory should not exceed t_{MBBmax}, otherwise it will differ from the real trajectories. In order to ensure this, the maximum timestamp t_{max} of the fake trajectory is calculated (line 3) as a function of the initial timestamp t_0 and the duration of the fake trajectory (i.e., $|fake\_trajectory| * T$). If ($t_{max} > t_{MBBmax}$) then a line simplification procedure is applied to reduce the number of line segments (lines 5–9). Douglas-Peucker algoriothm [6] compresses the generated segments by using a polyline representation and a parameter f that corresponds to a distance threshold, defined as a percentage of the trajectory's length (line 6).

The compression procedure is repeated until $t_{max} < t_{MBBmax}$ and at each iteration parameter f is halved.

Having calculated the initial timestamp, the algorithm adjusts the maximum length l_{max} of the segments that have been generated (lines 10–13) in order to manipulate long segments that will lead to the generation of non-realistic fake trajectories. Specifically, if l_{max} is greater than twice the average length l_{avg}, then l_{max} is being recalculated as a random value between l_{avg} and the twice of l_{avg}. Otherwise, the algorithm sets l_{max} randomly between l_{avg} and l_{max}. Then, the algorithm enters a loop (line 14) and assigns the time dimension to each line segment of the fake trajectory. The initial timestamp t_0 of the first line segment has been calculated in previous steps. The timestamp of the ending point of this segment equals to t_0 increased by the sampling rate's duration, i.e., is equal to $t_0 + Timestep$. The ending timestamp of the initial segment will be the starting timestamp of the next segment. Generally, for each line segment it holds that $t_{i+1} = t_i + Timestep$, where $0 \leq i < |fake\_trajectory|$.

Algorithm 11.1 Fake trajectory generation

---

function FAKE-GEN(line segments S_i, minimum number of points $MinLns$, smoothing parameter γ, time step of sampling rate $Timestep$, MBB(t_{MBBmin}, t_{MBBmax}), d_{min}, d_{max}, l_{min}, l_{max}, l_{avg}, $avgU_{min}$, $avgU_{max}$

1:    $fake\_trajectory \leftarrow RTG\ (S_i, MinLns, \gamma)$
2:    calculate initial timestamp t_0 of the fake trajectory
3:    $t_{max} \leftarrow t_0 + |fake\_trajectory| * Timestep$
4:    **if** $t_{max} > t_{MBBmax}$ **then**
5:        **repeat**
6          $Douglas\_Peucker(fake\_trajectory, f)$
7:            $f \leftarrow f/2$
8:            $t_{max} \leftarrow t_0 + |fake\_trajectory| * Timestep$
9:        **until** $t_{max} < t_{MBBmax}$
10:   **if** $l_{max} > 2 * l_{avg}$ **then**
11:       $l_{max} \leftarrow random\ (l_{avg}, 2 * l_{avg})$
12:   **else**
13:       $l_{max} \leftarrow l_{avg} * random\ (1, l_{max}/l_{avg})$
14:   **for each** $p_i \in fake\_trajectory$ **do**
15:       set timestamps of the initial and final point of p_i
16:       calculate speed U_i of p_i
17:       **if** $U_i < avgU_{min}$ or $U_i > avgU_{max}$ **then**
18:           **repeat**
19:               $l \leftarrow random\ (l_{min}, l_{max})$
20:               calculate new speed U_i of p_i
21:           **until** $U_{min} < U_i < U_{max}$
22:       calculate angle φ_i
23:       define coords of new ending point based on l
24:       map match $fake\_trajectory$
25:   **return** $fake\_trajectory$

After assigning the time dimension to the current segment p_i (line 15), the algorithm proceeds to calculate the speed U_i for each segment p_i (line 16) and checks if it lies within $avgU_{min}$ and $avgU_{max}$ (lines 17–21). If it is outside this range, the algorithm calculates a random segment length l, between l_{min} and l_{max}, such that the speed U_i of the specific segment is within the limits. As a final step, the coordinates of the new ending point are identified based on the length of segment l that was calculated before (lines 22–23).

Depending on the direction of the segment and its angle ϕ_i with x-axis, the fake trajectory generation algorithm calculates the new coordinates (x_{t+1}, y_{t+1}), according to the following formulas (l is the length of the line segment):

$$\phi_i = \arctan 2(y_{t+1} - y_t, x_{t+1} - x_t)$$

$$x_{t+1} = x_t + l * \cos(\phi), \ y_{t+1} = y_t + l * \sin(\phi)$$

In the case that trajectory data are related to an underlying road network, the fake trajectory generation algorithm map-matches the generated fake trajectory with the specific road network (line 24) by employing a state-of-the-art map-matching algorithm [5]. This functionality of the algorithm can lead to a more realistic representation of the fake trajectory. After calculating the new coordinates, the algorithm proceeds to the next segment and the procedure continues until all line segments are examined. Finally, the generated fake trajectory is returned (line 25).

11.3.3 Query Auditing

The main goal of Hermes++ query engine is to prevent the potential attacks that may occur while a malevolent user query the database. User identification attack is possible when the query engine answers a query involving a spatial (or spatiotemporal) region and then another, more specific query, involving part of this region. In this case, the attacker can breach the enforced privacy model by identifying the differences between the created fake trajectories which, in turn, increases her confidence regarding information about the corresponding real trajectories. To block this type of attack, Hermes++ uses auditing to track the queries initiated by each end-user in the system and denies answering overlapping queries.

Sensitive location tracking attack allows malevolent users to learn sensitive locations that real users have visited, and (possibly) reveal the identity of these users. To block these attacks, Hermes++ protects the starting and the ending location of trajectories, as well as any other (owner-specified) location in the course of the user trajectory that can be considered as sensitive for the user. As an example of this type of attack, assume a query that involves region Q_4, illustrated in Fig. 11.4. Since in this region the trajectory has its end point to a sensitive location, the attacker can map-match this location and reveal the user's identity. The attack can succeed even if fake trajectories are generated in this region by collecting more precise information about the real trajectories on every focused query, which in turn increases her confidence. To block the sensitive location tracking attack, the auditing approach

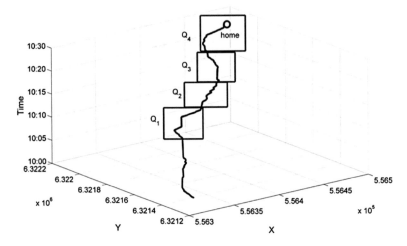

Fig. 11.4 Sensitive location tracking and sequential tracking attacks to user privacy [22]

identifies sensitive locations of trajectories that appear in the query window and proceeds to dislocate them so that the sensitive location is not disclosed.

Finally, in the sequential tracking attack an attacker attempts to "follow" a user trajectory in the system by using a set of focused queries involving spatiotemporal regions that are adjacent to each other. To block this attack, the auditing algorithm takes the necessary measures to smoothly continue the movement of fake trajectories from neighboring regions (returned as part of previous queries of the user) to the current region.

The query auditing approach for shielding the database against malevolent users (to be presented in Algorithm 11.3) is based on the *Hide Sensitive Location Algorithm*, originally presented in [22], that is discussed first. This algorithm (listed in Algorithm 11.2) takes as input a set of sensitive locations *SL*, a set of trajectories *T* and the *MBB* formed by user's query. Initially, the algorithm selects all sensitive locations SL' that lie inside the *MBB* (line 1). For each trajectory of the given set *T*, it defines those sensitive locations, SL'_i, that correspond to the current trajectory (lines 2–3). For every sensitive location, $SL'_{i,j}$, it examines if fake sub-trajectories that hide the sensitive locations have been previously computed for this trajectory and retrieves them from History (lines 4–6). Otherwise, it computes a new synthetic (fake) trajectory that is then stored for future reference (lines 8–13).

Algorithm 11.2 produces fake (synthetic) sub-trajectories by applying a variant of the GSTD trajectory synthesizer, called GSTD*, proposed by Pelekis et al. [24]. GSTD* produces trajectories following complex mobility patterns based on a given distribution of spatiotemporal focal points, to be visited by each trajectory in a specific order. The general idea behind GSTD* is to use the focal points so as to attract each trajectory's movement. When a particular trajectory has reached the area around a focal point, having at the same time completed the respective temporal predicate, the generation algorithm changes the attracting point to the next focal point in the list, and so on, until no focal points are left unvisited.

Fig. 11.5 Protecting
sensitive locations of user
trajectories [22]

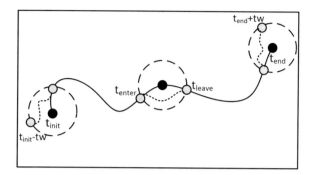

The idea of hiding sensitive locations of a trajectory by misplacing its route
is illustrated in Fig. 11.5. The algorithm discovers the intersection points of the
trajectory with a circle that is formed around a sensitive location by taking as radius
the distance between the sensitive location from a point where the object would have
been moved after a certain period of time *tw* (i.e., *tw* is a temporal window), if it was
moving with its current speed. The idea is to use these intersection points as the focal
points in GSTD* (line 8) (see the filled gray circles in Fig. 11.5). If the number of
focal points is greater than two (i.e. the object enters and/or leaves the circle more
than two times), the algorithm utilizes the first (entering) and the last (leaving) one.
In case where the sensitive location is either the initial or the ending point causing
the creation of only one focal point, the algorithm randomly selects another random
focal point in the perimeter of the circle (lines 9–10). After determining focal points
it produces a synthetic (fake) trajectory by applying GSTD* between the two chosen
focal points as illustrated in the figure with the dotted line (line 11). The algorithm
returns the set of trajectories that does not any longer contain sensitive locations
(line 14).

Algorithm 11.2 Hide sensitive locations

---

function HideSensitiveLocations(set of sensitive locations *SL*, set of trajectories *T*, user's query
MBB, temporal window *tw*)
1:   $SL' \leftarrow SL$ inside *MBB*
2:   **for each** ($T_i \in T$) **do**
3:      $SL_i' \leftarrow$ select the subset of SL' that correspond to T_i
4:      **for each** (sensitive location of T_i, $SL_{i,j}' \in SL_i'$) **do**
5:         **if** (fake sub-trajectory computed in the past for this $SL_{i,j}'$) **then**
6:            Retrieve the fake sub-trajectory from *History* and update T_i
7:         **else**
8:            $focal_{points} \leftarrow$ Intersection(T_i, buffer($SL_{i,j}'$, tw))
9:         **if** ($|focal_{points}| = 1$) **then**
10:            $focal_{points} \leftarrow$ AddRandomPointOnSurface(buffer($SL_{i,j}'$, tw))
11:            Produce a fake trajectory by applying *GSTD** on $focal_{points}$
12:            Update the part of T_i with the fake sub-trajectory
13:            UpdateHistory
14:   **return** (T)

Algorithm 11.3, originally presented in [22], describes the query auditing mechanism. When a new query is submitted to the engine, the auditing algorithm first examines if this query involves an area that (partially) overlaps with that of a previous query, submitted by the same end-user. If this is the case, then it denies serving the query (lines 1–2) to block a potential user identification attack. If the previous test is negative, the auditing mechanism executes the actual query of the user and retrieves the result set (line 3). In order to prohibit the identification of an individual by an adversary that is able to link sensitive locations that are visited by a user (e.g., the home of the user) with trajectories that belong to the specific query, the *Hide Sensitive Location Algorithm presented* earlier is invoked (line 4).

Having protected the sensitive locations of the trajectories in the querying region, Algorithm 11.3 commands the generation of the necessary fake trajectories for this region (lines 11–21). To generate the requested number of fake trajectories, the algorithm calculates a set of basic statistics (line 11) that are needed by the fake trajectory generation approach (Algorithm 11.1), while trying to find trajectories that follow more or less the same direction in the query region (lines 12–20). Specifically, a step dir_{step} (in degrees) is randomly selected (line 13) in the range of $(0, dir_{step_{max}})$, with $dir_{step_{max}}$ being an input parameter that defines the size of an angular range used to divide the Cartesian plane. As illustrated in Fig. 11.6, the algorithm selects those segments from the real trajectories that belong to the range (dir_{min}, dir_{max}) (see the solid lines in the figure), which are set by randomly assigning dir_{min} and then setting dir_{max} equal to $dir_{min} + dir_{step}$. Subsequently, it calls Algorithm 11.1 on these segments and passes the query window to create one new fake trajectory. The same process is repeated for the next range of directions, which leads to the generation of another fake trajectory, until the 360° range is exceeded. Then, the algorithm selects a new dir_{step} and repeats the same process, until the requested number of fake trajectories is generated (line 21). Note that the filtering approach on the directional property of the segments guarantees that the fake generation algorithm will produce nice representative trajectories of the query result, as it acts as a simple clustering methodology on the overall set of available segments.

After generating the fake trajectories, Algorithm 11.3 takes the necessary measures to protect the privacy of the users whose movement is depicted in the

Fig. 11.6 Selecting segments
from real trajectories [22]

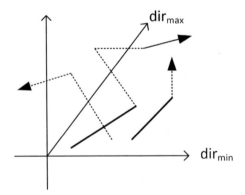

query window by smoothly continuing the movement of the fake trajectories from neighboring regions, returned as part of previous queries posed by the end-user, to the current one. Specifically, the algorithm examines if the query posed by the end-user has a nearby query made by the same end-user in the past, which does not exceed a spatial s_{thr} and a temporal t_{thr} threshold. In case that the query has only one such neighbor, the algorithm performs a one-by-one matching (line 23) between the fake trajectories of MBB and MBB_{hist} (i.e. the nearby query saved in $History$). In detail, it first finds the MBB with the minimum number of fake trajectories and then it randomly matches each one of them with fakes from the other query, by producing pairs P_i of fake trajectories. For each pair, it examines if MBB touches MBB_{hist} or if they are apart. In the first case, illustrated in Fig. 11.7a, a space time translation is performed to connect the two fake trajectories. The fake trajectory is transferred in the x and y axes, if necessary.

Algorithm 11.3 Query auditing algorithm

---

function TrajAuditor (user's query MBB, number of generated fake trajectories N, lower bound threshold L, spatial threshold s_{thr}, temporal threshold t_{thr}, maximum direction step $dir_{step_{max}}$, set of sensitive locations SL, temporal window tw, $MinIns$, γ, $Timestep$)

1: **if** CheckHistory(user posed in the past an overlapping query w.r.t. MBB) = true **then**
2:      **Privacy threat:** Overlapping queries
3:      $TR \leftarrow$ SpatioTemporalRangeQuery(MBB)
4:      $TR \leftarrow$ HideSensitiveLocations(SL, TR, MBB, tw)
5: **if** (CheckHistory(user posed in the past a nearby query w.r.t. s_{thr}, t_{thr}) = true) **then**
6:      **Privacy threat:** Sequential tracking attack
7: **else**
8:      **if** $|TR| \leq L$) **then**
9:          **Privacy threat:** Lower bound threshold violation
10:         **else**
11:             CalculateStatistics (d_{min}, d_{max}, l_{min}, l_{max}, l_{avg}, $avgU_{min}$, $avgU_{max}$)
12:             **repeat**
13:                 $dir_{step} \leftarrow random(0, dir_{step_{max}})$
14:                 $dir_{min} = random(0, 360)$; $dir_{max} = dir_{min} + dir_{step}$
15:                 **repeat**
16:                     $S_i \leftarrow$ FilterbyDirection(dir_{min}, dir_{max}, TR)
17:                     $FT \leftarrow FT \cup$ Fake_Gen(S_i, $MinIns$, γ, $Timestep$, MBB, $Statistics$)
18:                     $dir_{min} \leftarrow dir_{min} + dir_{step}$
19:                     $dir_{max} \leftarrow dir_{max} + dir_{step}$
20:                 **until** $dir_{max} > 360$
21:             **until** $|FT| = N$
22:         Retrieve from $History$ all fakes FT_{hist} from a nearby query of the user w.r.t. s_{thr}, t_{thr}
23:         $P_{match} \leftarrow$ MinRandomMax(FT, FT_{hist})
24:     **for each pair** $P_i(T_j, T_k) \in P_{match}$ **do**
25:         **if** (MBB touches a historic query of the user) **then**
26:             SpaceTimeTranslation(P_i)
27:         **else**
28:             $focal_{points} \leftarrow (T_{j_{end}}, T_{k_{start}})$
29:             GSTD* ($focal_{points}$)
30:         Update in FT the fake trajectory that corresponds to P_i
31:     $FT \leftarrow$ HideSensitiveLocations(SL, FT, MBB, tw)
32:     UpdateHistory
33:     **return** ($TR \cup FT$)

Fig. 11.7 Prohibiting
sequential tracking: (**a**) case I,
(**b**) case II [22]

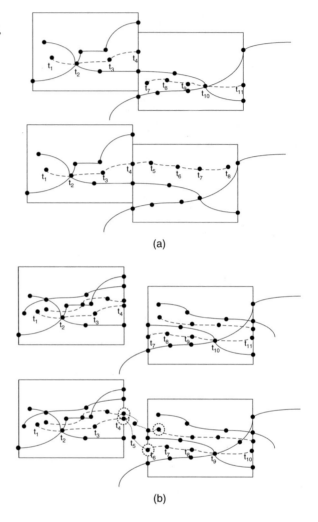

Then, the algorithm checks the time dimension to assure that there is no temporal
gap. If such a gap exists, the algorithm recalculates the timestamp of each point of
the fake trajectory. In the second case (illustrated in Fig. 11.7b), where a spatial
and/or a temporal gap exists between *MBB* and *MBB$_{hist}$*, Algorithm 11.3 generates
a connection-trajectory (see the dotted lines) between them using GSTD*. Focal
points are the ending point of the one trajectory with the starting point of its
matching trajectory in P_i. After generating the fake trajectories, the algorithm
applies the hiding process of the sensitive locations also for these trajectories
(line 31), to conceal the fact that they are fakes.

As a last remark, *TrajAuditor* (Algorithm 11.3) commands the generation of the
necessary number of fake trajectories based on the parts of the real trajectories
that appear inside the query window. An alternative approach (henceforth called

TrajFaker) would be to generate wide fake trajectories that exceed the limits of the window the user submitted. In this case, auditing would still be applicable but not forced, contrary to the case of *TrajAuditor*. *TrajFaker* differs from *TrajAuditor* in the following steps. When a user executes a query, *TrajFaker* finds the trajectories that are contained in the specific spatiotemporal window (or the k nearest neighbors in case of $k - NN$ queries) and then retrieves the whole trajectories and not the parts of them that lie inside the window. Subsequently, it generates fake trajectories by employing Algorithm 11.1 on the whole trajectories. Each generated fake trajectory is examined to see whether it crosses the spatiotemporal window of the query and, if so, it is included to the returning set. Otherwise, the trajectory is discarded and the same process is repeated. All generated fake trajectories are stored in order to participate to the generation of other fake trajectories. Finally, there are no privacy threats with respect to sequential tracking as before, since the generated fake trajectories are based on the whole trajectories and not parts of them, and are stored. If an adversary tries to execute overlapping or sequential queries, the fakes will appear in all of these queries' answers.

11.4 Private-Hermes Benchmark Framework

Building on top of Hermes++, Private-Hermes, developed by Pelekis et al. [23], integrates algorithms that enable the privacy-aware publishing of personal mobility data under a common, benchmark-oriented framework and gives the ability to users to evaluate the utility either of the fake or the sanitized trajectories via a variety of well-known mobility data mining algorithms, i.e. various types of clustering, frequent sequential patterns, etc. The idea is that by adding fake trajectories (that affect the cardinality of the MOD), as well as perturbating original ones (that affects the shape of the MOD) should not destroy the patterns hidden in the original MOD. Such an evaluation can be done by using clustering and frequent pattern mining techniques, appropriate for mobility data. Private-Hermes incorporates the following state-of-the-art algorithms:

- *Clustering*: Private-Hermes supports TRACLUS [12], T-Optics [14], and CenTR-I-FCM [24]. Two traditional clustering techniques, namely K-medoids [11] and Bisecting K-medoids [28], are also included with the special feature that the user can choose different distance functions between the trajectories (i.e. grouping only by their starting or destination point, without taking into account the whole route) [20].
- *Trajectory representatives*: related to cluster analysis, a useful requirement is to extract a compact representation of a set of trajectories (e.g. a cluster found through cluster analysis), in terms of "representative" trajectory. To this end, Private-Hermes supports CenTra "centroid" trajectories [24] and TRACLUS "typical" trajectories [12].

- *Frequent pattern mining*: Private-Hermes incorporates the T-pattern mining technique [7], which models sequences of visited regions, frequently visited in the specified order with similar transition times, out of trajectory databases.
- *Sampling*: Private-Hermes supports a state-of-the-art trajectory sampling technique proposed in [25].
- *Trajectory anonymization*: Private-Hermes incorporates NWA [1] and W4M [2].

The above-presented functionality is integrated in the Hermes MOD engine [21] by appropriately extending the query language with new constructs, in a fashion origi-nally proposed by Ortale et al. [19]. This allows users to progressively analyze the MOD and interchange between querying and mining operations. In detail, Pivate-Hermes users are given the ability to perform:

- *Querying and mining operations on Hermes*: the platform is capable of executing range and $k - NN$ queries on Hermes as well as mining operations using the algorithms listed above. Queries and mining operations are posed via Private-Hermes GUI, which provides essential capabilities, including query predicate selection, parameters selection and results projection. Graphical map user-interaction for predicate definition is also supported.
- *Privacy-aware querying on Hermes++*: users are able to run range and $k - NN$ queries enabling Hermes++, which protects from privacy attacks. The data owner requires that at least a certain number of trajectories are returned to the end-users in response to their queries, for all different types of supported queries. The result consists of a set of carefully crafted, realistic fake trajectories aiming to preserve the trend of the original user trajectories.
- *Comparison/evaluation of anonymization algorithms*: as already mentioned, Private-Hermes integrates NWA and W4M anonymization algorithms. Both algorithms take as input trajectories which may have been extracted from a query posed to Hermes, and transform them into anonymous equivalents, subsequently stored in the MOD. An advantage of the platform is its ability to design and execute benchmarks that evaluate the results from the application of anonymization algorithms regarding the distortion over real user trajectories. The incorporated data mining techniques can be applied, and patterns steaming from original data with patterns resulting from anonymized data can be compared. This can be achieved by executing queries in the original and the anonymized data (or patterns), and comparing the results.
- *Profiling end-user's behavior to identify malevolent users*: The platform supports query auditing techniques [8], which can be used to monitor the behavior of the end-users and build user profiles. These user profiles can be subsequently analyzed by the data owner, as explained in [8], to help her identify suspicious behavior of end-users in the system.

Figures 11.8 and 11.9 illustrate representative snapshots of Private-Hermes GUI. More specifically, in Fig. 11.8a, a dataset has been extracted using a range query, while in Fig. 11.8b the dataset has been anonymized using NWA [1]. From these

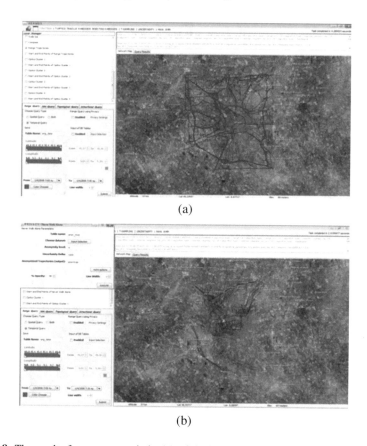

Fig. 11.8 The result of a range query in its (**a**) original vs. (**b**) NWA anonymized version [23]

outputs, a user can compare the distortion that has been caused to the dataset due to the anonymization algorithm. As a progressive analysis, Fig. 11.9a illustrates the result from the application of T-Optics [14] clustering on the original dataset (i.e. the one illustrated in Fig. 11.8a) in comparison with Fig. 11.9b, which presents the respective result when T-Optics is applied on the anonymized dataset (i.e. the one illustrated in Fig. 11.8b).

As for the technicalities of Private-Hermes components, illustrated in Fig. 11.10, the user interacts with a GUI with 3D rendering capabilities developed in Java and based on the Swing GUI widget toolkit [18]. The results from the operations that the program supports are visualized in the 3D globe provided by NASA World Wind [15]. To draw the charts reporting performance results, the JFreeChart library is

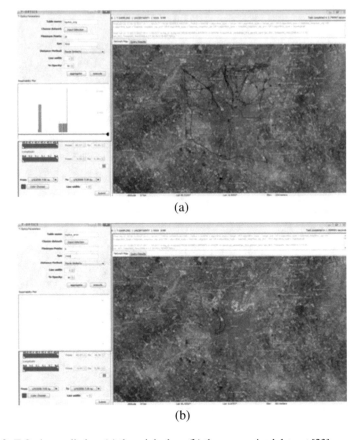

(a)

(b)

Fig. 11.9 T-Optics applied on (**a**) the original vs. (**b**) the anonymized dataset [23]

used [17]. Every component and library used during the development process is open source. Through the provided GUI, the user is able to setup his/her benchmark or, more generally, his/her analysis scenario. Private-Hermes retrieves the necessary data by calling the Hermes MOD engine.

The supported mobility data mining and anonymization algorithms have been incorporated as modules of the extensible DAEDALUS's MO-DMQL [19], while both of these sets of algorithms exchange data (i.e. real/fake/anonymized trajectories and mining models) directly with the database layer.

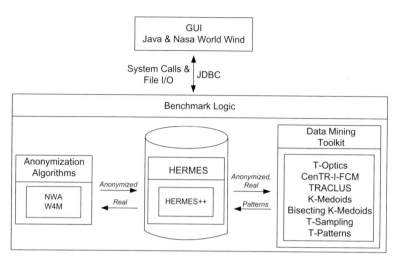

Fig. 11.10 Private-Hermes architecture [23]

11.5 HipStream: A Privacy-Preserving System for Managing Mobility Data Streams

The ability to build database systems able to connect individuals' information across different data repositories turns out to be simpler since individual information become more ubiquitous. Consequently, the privacy provisions incorporated into data collections and the privacy regulations that shield personal data, are debilitated. Data owners have no guarantee that the data that are donating are not misused for the sake of knowledge extraction since they have no control whether the privacy policies are enforced or not.

In order to deal with the lack of systemic control over the data use, the concept of Hippocratic management system was introduced by Agrawal et al. [4] to guarantee privacy and security of information they manage as a founding principle. Hippocratic databases extend the functionalities of traditional databases with privacy-preserving capabilities. The goal is to prevent disclosure of private information by placing data donor privacy as a main concern throughout data collection and management. Ten fundamental principles have been proposed that guide the behavior of Hippocratic stream data management:

- *Purpose Specification:* description of the purpose for which the data is collected needs to be collected and associated with the data itself.
- *Consent:* the purpose for which the data is collected has the consent of the user.
- *Limited collection:* collect the minimum amount of data from a user that satisfies the user's specified purposes
- *Limited use:* the purposes of the collected data should not be violated by operations carried out.
- *Limited Disclosure:* no personal information should be released to third parties without data owner's permission.

- *Limited Retention:* after the purpose of data collection is satisfied, user's data should be directly deleted.
- *Accuracy:* the information stored in the database system should be accurate and up-to-date.
- *Safety:* the adoption of security measures for protecting sensitive data from various types of attack.
- *Openness:* the individual whose data are recorded is allowed to access all information that is stored in the database and is related to herself.
- *Compliance:* data owners are able to validate that the privacy principles are conformed.

Wu et al. [31] developed a data management system, the so-called HipStream, which implements some of the aforementioned Hippocratic principles such as limited collection, limited use and limited disclosure. Data streams are collected and dropped dynamically in a system according to the data owner's policy. When data tuples arrive, the system is responsible to decide whether the data should be collected to serve the query or stored for analysis purposes in order to achieve limited collection. Controlling the access to the data w.r.t. data provider's preferences leads to limited disclosure. HipStream is able to preserve the privacy of the data streams that are shared between data providers and data users. The system guarantees not only that data providers' defined privacy specifications are enforced but also that the access to the data is limited. The idea behind HipStream is that service providers are allowed to access part of the data streams which are entirely controlled by data providers.

The architecture of HipStream is illustrated in Fig. 11.11. The basic components of the system are Security Manager, Privacy Controller, Query Management and Stream Manager. Through Web interface each end user has the ability to generate, retrieve and manage protected stream data.

A data owner is first registered to the system and specifies her privacy preferences such as who, for what purpose, under what conditions and which parts. Privacy policies are designed based on modelling users, data to be accessed and data accessing purposes with hierarchical categories. The preferences are then registered into the Policy Controller which is responsible for maintaining the privacy.

Prior to the registration of a data stream to the system, the Stream Registration acquires the purpose of the stream data and the consent of the stream owner. The Stream Manager receives the input stream and prepares it for further processing inside the system.

On the other hand, service providers may pose queries directly to the system while defining at the same time the query purpose. The query is forwarded through query getaway to the query rewriter. Query rewriter is then responsible to examine if the privacy preferences are satisfied. In case where the policies are not met, the query is rewritten. The query is dropped if its purpose is not in line with the authorised purpose on using the data. At this level, limited disclosure and limited use principles are enforced.

Next, the query is forwarded to the query manager and the stream filter. The stream filter is in charge of enforcing limited collection principle. The records that

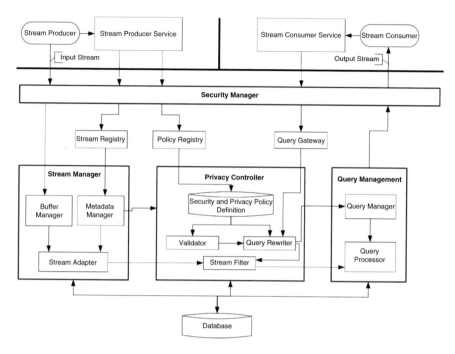

Fig. 11.11 HipStream architecture [31]

are participating to the answer set are maintained while the others are dropped. Moreover, the attributes that are not asked by any query are anonymized (i.e. replaced by null). Query processor executes the registered continuous queries and streams the result out to the query owner.

11.6 Conclusions

In this chapter, we presented techniques and Mobility Data Management Systems that have been proposed in the literature able to preserve the privacy of the users whose data are kept to the hosting organisation for analysis purposes. Hermes++ is a privacy-aware query engine that enables the remote analysis of user mobility data, supports a variety of popular spatial and spatiotemporal queries and uses auditing and fake trajectory generation techniques to identify and block, respectively, potential attacks to user privacy. On top of Hermes++, Private-Hermes is an integrated platform for applying data mining and privacy-preserving querying over mobility data. Finally, Hipstream, a data stream management system aiming at preserving users' privacy by enforcing Hippocratic principles was presented. Limited collection, limited use and limited disclosure of data are the main privacy requirements that the system implements.

References

1. Abul, O., Bonchi, F., and Nanni, M. (2008) Never walk alone: Uncertainty for anonymity in moving objects databases. In Proceedings of ICDE, pages 376–385.
2. Abul, O., Bonchi, F., and Nanni, M. (2010) Anonymization of moving objects databases by clustering and perturbation. Information Systems, 35(8), pages 884–910.
3. Adam, N.R. and Worthmann, J. C. (1989) Security-control methods for statistical databases: A comparative study. ACM Computing Surveys, 21(4), pages 515–556.
4. Agrawal, R., Kiernan, J., Srikant, R., & Xu, Y. (2002) Hippocratic databases. In Proceedings of VLDB Endowment, pages 143–154
5. Brakatsoulas, S., Pfoser, D., Salas, R. and Wenk, C. (2005) On map-matching vehicle tracking data. In Proceedings of VLDB Endowment, pages 853–864.
6. Douglas, D. and Peucker, T. (1973) Algorithms for the reduction of the number of points required to represent a digitized line or its caricature. Cartographica: The International Journal for Geographic Information and Geovisualization, 10(2), pages 112–122.
7. Giannotti, F., Nanni, M., Pedreschi, D., and Pinelli, F. (2007) Trajectory Pattern Mining. In Proceedings of SIGKDD, pages 330–339.
8. Gkoulalas-Divanis, A. and Verykios, V. S. (2008) A privacy–aware trajectory tracking query engine. ACM SIGKDD Explorations Newsletter, 10(1), pages 40–49.
9. Hoh, B. and Gruteser, M. (2005) Protecting location privacy through path confusion. In SECURECOMM, pages 194–205.
10. Hoh, B., Gruteser, M., Xiong, H. and Alrabady, A. (2007) Preserving privacy in GPS traces via uncertainty-aware path cloaking. In Proceedings of CCS, pages 161–171.
11. Kaufman, L., Rousseeuw, P. J. (1990) Finding Groups in Data: An Introduction to Cluster Analysis. Wiley, NY, Vol. 334.
12. Lee, J. G., Han, J., and Whang, K. Y. (2007) Trajectory clustering: a partition-and-group framework. In Proceedings of SIGMOD, pages 593–604.
13. LeFevre, K., DeWitt, D. and Ramakrishnan, R. (2006) Mondrian multidimensional k-anonymity. In Proceedings of ICDE, page 25.
14. Nanni, M. and Pedreschi, D. (2006) Time-focused clustering of trajectories of moving objects. Journal of Intelligent Information Systems, 27(3), pages 267–289.
15. NASA, World Wind Java SDK. URL: http://worldwind.arc.nasa.gov/java. (accessed: 6 Oct. 2011).
16. Nergiz, M. E., Atzori, M. and Saygin, Y. (2008) Towards trajectory anonymization: A generalization-based approach. In Proceedings of the SIGSPATIAL, pages 52–61.
17. Object Refinery, the JFreeChart project. URL: http://www.jfree.org/jfreechart. (accessed: 6 Oct. 2011)
18. Oracle, The Swing Tutorial. URL: http://download.oracle.com/javase/tutorial/uiswing. (accessed: 6 Oct. 2011).
19. Ortale, R., Ritacco, E., Pelekis, N., Trasarti, R., Costa, G., Giannotti, F., Manco, G., Renso, C., and Theodoridis, Y. (2008) The DAEDALUS Framework: Progressive Querying and Mining of Movement Data. In Proceedings of ACM SIGSPATIAL, page 52.
20. Pelekis, N., Andrienko, G., Andrienko, N., Kopanakis, I., Marketos, G., and Theodoridis, Y. (2011) Visually Exploring Movement Data via Similarity-based Analysis. Journal of Intelligent Information Systems, 38(2), pages 343–391.
21. Pelekis, N., Frentzos, E., Giatrakos, N., and Theodoridis, Y. (2008) Hermes: Aggregative LBS via a trajectory DB engine. In Proceedings of SIGMOD, pages 1255–1258.
22. Pelekis, N., Gkoulalas-Divanis, A., Vodas, M., Kopanaki, D., and Theodoridis, Y. (2011). Privacy-aware querying over sensitive trajectory data. In Proceedings of CIKM, pages 895–904.
23. Pelekis, N., Gkoulalas-Divanis, A., Vodas, M., Plemenos, A., Kopanaki, D., and Theodoridis, Y. (2012) Private-Hermes: A Benchmark Framework for Privacy-Preserving Mobility Data Querying and Mining Methods. In Proceedings of EDBT, pages 598–601.

24. Pelekis, N., Kopanakis, I., Kotsifakos, E., Frentzos, E. and Theodoridis, Y. (2011) Clustering uncertain trajectories. Knowledge and Information Systems, 28(1), pages 117–147.
25. Pelekis, N., Panagiotakis, C., Kopanakis, I., and Theodoridis, Y. (2010) Unsupervised trajectory sampling. In Proceedings of ECML PKDD, pages 17–33.
26. Pfoser, D., Jensen, C. S., and Theodoridis, Y. (2000) Novel approaches to the indexing of moving object trajectories. In Proceedings of VLDB, pages 395–406.
27. Samarati, P. (2001) Protecting respondents' identities in microdata release. Transactions on Knowledge and Data Engineering, 13(6), pages 1010–1027.
28. Steinbach, M., Karypis, G., Kumar, V. (2000). A comparison of document clustering techniques. In Proceedings of KDD Workshop on Text Mining, 400(1), pages 525–526.
29. Sweeney, L. (2002) k-anonymity: A model for protecting privacy. International Journal on Uncertainty, Fuzziness and Knowledge Based Systems, 10(5), pages 557–570.
30. Terrovitis, M. and Mamoulis, N. (2008) Privacy preservation in the publication of trajectories. In Proceedings of Mobile Data Management, pages 65–72.
31. Wu, H., Xiang, S., Ng, W. S., Wu, W., & Xue, M. (2014) HipStream: A Privacy-Preserving System for Managing Mobility Data Streams. In Proceedings of Mobile Data Management, Vol. 1, pages 360–363.

Chapter 12
Privacy-Preserving Release of Spatio-Temporal Density

Gergely Acs, Gergely Biczók, and Claude Castelluccia

Abstract In today's digital society, increasing amounts of contextually rich spatio-temporal information are collected and used, e.g., for knowledge-based decision making, research purposes, optimizing operational phases of city management, planning infrastructure networks, or developing timetables for public transportation with an increasingly autonomous vehicle fleet. At the same time, however, publishing or sharing spatio-temporal data, even in aggregated form, is not always viable owing to the danger of violating individuals' privacy, along with the related legal and ethical repercussions. In this chapter, we review some fundamental approaches for anonymizing and releasing spatio-temporal density, i.e., the number of individuals visiting a given set of locations as a function of time. These approaches follow different privacy models providing different privacy guarantees as well as accuracy of the released anonymized data. We demonstrate some sanitization (anonymization) techniques with provable privacy guarantees by releasing the spatio-temporal density of Paris, in France. We conclude that, in order to achieve meaningful accuracy, the sanitization process has to be carefully customized to the application and public characteristics of the spatio-temporal data.

12.1 Introduction

Spatio-temporal, geo-referenced datasets are growing rapidly nowadays. With billions of location-aware devices in use worldwide, the large scale collection of space-time trajectories of people produces gigantic mobility datasets. Such datasets are invaluable for traffic and sustainable mobility management, or studying

G. Acs (✉) · G. Biczók
CrySyS Lab, Department of Networked Systems and Services, Budapest University
of Technology and Economics (BME-HIT), Budapest, Hungary
e-mail: acs@crysys.hu; biczok@crysys.hu

C. Castelluccia
INRIA, Montbonnot-Saint-Martin, France
e-mail: claude.castelluccia@inria.fr

© Springer Nature Switzerland AG 2018
A. Gkoulalas-Divanis, C. Bettini (eds.), *Handbook of Mobile Data Privacy*,
https://doi.org/10.1007/978-3-319-98161-1_12

accessibility to services. Even more, they can help understand complex processes, such as the spread of viruses or how people exchange information, interact, and develop social interactions. While the benefits provided by these datasets are indisputable, their publishing or sharing is not always viable owing to the danger of violating individuals' privacy, along with the related legal and ethical repercussions. This problem is socially relevant: companies and researchers are reluctant to publish any mobility data by fear of being held responsible for potential privacy breaches. This limits our ability to analyze such large datasets to derive information that could benefit the general public.

Unsurprisingly, personal mobility data reveals tremendous sensitive information about individuals' behavioural patterns such as health life or religious/political beliefs. Somewhat more surprisingly, such mobility data is also unique to individuals even in a relatively large population containing millions of users. For instance, only four spatio-temporal positions are enough to uniquely identify a user 95% of the times in a dataset of one and a half million users [13], even if the dataset is *pseudonymized*, i.e., identitifiers such as personal names, phone numbers, home address are suppressed. Moreover, the top 2 mostly visited locations of an individual is still unique with a probability of 10–50% [63] among millions of users. Notice that the most visited locations, such as home and working places, are easy to learn today from different social media where people often publicly reveal this seemingly harmless personal information. Therefore, publishing mobility datasets would put at risk our own privacy; if someone knows where we live and work could potentially find our record and learn all of our potentially sensitive location visits. Moreover, due to the large uniqueness of records, these datasets are regarded as personal information under several laws and regulations internationally, such as overall in the European Union. Therefore, their release prompt not only serious privacy concerns but also possible monetary penalties [18].

12.1.1 Privacy Implications of Aggregate Location Data

One might argue that publishing aggregate information, such as the number of individuals at a given location, is enough to reconstruct aggregate mobility patterns, and has no privacy implications. Indeed, aggregated information is usually related to large groups of individuals and is seemingly safe to disclose. However, this reasoning is flawed as shown next. First, an attack is described that can reconstruct even entire individual trajectories from aggregate location data, if aggregates are periodically and sufficiently frequently published (e.g., in every half an hour). We also illustrate the potential privacy threats of irregularly published aggregate location data, for example, when a querier (or the adversary) specifies the spatio-temporal points whose visits are then aggregated and released.

Consequently, aggregation *per se* do not necessarily prevent privacy breaches, and we need additional countermeasures to guarantee privacy for individuals even in a dataset of aggregate mobility data such as spatio-temporal densities.

12.1.1.1 Reconstruction from Periodically Published Aggregate Data

The attack described in [61] successfully reconstructed more than 70% of 100,000 trajectories merely from the total number of visits at 8000 locations, which were published every half an hour over a whole week in a large city. The attack exploits three fundamental properties of location trajectories:

Predictability: The current location of an individual can be accurately predicated from his previous location because consecutively visited locations are usually geographically close. This implies that trajectories can be well-separated in space; if two trajectories are far away in time t then they remain so in time $(t+1)$ assuming that t and $t+1$ are not too distant in time.

Regularity: Most people visit very similar (or the same) locations every day. Indeed, human mobility is governed by daily routines and hence periodic. For example, people go to work/school and return home at almost the same time every day.

Uniqueness: Every person visits quite different locations than any other person even in a very large population, which has already been demonstrated by several studies. For example, any four locations of an individual trajectory are unique to that trajectory with a probability of more than 95% for one and a half million individuals [13].

The attack has three main phases. In the first phase, it reconstructs every trajectory within every single day by exploiting the predictability of trajectories. This is performed by finding an optimal match of locations between consecutive time slots, where geographically close locations are more likely to be matched. After the first phase, we have the daily fragments of every trajectory, but we do not know which fragments belong to the same trajectory. Hence, in the second phase, complete trajectories are reconstructed by identifying their daily fragments. This is feasible due to the regularity and uniqueness properties of trajectories, i.e. every trajectory has similar daily fragments which are also quite different from the fragments of other trajectories. Similarity of fragments can be measured by the frequency of visits per location within a fragment. Finally, in the last phase, re-identification of individuals are carried out by using the uniqueness property again; a few locations of any individual known from external sources (e.g., social media) will single out the individual's trajectory [13]. As individual trajectories are regarded as personal data in several regulations internationally, the feasibility of this attack demonstrates that aggregate location data can also be regarded as personal data.

12.1.1.2 Reconstruction from Irregularly Published Aggregate Data

Another approach of releasing spatio-temporal density is to answer some counting queries executed on the location trajectories. The querier is interested in the number of people whose trajectories satisfy a specified condition (e.g., the number of trajectories which contain a certain hospital). Queries can be filtered instantly by an auditor, e.g. all queries which have too small support, say less than k (i.e., only k trajectories satisfy the condition), are simply refused to answer. However, this approach is not enough to prevent privacy breaches; if the support of two queries are both greater than k, their difference can still be 1. For instance, the first query may ask for the number of people who visited a hospital, and the second query for the number of people who visited the same hospital except locations L_1 and L_2. If the querier knows that L_1 and L_2 are unique to John then it learns whether John visited the hospital.

Defenses against such *differencing attacks* are not straightforward. For example, verifying whether the answers of two or more queries disclose any location visit can be computationally infeasible; if the query language is sufficiently complex there is no efficient algorithm to decide whether two queries constitute a differencing attack [30]. In Sect. 12.3.1, we show more principled techniques to recover individual location visits from the answers of a given query set.

12.1.2 Applications of Spatio-Temporal Density

Spatio-temporal density data, albeit aggregated in nature, can enable a wide variety of optimization use cases by providing a form of location awareness, especially in the context of the Smart City concept [46]. Depending on both its spatial and temporal granularity, such data can be useful for optimizing the (1) design and/or (2) operational phases of city management with regard to e.g., public transportation, local businesses or emergency preparedness. Obviously, spatial resolution determines the scale of such optimization, e.g., whether we can tell a prospective business owner to open her new cafe in a specific district or a specific street. On the other hand, it is the temporal granularity of density data that separates the application scenarios in terms of design and operational use cases.

In case of low temporal granularity (i.e., not more than a few data points per area per day), city officials can use the data for optimizing design tasks such as:

- planning infrastructure networks, such as new roads, railways or communication networks;
- advising on the location of new businesses such as retail, entertainment and food;
- developing timetables for public transportation;
- deploying hubs for urban logistics systems such as post, vehicle depos (e.g., for an urban bike rental system), electric vehicle chargers and even city maintenance personnel;

In case of high temporal granularity (i.e., several data points per area per hour) [33], spatio-temporal density data might enable on-the-fly operational optimization in the manner of:

- reacting to and forecasting traffic-related phenomena including traffic anomaly detection and re-routing;

- implementing adaptive public transportation timetables also with an increasingly autonomous vehicle fleet [52];
- scheduling maintenance work adaptively causing the least amount of disturbance to inhabitants;
- promoting energy efficiency by switching off unneeded electric equipment on-demand (cell towers, escalators, street lighting);
- location-aware emergency preparedness protocols in case of natural disasters or terrorist attacks [7].

These lists of application scenarios are not comprehensive. Interestingly, such an aggregated view on human mobility enables a large set of practical applications.

12.2 Privacy Models

Privacy has a multitude of definitions, and thus different privacy models have been proposed. In terms of privacy guarantee, we distinguish between syntactic and semantic privacy models. Syntactic models focus on syntactic requirements of the anonymized data (e.g., each record should appear at least k times in the anonymized dataset) without any guarantee on what sensitive information the adversary can exactly learn about individuals. As opposed to this, semantic models[1] are concerned with the private information that can be inferred about individuals using the anonymized data as well as perhaps some prior (or background) knowledge about them. The commonality of all privacy models is the inherent trade-off between privacy and utility: guaranteeing any meaningful privacy requires the distortion of the original dataset which yields imprecise, coarse-grained knowledge even about the population as a whole. There is no free lunch: perfect privacy with maximally accurate anonymized data is impossible. Each model has different privacy guarantees and hence provide different accuracy of the (same) data.

12.2.1 Syntactic Privacy Models

One of the most influential privacy model is k-anonymity, which was first introduced in computer science by Sweeney [53], albeit the same notion had already existed before in statistical literature. In general, for location data, k-anonymity guarantees that any record is indistinguishable with respect to spatial and temporal information

---

[1] In our context, semantic privacy is not analogous to semantic security used in cryptography, where ciphertexts must not leak any information about plaintexts. Anonymized data ("ciphertext") should allow partial information leakage about the original data ("plaintext"), otherwise any data release would be meaningless. Such partial leakage should include the release of useful population (and not individual specific) characteristics.

from at least $k - 1$ other records. Hence, an adversary who knows some attributes of an individual (such as few visited places) may not be able decide which record belongs to this person. Now, let us define k-anonymity more formally.

Definition 12.1 (k-**Anonymity** [53]) Let $\mathbb{P} = \{P_1, \ldots, P_{|\mathbb{P}|}\}$ be a set of public attributes, and $\mathbb{S} = \{S_1, \ldots, S_{|\mathbb{S}|}\}$ be a set of sensitive attributes. A relational table $R(\mathbb{P}, \mathbb{S})$ satisfies k-anonymity iff, for each record in r in R, there are at least $k - 1$ other records in R which have the same public attribute values as r.

k-anonymity requires (syntactic) indistinguishability of every record in the dataset from at least $k - 1$ other records with respect to their public attributes. Originally, public attributes included all (quasi)-identifiers of an individual (such as sex, ZIP code, birth date) which are easily learnable by an adversary, while the sensitive attribute value (e.g., salary, medical diagnosis, etc.) of any individual should not be disclosed. Importantly, the values of public attributes are likely to be unique to a person in a population [23], and hence can be used to link multiple records of the same individual across different datasets, if these datasets share common public attributes. In the context of location data, where a spatio-temporal point (L, t) corresponds to a binary attribute whose value is 1 if the individual visited location L at time t and 0 otherwise, such distinction of public and sensitive attributes is usually pointless. Indeed, the same location can be insensitive to one person while sensitive to another one (e.g., a hospital may be an insensitive place for a doctor, who works there, and sensitive for a patient). Therefore, in a location dataset, k-anonymity should require that each record (trajectory) must be completely identical to at least $k - 1$ other trajectories in the same dataset. Syntactically indistinguishable trajectories/records form a single anonymity group.

k-anonymity can be achieved by generalizing and/or suppressing the location visits of individuals in the anonymized dataset. Generalization can be performed by either forming clusters of similar trajectories, where each cluster has at least k trajectories, or by replacing the location and/or time information of trajectories with a less specific, but semantically consistent, one. For example, cities are represented by their county, whereas minutes or hours are represented by the time of day (morning/afternoon/evening/night).

A relaxation of k-anonymity, called k^m-anonymity, was first proposed in [54]. This model imposes an explicit constraint on the background knowledge of the adversary, and requires k-anonymity with respect to this specific knowledge. For example, if the adversary can learn at most m location visits of an individual, then, for any set of m location visits, there must be at least 0 or k records in the anonymized dataset which contain this particular set of visits. Formally:

Definition 12.2 (k^m-**Anonymity** [54]) Given a dataset D where each record is subset of items from a universe $\mathcal{U}$. D is k^m-anonymous iff for any m items from $\mathcal{U}$ there are 0 or at least k records which contain these items.

In our context, universe $\mathcal{U}$ represents all spatio-temporal points, and an individual's record has an item from $\mathcal{U}$ if the corresponding spatio-temporal is visited by the individual.

Table 12.1 Examples for k- and k^m-anonymity, where each row represents a record, public and sensitive attributes are not distinguished, and temporal information is omitted for simplicity

(a) Original		(b) 2-anonymous		(c) 2^2-anonymous	
No.	Locations	No.	Locations	No.	Locations
1	{LA}	1	{West US}	1	{LA}
2	{LA, Seattle}	2	{West US}	2	{LA, Seattle}
3	{NYC, Boston}	3	{NYC, Boston}	3	{West US}
4	{NYC, Boston}	4	{NYC, Boston}	4	{West US}
5	{LA, Seattle, NYC}	5	{LA, Seattle, West US}	5	{LA, Seattle, West US}
6	{LA, Seattle, NYC}	6	{LA, Seattle, West US}	6	{LA, Seattle, West US}
7	{LA, Seattle, NYC, Boston}	7	{LA, Seattle, West US}	7	{LA, Seattle, West US}

2^2-anonymity requires fewer generalizations and hence provides more accurate data at the cost of privacy

If m equals the maximum number of location visits per record, then k^m-anonymity boils down to standard k-anonymity. However, the rationale behind k^m-anonymity is that the adversary is usually incapable of learning more than a few locations visits per individual (e.g., most people publicly reveal only their home and working places on social media, in which case $m = 2$ if temporal data is disregarded). Clearly, requiring indistinguishability with respect to only m instead of all location visits of an individual requires less generalization and/or suppression thereby providing more accurate anonymized data. This is also illustrated in Table 12.1.

We must note that many more different syntactic privacy models (e.g., ℓ-diversity [39], t-closeness [37], (L, K, C)-privacy [42], etc.) have been proposed to mitigate the deficiencies of k-anonymity. We refer the interested reader to [21] and [56] for more details on privacy models and their usage. In this chapter, we only consider syntactic anonymization schemes which rely on k- or k^m-anonymity.

12.2.2 Semantic Privacy Models

Most syntactic privacy models, such as k-anonymity, aim to mitigate only identity disclosure, when the adversary re-identifies a record in the dataset (i.e., infer the exact identity of the record owner). Although re-identification is clearly undesirable and explicitly addressed by most legal regulations worldwide, it is not a necessary condition of privacy violations. That is, locating the anonymity group of a person (e.g., using his home and working places), the group itself can still leak a person's visited places no matter how large the group is. For instance, each of the k trajectory may contain the same sensitive place, which means that the person also passed this place. The real culprit is the lack of uncertainty about the individuals' presence in the *anonymized* dataset; even a knowledgeable adversary, who may know that a person's record is part of the original dataset, should not be able learn if this record was indeed used to generate the anonymized data. Another common pitfall of syntactic privacy models is the lack of *composability*; the privacy of independent releases of the same or correlated datasets should not collapse but rather "degrade gracefully". However, this does not hold for k-anonymity: the composition of k-anonym datasets, where k can be arbitrarily large, can only be 1-anonym (i.e., the anonymity guarantee completely collapses), which is also demonstrated in [22]. Composability is a natural requirement of any privacy model in the era of Big Data where many different pieces of personal data get anonymized and published about people by many different stakeholders independently. These different pieces may be gathered and combined by a knowledgeable adversary in order to breach individuals' privacy. Next, we present a model which addresses these concerns.

Intuitively, differential privacy [15] requires that the outcome of any computation be insensitive to the change of any single record inside and outside the dataset. It allows a party to privately release a dataset: with perturbation mechanisms, a function of an input dataset is modified, prior to its release, so that any information which can discriminate a record from the rest of the dataset is bounded [16].

Definition 12.3 (Differential Privacy [16]) A privacy mechanism $\mathscr{A}$ guarantees (ε, δ)-differential privacy if for any database D and D', differing on at most one record, and for any possible output $S \subseteq Range(\mathscr{A})$,

$$Pr[\mathscr{A}(D) \in S] \leq e^{\varepsilon} \times Pr[\mathscr{A}(D') \in S] + \delta$$

or, equivalently, $Pr_{O \sim \mathscr{A}(D)}\left[\log\left(\frac{Pr[\mathscr{A}(D)=O]}{Pr[\mathscr{A}(D')=O]}\right) > \varepsilon\right] \leq \delta$.

Here, ε is typically a modest value (i.e., less than 1), and δ is a negligible function of the number of records in D (i.e., less then $1/|D|$) [16].

We highlight two consequences of the above definition which are often overlooked or misinterpreted. First, differential privacy guarantees plausible deniability to every individual *inside as well as outside* of the dataset, as an adversary, provided with the output of $\mathscr{A}$, can draw almost the same conclusions about any individual no matter if this individual is included in the input of $\mathscr{A}$ or not [16]. Specifically, Definition 12.3 guarantees that every output of algorithm $\mathscr{A}$ is almost equally likely (up to ε) on datasets differing in a single record except with probability at most δ. This implies that *every possible* binary inference (i.e., predicate) has almost the same probability to be true (false) on neighboring datasets [15]. For example, if an adversary can infer from $\mathscr{A}(D)$ that an individual, say John, visited a hospital with probability 0.95, where D excludes John's record, then the same adversary infers the same from $\mathscr{A}(D')$ with probability $\approx e^{\pm\varepsilon} \times 0.95 + \delta$, where $D' = D \cup \{$John's record$\}$. This holds for *any* adversary and inference irrespective of the applied inference algorithm and prior (background) knowledge.[2] That is, the

---

[2]The inference algorithm and background knowledge influences only the probability of the conclusion, which is 0.95 in the current example.

privacy measure ε and δ are "agnostic" to the adversarial background knowledge and inference algorithm.

Second, Definition 12.3 *does not* provide any guarantee about the (in)accuracy of any inference. There can be inferences (adversaries) which may predict the hospital visit of John quite accurately, e.g., by noticing that all records, which are very similar to John's record (such as the records having the same age and profession as John), also visited a hospital [11], while other inferences may do a bad job of prediction as they cannot reliably sort out the records being similar (correlated) to John's record. Definition 12.3 guarantees that the accuracy of *any* inferences, no matter how sensitive are, remain unchanged (up to ε and δ) if John's own record is included in the anonymized data. In other words, differential privacy allows to learn larger statistical trends in the dataset, even if these trends reveal perhaps sensitive information about each individual, and protects secrets about individuals which can only be revealed with their participation in the dataset.[3] Learning such trends (i.e., inferences which are generalizable to a larger population in interest) is the ultimate goal of any data release in general.

Therefore, the advantage of differential privacy, compared to the many other models proposed in the literature, is twofold. First, it provides a formal and measurable privacy guarantee regardless what other background information or sophisticated inference technique the adversary uses even in the future. Second, following from Definition 12.3, it is closed with respect to sequential and parallel composition, i.e., the result of the sequential or parallel combination of two differential private algorithms is also differential private.

Theorem 12.1 ([40]) *If each of $\mathscr{A}_1, \ldots, \mathscr{A}_k$ is (ε, δ)-differential private, then their k-fold adaptive composition*[4] *is $(k\varepsilon, k\delta)$-differential private.*

Composition property has particular importance in practice, since it does not only simplify the design of anonymization (sanitization) solutions, but also allows to measure differential privacy when a given dataset, or a set of correlated datasets, is anonymized (and released) several times, possibly by different entities.

There are a few ways to achieve DP and all of them are based on the randomization of a computation whose result ought to be released. Most of these techniques are composed of adding noise to the true output with zero mean and variance calibrated to desired privacy guarantee which is measured by ε and δ. A fundamental concept of these techniques is the *global sensitivity* of the computation (function) [16] whose result should be released:

Definition 12.4 (Global L_p-Sensitivity)   For any function $f : \mathscr{D} \to \mathbb{R}^d$, the L_p-sensitivity of f is $\Delta_p f = \max_{D,D'} \|f(D) - f(D')\|_p$, for all D, D' differing in at most one record, where $\| \cdot \|_p$ denotes the L_p-norm.

---

[3]These secrets are the *private* information which discriminate the individual from the rest of the dataset and should be protected.

[4]Adaptive composition means that the output of $\mathscr{A}_{i-1}$ is used as an input of $\mathscr{A}_i$, that is, their executions are not necessarily independent except their coin tosses.

The Gaussian Mechanism [16] consists of adding Gaussian noise to the true output of a function. In particular, for any function $f : D \rightarrow \mathbb{R}^d$, the mechanism is defined as $\mathcal{G}(D) = f(D) + \langle \mathcal{N}_1(0, \sigma), \ldots, \mathcal{N}_d(0, \sigma) \rangle$, where $\mathcal{N}_i(0, \sigma)$ are i.i.d. normal random variables with zero mean and with probability density function $g(z|\sigma) = \frac{1}{\sqrt{2\pi\sigma^2}} e^{-z^2/2\sigma^2}$. The variance σ^2 is calibrated to the L_2-sensitivity of f which is shown by the following theorem.

Theorem 12.2 ([16]) *For any function* $f : \mathcal{D} \rightarrow \mathbb{R}^d$, *the mechanism* $\mathcal{A}$

$$\mathcal{A}(\mathcal{D}) = f(\mathcal{D}) + \langle \mathcal{G}_1(\sigma), \ldots, \mathcal{G}_d(\sigma) \rangle$$

gives (ε, δ)-*differential privacy for any* $\varepsilon < 1$ *and* $\sigma^2 \geq 2(\Delta_2 f)^2 \ln(1.25/\delta)/\varepsilon^2$, *where* $\mathcal{G}_i(\sigma)$ *are i.i.d Gaussian variables with variance* σ^2.

For example, if there are d possible locations and f returns the number of visits per location (i.e., the spatial density), then $\Delta_1 f$ equals the maximum number of all visits of any single individual *in any input dataset*, where $\Delta_2 f \leq \Delta_1 f$. If $\Delta_2 f$ is "too" large or ε and/or δ are "too" small, large noise is added providing less accurate visit counts. Also notice that the noise variance is calibrated to the worst-case contribution of any single individual to the output of f, which means that the count of popular locations visited by many individuals can be more accurately released than less popular locations with smaller counts. Indeed, all location counts are perturbed with the same magnitude of noise, hence the signal-to-noise ratio is higher for larger counts providing smaller relative error.

12.3 Releasing Spatio-Temporal Data

Suppose a geographical region which is composed of a set $\mathbb{L}$ of locations visited by N individuals over a time of interest with T discretized epochs.[5] These locations may represent a partitioning of the region (e.g., all districts of the metropolitan area of a city). The mobility dataset D of N users is a binary data cube with size $N \cdot |\mathbb{L}| \cdot T$, where $D_{i,L,t} = 1$ if individual i visited location L in epoch t otherwise $D_{i,L,t} = 0$. That is, each individual's record (or trajectory) is represented by a binary vector with size $|\mathbb{L}| \times T$. The spatio-temporal density of locations $\mathbb{L}$ is defined by the number of individuals who visited these locations as a function of time. More precisely, there is a time series $\mathbf{X}^L = \langle X_0^L, X_1^L, \ldots, X_{T-1}^L \rangle$ for any location $L \in \mathbb{L}$, where $X_t^L = \sum_{i=1}^N D_{i,L,t}$ and $0 \leq t < T$. $\mathbf{X}^{\mathbb{L}}$ denotes the set of time series of all locations $\mathbb{L}$ and is referred to as the spatio-temporal density of locations $\mathbb{L}$ in the sequel.

In general, any data release is modelled by the execution of data queries. For example, if the querier is interested in the spatio-temporal density of locations

---

[5]An epoch can be any time interval such as a second, a minute, an hour, etc.

$S_L \subseteq \mathbb{L}$ at time $S_T \subseteq \{0, 1, \ldots, T - 1\}$, then the query $Q(S_L, S_T)$ is computed as $Q(S_L, S_T) = \sum_{L \in S_L, t \in S_T} \sum_{i=1}^{N} D_{i,L,t} = \sum_{L \in S_L, t \in S_T} X_t^L$. This gives rise to at least three approaches for the privacy-preserving release of spatio-temporal density:

Approach 1: compute any query Q on the original data D (or $\mathbf{X}^{\mathbb{L}}$) and release only the anonymized query result $\hat{Q}(S_L, S_T)$;

Approach 2: anonymize the mobility dataset D into $\hat{D}$, then release $\hat{D}$ which can be used to answer any query Q as $\hat{Q}(S_L, S_T) = \sum_{L \in S_L, t \in S_T} \sum_{i=1}^{N} \hat{D}_{i,L,t}$;

Approach 3: compute the density $\mathbf{X}^{\mathbb{L}}$ from the original mobility data D as $X_t^L = \sum_{i=1}^{N} D_{i,L,t}$, and release the anonymized $\hat{\mathbf{X}}^{\mathbb{L}}$, where $\hat{\mathbf{X}}^{\mathbb{L}}$ can be used to answer any query Q.

In Approach 1, a querier can adaptively (i.e., interactively) choose its queries depending on the result of previously answered queries. By contrast, in Approach 2 and 3, the released data are used to answer arbitrary number and type of queries non-interactively (i.e., the queries are independent of each other). In fact, Approach 1, 2 and 3 only differ in their adversary models: Approach 2 and 3 are instantiations of Approach 1 in the non-interactive setting where the possibly adversarial querier must fix all queries before learning any of its results. Specifically, Approach 2 is simply consists of answering $N \cdot |\mathbb{L}| \cdot T$ binary queries at once, where a query returns an element of the cube D. Similarly, in Approach 2, $|\mathbb{L} \cdot T|$ queries can represent the elements of every time series, where all queries are answered together. As detailed in the sequel, the decreased number of queries as well as the non-interactive answering mechanism is the reason that Approach 3 usually outperforms Approach 1 and 2 in practice as long as the only goal is to release $\mathbf{X}^{\mathbb{L}}$ as accurately as possible meanwhile preserving the privacy of individuals. Hence, we will detail a specific solution of Approach 3 in Sect. 12.3.3 and briefly review the rest in Sects. 12.3.1 and 12.3.2.

12.3.1 Approach 1: Anonymization of Specific Query Results

12.3.1.1 Syntactic Anonymization

Privacy breaches may be alleviated by query auditing which requires to maintain all released queries. The database receives a set of counting queries $Q_1(S_{L_1}, S_{T_1}), \ldots, Q_n(S_{L_n}, S_{T_n})$, and the auditor needs to decide whether the queries can be answered without revealing any single visit or not. Specifically, the goal is to prevent the *full disclosure* of any single visit of any spatio-temporal point in the dataset.

Definition 12.5 (Full Disclosure) $D_{i,L,t}$ is fully disclosed by a query set $\{Q_1(S_{L_1}, S_{T_1}), \ldots, Q_n(S_{L_n}, S_{T_n})\}$ if $D_{i,L,t}$ can be uniquely determined, i.e., in all possible data sets D consistent with the answers $\mathbf{c} = (c_1, \ldots, c_n)$ to queries $Q_1, \ldots, Q_n, D_{i,L,t}$ is the same.

As each query corresponds to a linear equation on location visits, the auditor can check whether any location visit can be uniquely determined by solving a system of linear equations specified by the queries. To ease notation, let $\mathbf{x} = (x_1, \ldots, x_{N \cdot |\mathbb{L}| \cdot T})$ denote the set of all location visits, i.e., there is a bijection $\alpha : [1, N] \times \mathbb{L} \times [1, T] \to [1, N \cdot |\mathbb{L}| \cdot T]$ such that $x_{\alpha(i,L,t)} = D_{i,L,t}$. Let $\mathbf{Q}$ be a matrix with n rows and $N \cdot |\mathbb{L}| \cdot T$ columns. Each row in $\mathbf{Q}$ corresponds to a query, which is represented by a binary vector, indexing the visits that are covered by the query. The system of linear equations is described in matrix form as $\mathbf{Qx} = \mathbf{c}$. Hence, the auditor checks whether any x_i can be uniquely determined by solving the following system of equations:

$$\mathbf{Qx} = \mathbf{c}$$
$$\text{subject to} \quad x_i \in \{0, 1\} \quad \text{for } 1 \le i \le N \cdot |\mathbb{L}| \cdot T \tag{12.1}$$

In general, this problem is coNP-hard as the variables x_i have boolean values [34]. However, there exists an efficient polynomial time algorithm in the special case when the queries are 1-dimensional, i.e. there is a permutation of $\mathbf{x}$ where each query covers a subsequence of the permutation. Typical examples include range queries. For instance, if locations are ordered according to their coordinates on a space-filling Hilbert curve, then range queries can ask for the total number of visits of locations (over all epochs) that are geographically also close. In the case of 1-dimensional queries, the auditor has to determine the integer solutions of the following system of equations and inequalities:

$$\mathbf{Qx'} = \mathbf{c}$$
$$\text{subject to} \quad 0 \le x_i' \le 1 \quad \text{for } 1 \le i \le N \cdot |\mathbb{L}| \cdot T \tag{12.2}$$

Notice that the variables in Eq. (12.2) are no longer over boolean data and hence Eq. (12.2) can be solved in polynomial time with any LP solver [55]. The integer solutions of Eq. (12.2) equals the solutions of Eq. (12.1) for 1-dimensional location queries [34].

In the general case, when the queries are multi-dimensional, the auditor can also solve Eq. (12.2), and the final solutions are obtained by rounding: $\hat{x}_i = 1$ if $x_i' > 1/2$ and $\hat{x}_i = 0$ otherwise. In that case, $\hat{\mathbf{x}} \approx \mathbf{x}$ for sufficiently large number of queries [14]. In particular, if each query covers a visit with probability $1/2$, then $O(|\mathbf{x}| \log^2 |\mathbf{x}|)$ queries are sufficient to recover almost the whole $\mathbf{x}$ (i.e., dataset D). Even more, only $|\mathbf{x}|$ number of deterministically chosen queries are enough to recover almost the entire original data [17]. In fact, these reconstruction techniques are the best known attacks against a database curator who answers only aggregate counting queries over boolean data.

Therefore, equipped with the original data $\mathbf{x}$, the auditor can check whether any of the above attacks would be successful by comparing $\mathbf{x}$ with the reconstructed values $\hat{\mathbf{x}}$ (or $\mathbf{x'}$). If so, the auditor refuses to answer any of the n queries.

The above query auditing techniques have several problems. First and foremost, refusing to answer a query itself can leak information about the underlying dataset (i.e., D) [44]. This would not be the case if refusal was independent of the underlying dataset (e.g., auditing is carried out without accessing the true answers c). Second, they can be computationally expensive. Indeed, using the solver in [55] the worst-case running time is $O(n|\mathbf{x}|^4)$ if $|\mathbf{x}| \gg n$. Finally, most query auditing schemes assume that the adversary has either no background knowledge about the data, or it is known to the auditor. These are impractical assumptions which is also demonstrated in Sect. 12.1.1.1, where the adversary reconstructed complete trajectories from aggregate location counts exploiting some inherent characteristics of human mobility.

12.3.1.2   Semantic Anonymization

An alternative approach to query auditing perturbs each query result with some random noise and releases these noisy answers. In order to guarantee (ε, δ)-differential privacy, the added noise usually follows a Laplace or Gaussian distribution. If the noise is added independently to each query answer, then the error is $O(\sqrt{n \log(1/\delta)}/\varepsilon N)$ [16], where N is the number of individuals and n is the number of queries. This follows from the advanced composition property of differential privacy [16]. Therefore, $\tilde{\Omega}(N^2)$ queries can be answered using this approach with non-trivial error (i.e., it is less than the magnitude of the answer). We note that at least $\Omega(\sqrt{N})$ noise is needed per query in order to guarantee any reasonable notion of privacy [14, 16]. There also exist better techniques that add correlated noise to the answers. For instance, the private multiplicative weight mechanism [26] can answer exponentially many queries in N with non-trivial error, where the added noise scales with $O(\sqrt{\log(T|\mathbb{L}|)} \cdot \log(1/\delta) \cdot \log(n)/\varepsilon N)^{1/2}$.

In contrast to query auditing described in Sect. 12.3.1, the above mechanisms can answer queries in an on-line fashion (i.e., each query is answered as it arrives) and run in time $\text{poly}(N, T|\mathbb{L}|)$ per query. Moreover, the privacy guarantee is independent of the adversarial background knowledge (see Sect. 12.2.2). On the other hand, they distort (falsify) the data by perturbation, which may not be desirable in some practical applications of spatio-temporal density. Another drawback is that they are data agnostic and may not exploit some inherent correlation between query results which are due to the nature of the location data. For example, query results usually follow a publicly known periodic trend, and adding noise in the frequency domain can provide more accurate answers [5].

12.3.2   Approach 2: Anonymization of the Mobility Dataset

12.3.2.1   Syntactic Anonymization

In general, anonymizing location trajectories (i.e., the whole cube D) while preserving practically acceptable utility is challenging. This is due to the fact that

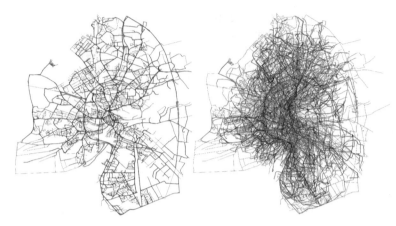

Fig. 12.1 Never-Walk-Alone anonymization. Original dataset (city of Oldenburg in Germany) with 1000 trajectories (left) and its anonymized version (NWA from [2]) with $k = 3$ where the distance between any points of two trajectories within the same cluster is at most 2000 m (right) (image courtesy of Gábor György Gulyás)

location data is typically high-dimensional and sparse, that is, any individual can visit a large number of different locations, but most of them typically visit only a few locations which are quite different per user. This has devastating effect on the utility of anonymized datasets: most k-anonymization schemes generalize multiple trajectories into a single group (or cluster) and represent each trajectory with the centroid of their cluster [2, 43, 47]. Hence, every record becomes (syntactically) indistinguishable from other records within its cluster. This generalization is often implemented by some sophisticated clustering algorithm, where the most similar trajectories are grouped together with an additional (privacy) constraint: each cluster must contain at least k trajectories. Unfortunately, such approaches fail to provide sufficiently useful anonymized datasets because of the *curse of dimensionality* [6]: any trajectory exhibits almost identical similarity to any other trajectory in the dataset. This implies that the centroid of each cluster tend to be very dissimilar from the cluster members implying weak utility. Moreover, as the distribution of the number of visits of spatio-temporal points are typically heavy-tailed [45], projection to low dimensions and then clustering in low dimension also loses almost all information about the trajectories. This is illustrated by Fig. 12.1 which shows the result of a state-of-the-art anonymization scheme, referred to as Never-Walk-Alone (NWA) [2], on a synthetic dataset with 1000 trajectories.[6] This scheme groups k co-localized trajectories within the same time period to form a k-anonymized aggregate trajectory, where k was set to 3 in our experiment and the greatest difference between any spatial point of two members of the same cluster is set to 2000 m. Figure 12.1

---

[6]We used a subset of a larger synthetic trajectory dataset available on https://iapg.jade-hs.de/personen/brinkhoff/generator/.

shows that even with modest values of k, the anonymized dataset provides quite imprecise spatial density of the city.

To improve utility while relaxing privacy requirements, k^m-anonymity has also been considered to anonymize location trajectories in [48]. However, most anonymization solutions guaranteeing k^m-anonymity has a computational cost which is exponential in m in the worst-case, hence this approach is only feasible if m is small. This drawback is alleviated in [3], where a probabilistic relaxation of k^m-anonymity is proposed to release the location visits of individuals without temporal information. In theory, temporal data can also be released along with the location information if the m items are composed of pairs of spatial and temporal positions. However, care must be taken as the background knowledge of a realistic adversary cannot always be represented by m items (e.g., it perhaps also knows the frequency of m items of a targeted individual).

Another approach improving on k-anonymization is p-confidentiality [10]; instead of grouping the trajectories, the underlying map is anonymized, i.e., points of interest are grouped together creating obfuscation areas around sensitive locations. More precisely, given the path of a trajectory, p bounds the probability that the trajectory stops at a sensitive node in any group. Supposing that (1) the background knowledge of the adversary consists of stopping probabilities for each location in a single path and (2) sensitive locations are pre-specified by data owners, groups of locations are formed in such a way that the parts of trajectories entering the groups do not increase the adversary's belief in violating the p-confidentiality. Trajectories are then anonymized based on the above map anonymization. The efficiency and utility of this solution is promising, however, in cases where the adversarial background knowledge cannot be approximated well (or at all), semantic privacy models such as differential privacy is preferred.

12.3.2.2   Semantic Anonymization

A more promising approach is to publish a synthetic (anonymized) mobility dataset resembling the original dataset as much as possible, while achieving provable guarantees w.r.t. the privacy of each individual. The records in both datasets follow similar underlying distributions, i.e., after modeling the generator distribution of the original dataset, random samples (records) are drawn from a noisy version of this distribution. A few solutions exist in literature where the generator distribution is modeled explicitly and noised to guarantee differential privacy. For example, DP-WHERE [41] adds noise to the set of empirical probability distributions which is derived from CDR (Call-Detail-Record) datasets, and samples from these distributions to generate synthetic CDRs which are differential private. Although this synthetic dataset can also be used to compute spatio-temporal density, it is usually not as accurate as perturbing the generator distribution of the spatio-temporal density exclusively [4]. Indeed, the accurate model of more complex data (such as the original mobility data) is also more complex in general (i.e., have larger number of parameters), which usually requires increased perturbation.

Some other works generate synthetic sequential data using more general data generating models such as different Markov models [8, 9, 29]. These approaches have wide applicability but they are usually not as accurate as a specific model tailored to the *publicly known* characteristics of the dataset to be anonymized. We illustrate this important point by the following example. DP-WHERE is designed for CDR datasets, and provides more accurate anonymized CDR data than a simple n-gram model [8]. For example, DP-WHERE models the distribution of commute distances per home location and then generates a pair of home and working places as follows. First, a home location is selected, which is followed by picking a distance from the (noisy) distribution of commute distances. Finally, a working place is selected which has this distance from the selected home location. This approach results in more accurate representation of home and working places than using the noisy occurrence counts of different pairs of home and working places like in [8]. This is because commute distances are modeled by an exponential distribution [41], and its single rate parameter can be estimated by the median of the empirical data (i.e., commute distances). Therefore, in DP-WHERE, the probability of a particular pair of home and working location depends on their distance, while in an n-gram model, it depends on the occurrence count of this pair in the original dataset. For instance, New York, as a home location, occurs equally likely with LA and Philadelphia, as working places, in an n-gram model, if these pairs have the same frequency in the original dataset. By contrast, in DP-WHERE, New York is much more likely to co-occur with the geographically closer Philadelphia than with LA. The moral of the story is that achieving the best performance requires to find the most faithful model of the data whose accuracy does not degrade significantly due to additional perturbation.

12.3.3   Approach 3: Anonymization of Spatio-Temporal Density

A simple k-anonymization of time series $\mathbf{X}^{\mathbb{L}}$ releases X_t^L only if $X_t^L \geq k$. However, as it is detailed in Sect. 12.1, this still allows privacy violations through various reconstruction attacks. Hence, releasing spatio-temporal density with provable privacy guarantees, such as differential privacy, is preferred in many practical scenarios.

Within the literature of differential privacy, a plethora of techniques have been proposed to release 1- and 2-dimensional range queries (or histograms) while preserving differential privacy [5, 12, 26, 28, 35, 36, 38, 49, 59, 60, 62, 64] and they are also systematically compared in [27]. Indeed, interpreting query results (or bin counts in a histogram) as location counts, these techniques are directly applicable to release spatial density without temporal data. In theory, low-dimensional embedding, such as Locality-sensitive hashing (LSH) [50], may allow to use any of the above techniques to release spatio-temporal density.

Another line of research addresses the release of time series data with the guarantees of differential privacy. This is challenging as time series are large dimensional data whose global sensitivity is usually so large that the magnitude of the added noise is greater than the actual counts of the series for stringent privacy requirement (i.e., $\varepsilon < 1$ and $\delta \leq 1/|N|$ where N is the number of records). Consequently, naively adding noise to each count of a time series often results in useless data. Several more sophisticated techniques [19, 31, 51] have been proposed to release time series data meanwhile guaranteeing differential privacy. Most of these methods reduce the global sensitivity of the time series by using standard lossy compression techniques borrowed from signal processing such as sampling, low-pass filtering, Kalman filtering, and smoothing via averaging. The main idea that the utility degradation is decomposed into a reconstruction error, which is due to lossy compression, and a perturbation error, which is due to the injected Laplace or Gaussian noise to guarantee differential privacy. Although strongly compressed data is less accurate, it also requires less noise to be added to guarantee privacy. The goal is to find a good balance between compression and perturbation to minimize the total error.

There are only a few existing papers addressing the release of spatio-temporal density specifically. Although data sources (and hence the definition of spatio-temporal density) vary to a degree in these papers, the commonality is the usage of domain-specific knowledge, i.e., the correlation of data points at hand in both the spatial and the temporal dimension. This domain-specific knowledge helps overcome several challenges including high perturbation error, data sparsity in the spatial domain, and (in some of the cases) real-time data publication. In the context of releasing multi-location traffic aggregates, road network and density are utilized to model the auto-correlation of individual regions over time as well as correlation between neighboring regions [20]. Temporal estimation establishes an internal time series model for each individual cell and performs posterior estimation to improve the utility of the shared traffic aggregate per time stamp. Spatial estimation builds a spatial indexing structure to group similar cells together reducing the impact of data sparsity. All computations are lightweight enabling real-time data publishing. Drawing on the notion of w-event privacy [32], RescueDP studies the problem of the real-time release of population statistics per regions [57]. Such w-event privacy protects each user's mobility trace over any successive w time stamp inside the infinite data grouping algorithm that dynamically aggregates sparse regions together. The criterion for regions to be grouped is that local population statistics should follow a similar trend. Finally, a practical scheme for releasing the spatio-temporal density of a large municipality based on a large CDR dataset is introduced in [4]. Owing to the complexity of its scenario and the innovative techniques used, we present this work in detail in Sect. 12.4.

12.4 A Case-Study: Anonymizing the Spatio-Temporal Density of Paris

In this section, we present an anonymization (or sanitization) technique in order to release the spatio-temporal density with provable privacy guarantees. Several optimizations are applied to boost accuracy: time series are compressed by sampling,

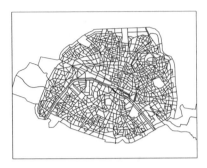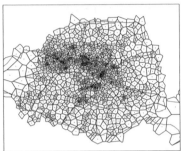

Fig. 12.2 IRIS cells of Paris (left) and Voronoi-tesselation of tower cells (right)

clustering and low-pass filtering. The distortion of the perturbation is attenuated via further optimization and post-processing algorithms. A striking demonstration shows that the achieved performance is high and can be practical in real-world applications: the spatio-temporal density of the city of Paris in France, covering roughly two million people over $105\,\mathrm{km}^2$, is anonymized using the proposed approach.

The specific goal is to release the spatio-temporal density of 989 non-overlapping areas in Paris, called IRIS cells. Each cell is defined by INSEE[7] and covers about 2000 inhabitants. $\mathbb{L}$ denotes the set of all IRIS cells henceforth, and are depicted in Fig. 12.2 based on their contours.[8] We aim to release the number of all individuals who visited a specific IRIS cell in each hour over a whole week. Since human mobility trajectories exhibit a high degree of temporal and spatial regularity [24], 1 week long period should be sufficient for most practical applications. Therefore, we are interested in the time series $\mathbf{X}^L = \langle X_0^L, X_1^L, \ldots, X_{167}^L \rangle$ of any IRIS cell $L \in \mathbb{L}$, where X_t^L denotes the number of individuals at L in the $(t+1)$th hour of the week, such that any single individual can visit a tower only once in an hour. We will omit t and L in the sequel, if they are unambiguous in the given context. $\mathbf{X}^{\mathbb{L}}$ denotes the set of time series of all IRIS cells in the sequel.

To compute $\mathbf{X}^{\mathbb{L}}$, we use a CDR (Call Detail Record) dataset provided by a large telecom company. This CDR data contains the list of events of each subscriber (user) of the operator, where an event is composed of the location (GPS coordinate of the cell tower), along with a timestamp, where an incoming/outgoing call or message is sent to/from the individual. The dataset contains the events of $N = 1{,}992{,}846$ users at 1303 towers within the administrative region of Paris (i.e., the union of all IRIS cells) over a single week (10/09/2007–17/09/2007). Within this interval, the average number of events per user is 13.55 with a standard deviation of 18.33 (assuming that

---

[7]National Institute of Statistics and Economics: http://www.insee.fr/fr/methodes/default.asp?page=zonages/iris.htm.

[8]Available on IGN's website (National Geographic Institute): http://professionnels.ign.fr/contoursiris.

an individual can visit any tower cell only once in an hour) and with a maximum at 732. Similarly to IRIS cells, we can create another set of time series $\mathbf{X}^{\mathbb{C}}$, where X_t^C denotes the number of visits of tower C in the $(t + 1)$th hour of the week.

To map the counts in $\mathbf{X}^{\mathbb{C}}$ to $\mathbf{X}^{\mathbb{L}}$, we compute the Voronoi tessellation of the towers cells $\mathbb{C}$ which is shown in Fig. 12.2. Then, we calculate the count of each IRIS cell in each hour from the counts of its overlapping tower cells; each tower cell contributes with a count which is proportional to the size of the overlapping area. More specifically, if an IRIS cell L overlaps with tower cells $\{C_1, C_2, \ldots, C_c\}$, then

$$X_t^L = \sum_{i=1}^{c} X_t^{C_i} \times \frac{\text{size}(C_i \cap L)}{\text{size}(C_i)} \tag{12.3}$$

at time t.

Algorithm 12.1 Anonymization scheme

Input: $\mathbf{X}^{\mathbb{T}}$ - input time series (from CDR), (ε, δ)-privacy parameters, $\mathbb{L}$ - IRIS cells, ℓ - maximum visits per user
Output: Noisy time series $\hat{\mathbf{X}}^{\mathbb{L}}$

1: Create $\overline{\mathbf{X}}^{\mathbb{C}}$ by sampling at most ℓ visits per user from $\mathbf{X}^{\mathbb{C}}$
2: Compute the IRIS time series $\overline{\mathbf{X}}^{\mathbb{L}}$ from $\overline{\mathbf{X}}^{\mathbb{C}}$ using Eq. (12.3)
3: Perturb $\overline{\mathbf{X}}^{\mathbb{L}}$ into $\hat{\mathbf{X}}^{\mathbb{L}}$ //see Algorithm 12.2
4: Apply smoothing on $\hat{\mathbf{X}}^{\mathbb{L}}$

The rationale behind this mapping is that users are usually registered at the geographically closest tower at any time. Notice that this mapping technique might sometimes be incorrect, since the real association of users and towers depends on several other factors such as signal strength or load-balancing. Nevertheless, without more details of the cellular network beyond the towers' GPS position, there is not any better mapping technique.

12.4.1 Outline of the Anonymization Process

The aim is to transform the time series of all IRIS cells $\mathbf{X}^{\mathbb{L}}$ to a sanitized version $\hat{\mathbf{X}}^{\mathbb{L}}$ such that $\hat{\mathbf{X}}^{\mathbb{L}}$ satisfies Definition 12.3. That is, the distribution of $\hat{\mathbf{X}}^{\mathbb{L}}$ will be insensitive (up to ε and δ) to all the visits of any single user during the whole week, meanwhile the error between $\hat{\mathbf{X}}^{\mathbb{L}}$ and $\mathbf{X}^{\mathbb{L}}$ is small.

The anonymization algorithm is sketched in Algorithm 12.1. First, the input dataset is pre-sampled such that only ℓ visits are retained per user (Line 1). This ensures that the global L_1-sensitivity of all the time series (i.e., $\mathbf{X}^{\mathbb{L}}$) is no more than ℓ. Then, the pre-sampled time series of each IRIS cell is computed from that of the tower cells using Voronoi-tesselation (Line 2), which is followed by the perturbation of the time series of all IRIS cells to guarantee privacy (Line 3). In order to mitigate the distortion of the previous steps, smoothing is applied on the perturbed time series as a post-processing step (Line 4).

12.4.2 Pre-sampling

To perturb the time series of all IRIS cells, we first compute their sensitivity, i.e., $\Delta_1(\mathbf{X}^{\mathbb{L}})$. To this end, we first need to calculate the sensitivity of the time series of all tower cells, i.e., $\Delta_1(\mathbf{X}^{\mathbb{C}})$. Indeed, Eq. (12.3) does not change the L_1-sensitivity of tower counts, and hence, $\Delta_1(\mathbf{X}^{\mathbb{C}}) = \Delta_1(\mathbf{X}^{\mathbb{L}})$.

$\Delta_1(\mathbf{X}^{\mathbb{C}})$ is given by the maximum *total* number of (tower) visits of a single user in *any* input dataset. This upper bound must universally hold for all possible input datasets, and is usually on the order of few hundreds; recall that the maximum number of visits per user is 732 in our dataset. This would require excessive noise to be added in the perturbation phase. Instead, each record of any input dataset is truncated by considering at most one visit per hour for each user, and then at most ℓ of such visits are selected per user uniformly at random over the whole week. This implies that a user can contribute with at most ℓ to all the counts in total regardless of the input dataset, and hence, the L_1-sensitivity of the dataset always becomes ℓ. The pre-sampled dataset is denoted by $\overline{\mathbf{X}}$, and $\Delta_1(\overline{\mathbf{X}}^{\mathbb{C}}) = \Delta_1(\overline{\mathbf{X}}^{\mathbb{L}}) = \ell$.

In order to compute the L_2-sensitivity $\Delta_2(\mathbf{X}^{\mathbb{L}})$, observe that, for any t, there is only a single tower whose count can change (by at most 1) by modifying a single user's data. From Eq. (12.3), it follows that the total change of all IRIS cell counts is at most 1 at any t, and hence $\Delta_2(\overline{\mathbf{X}}^{\mathbb{L}}) \leq \Delta_2(\overline{\mathbf{X}}^{\mathbb{C}}) = \sqrt{\ell}$ based on the definition of L_2-norm.

12.4.3 Perturbation

The time series $\overline{\mathbf{X}}^{\mathbb{L}}$ can be perturbed by adding $\mathscr{G}(\sqrt{2\ell \ln(1.25/\delta)}/\varepsilon)$ to each count in all time series (see Theorem 12.2) in order to guarantee (ε, δ)-DP. Unfortunately, this naive method provides very poor results as individual cells have much smaller counts than the magnitude of the injected noise; the standard deviation of the Gaussian noise is 95 with $\varepsilon = 0.3$ and $\delta = 2 \cdot 10^{-6}$, which is comparable to the mean count in $\overline{\mathbf{X}}^{\mathbb{L}}$.

A better approach exploits (1) the similarity of geographically close time series, as well as (2) their periodic nature. In particular, nearby less populated cells are first clustered until their aggregated counts become sufficiently large to resist noise. The key observation is that the time series of close cells follow very similar trends, but their counts usually have different magnitudes. Hence, if we simply aggregate (i.e., sum up) all time series within such a cluster, the aggregated series will have a trend close to its individual components yet large enough counts to tolerate perturbation. To this end, the time series of individual cells are first accurately approximated by normalizing their aggregated time series (i.e., the aggregated count of each hour is divided with the total number of visits inside the cluster), and then scaled back with the (noisy) total number of visits of individual cells.

Algorithm 12.2 Perturbation

Input: Pre-sampled time series $\overline{\mathbf{X}}^{\mathbb{L}}$, Privacy budget ε, δ, Sensitivity $\Delta_1(\overline{\mathbf{X}}^{\mathbb{L}}) = \ell$
Output: Noisy time series $\hat{\mathbf{X}}^{\mathbb{L}}$

1: $\hat{S}_i := \sum_{t=0}^{167} \overline{X}_t^i + \mathscr{G}(2\sqrt{2\ell \ln(2.5/\delta)}/\varepsilon)$ for each $i \in \mathbb{L}$
2: $\mathbb{E} := \text{Cluster}(\mathbb{L}, \hat{S})$
3: **for** each cluster $E \in \mathbb{E}$ **do**
4:     $\overline{\mathbf{X}}^E := \langle \sum_{i \in E} \overline{X}_0^i, \sum_{i \in E} \overline{X}_1^i, \ldots, \sum_{i \in E} \overline{X}_{167}^i \rangle$
5:     $\hat{\mathbf{X}}^E := \text{FourierPerturb}(\overline{\mathbf{X}}^E, \varepsilon/2, \delta)$
6:     **for** each cell $i \in E$ **do**
7:         $\hat{X}^i := \hat{S}_i \cdot (\hat{X}_t^E / \|\hat{\mathbf{X}}^E\|_1)$
8:     **end for**
9: **end for**

In order to guarantee differential privacy (DP), the aggregated time series are perturbed before normalization. To do so, their periodic nature is exploited and a Fourier-based perturbation scheme [5, 51] is applied: Gaussian noise is added to the Fourier coefficients of the aggregated time series, and all high-frequency components are removed that would be suppressed by the noise. Specifically, the low-frequency components (i.e., largest Fourier coefficients) are retained and perturbed with noise $\mathscr{G}(\sqrt{2\ell \ln(1.25/\delta)}/\varepsilon)$, while the high-frequency components are removed and padded with 0. As only (the noisy) low-frequency components are retained, this method preserves the main trends of the original data more faithfully than simply adding Gaussian noise to $\mathbf{X}^{\mathbb{L}}$, while guaranteeing the same $(\varepsilon/2, \delta/2$-DP. Further details of this technique can be found in [4].

The whole perturbation process is summarized in Algorithm 12.2. First, the noisy total number of visits of each cell in $\mathbb{L}$ is computed by adding noise $\mathscr{G}(2\sqrt{2\ell \ln(2.5/\delta)}/\varepsilon)$ to $\sum_{t=0}^{167} \overline{X}_t^i$ for cell i (Line 1). These noisy total counts are used to cluster similar cells in Line 2 by invoking any clustering algorithm aiming to create clusters with large aggregated counts overall (i.e., the sum of all cells' time series within the cluster has large counts) using only the noisy total number of visits $\hat{S}_i$ as input. The output $\mathbb{E}$ of this clustering algorithm is a partitioning of cells $\mathbb{L}$. When clusters $\mathbb{E}$ are created, their aggregated time series (i.e., the sum of all cells' time series within the cluster) is perturbed with a Fourier-based perturbation scheme [5] in Line 5. Finally, the perturbed time series of each cell i in $\mathbb{L}$ is computed in Line 7 by scaling back the normalized aggregated time series with the noisy total count cell i (i.e., with $\hat{S}_i$). Since Line 1 guarantees $(\varepsilon/2, \delta/2)$-DP to the total counts $(\Delta_1(\overline{\mathbf{X}}^{\mathbb{L}}) = \sqrt{\ell})$, it follows from Theorem 12.1 that Algorithm 12.2 is (ε, δ)-DP as the Fourier perturbation of time-series is $(\varepsilon/2, \delta/2)$-DP in Line 5 [4].

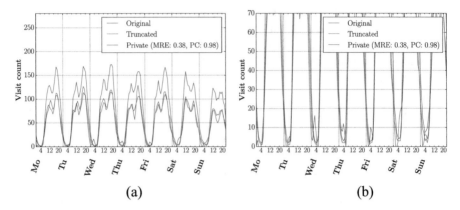

Fig. 12.3 Algorithm 12.1 before improvements ($\varepsilon = 0.3$, $\delta = 2 \cdot 10^{-6}$, $\ell = 30$). (**a**) Large counts. (**b**) Small counts around local minimas

12.4.4 Improvements: Scaling and Smoothing

The result of the above perturbation technique, which is illustrated in Fig. 12.3, still suggests a large error on average. The difference between $\hat{\mathbf{X}}$ and $\mathbf{X}$ is the result of two errors: the sampling error (between $\overline{\mathbf{X}}$ and $\mathbf{X}$) is attributed to pre-sampling, whereas the perturbation error (between $\hat{\mathbf{X}}$ and $\overline{\mathbf{X}}$) is due to the perturbation scheme presented in Algorithm 12.2. Indeed, since $\overline{\mathbf{X}}^i$ is the pre-sampled time series of cell i, $\hat{\mathbf{X}}^i$ (Line 6 of Algorithm 12.2) will be a scaled down version of the original time series $\mathbf{X}^i$ due to the fact that the ℓ visits per individual are sampled *uniformly* at random.

As illustrated by Fig. 12.3a, sampling error mainly distorts large counts: although the noisy counts are close to the counts of the *truncated* (pre-sampled) time series between 9:00 AM and 11:00 PM, it is still far from the original count values. This significantly increases the mean relative error. In addition, as Fig. 12.3b also shows, noisy counts also deviate from pre-sampled as well as from original counts around the local minimas (close to 4:00 AM every day), which further deteriorates the relative error.

To alleviate these errors, two further improvements are proposed in [4], which are also illustrated in Fig. 12.4: first, the perturbation of total cell counts (Line 1 in Algorithm 12.2) is improved, which is used in cell clustering (Line 2 in Algorithm 12.2) and scaling (Line 6 in Algorithm 12.2). The main idea is that the real scaling factor $\sum_{t=0}^{167} X_t^i$ (in Line 1 of Algorithm 12.2) is approximated by a more accurate technique: the relative frequency of each tower is first estimated by sampling only a single visit per user, then the perturbed relative frequencies are multiplied with the (perturbed) total number of visits of the original data $\mathbf{X}$ to obtain an estimation of $\sum_{t=0}^{167} X_t^i$. The relative frequencies have L_2-sensitivity 1, while the L_2-sensitivity of the total number of visits is $\sqrt{753} < 27.44$. Hence, the relative error of this new estimation becomes small, as the relative frequencies of towers

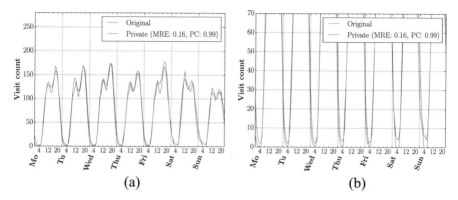

Fig. 12.4 Algorithm 12.1 after improvements ($\varepsilon = 0.3, \delta = 2 \cdot 10^{-6}, \ell = 30$). (**a**) Scaling. (**b**) Smoothing

require very small noise, while the total number of visits is incomparably larger than its L_2-sensitivity. The result of scaling with this more accurate scaling factor is shown in Fig. 12.4a. Finally, in order to diminish perturbation error of small counts, counts between 0:00 and 6:00 AM are smoothed out through non-linear least-square fitting as a post-processing step (Fig. 12.4b).

12.4.5 Time Complexity

The pre-sampling step has a complexity of $O(\ell N)$ and the computation of $\mathbf{X}^{\mathbb{L}}$ (see Eq. (12.3)) needs $O(T|\mathbb{C}||\mathbb{L}|)$ steps in the worst case. In the perturbation algorithm (Algorithm 12.2), the clustering of time-series runs in $O(T|\mathbb{L}|^2)$ and the Discrete Cosine Transform can be implemented with Fast Fourier Transform that has a complexity of $O(T \log T)$. Therefore, the overall complexity is $O(|\mathbb{L}|T \log T + T|\mathbb{L}|^2 + T|\mathbb{C}||\mathbb{L}| + \ell N)$ disregarding the post-processing step (in Line 4 of Algorithm 12.1).

12.4.6 Results

The error between the anonymized and original time series is measured by two metrics: the mean relative error (MRE) and the Pearson Correlation (PC), where $\text{MRE}(\mathbf{X}, \hat{\mathbf{X}}) = (1/n) \sum_{i=0}^{n-1} \frac{|\hat{X}_i - X_i|}{\max(\gamma, X_i)}$.[9] The Pearson correlation measures the linear correlation between the noisy and the original time series (i.e., whether they have similar trends), and it always falls between -1 and 1.

---

[9]The sanity bound γ mitigates the effect of very small counts and is adjusted to 0.1% of $\sum_{i=0}^{n-1} X_i$ [58].

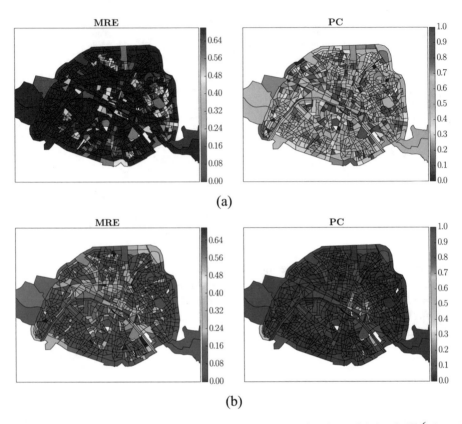

Fig. 12.5 Mean relative error and Pearson correlation of each IRIS cell ($\varepsilon = 0.3$, $\delta = 2 \cdot 10^{-6}$, $\ell = 30$). (**a**) Naive Gaussian Perturbation (Avg. MRE: 1.01, PC: 0.47). (**b**) Algorithm 12.1 (Avg. MRE: 0.17, PC: 0.96)

The MRE and PC of individual IRIS cells are illustrated by color maps in Fig. 12.5. This figure shows that the presented anonymization (Fig. 12.5b) scheme outperforms the naive Gaussian Perturbation Algorithm (Fig. 12.5a) when $\mathscr{G}(\sqrt{2\ell \ln(1.25/\delta)}/\varepsilon)$ is added to each count in $\overline{\mathbf{X}}^{\mathbb{L}}$ without any further optimization. Moreover, Algorithm 12.1 can also provide practical utility for most cells with strong privacy guarantee. Specifically, the average MRE over all cells is only 0.17 with $\varepsilon = 0.3$, $\delta = 2 \cdot 10^{-6}$ and $\ell = 30$.

12.5 Summary and Conclusions

In this chapter, we gave an overview of the privacy models and anonymization/sanitization techniques for releasing spatio-temporal density in a privacy-preserving manner. We first illustrated the privacy threats of releasing spatio-temporal density and described two attacks that can recover individual visits or even complete trajectories merely from spatio-temporal density. Then, we reviewed the mainstream privacy models, and distinguished syntactic models (such as k-anonymity) and semantic models (such as differential privacy). As spatio-temporal density is a function of the raw mobility data, we identified three main approaches to anonymize spatio-temporal density: (1) anonymize and release the results of queries executed on the original mobility data, (2) anonymize and release the original mobility data (i.e., location trajectories) used to compute the spatio-temporal density, and (3) anonymize and release the spatio-temporal density directly which is computed from the original mobility data.

The first approach relies on query auditing, or query perturbation using differential privacy. Query auditing is computationally expensive, and disregards the background knowledge of the adversary. Although query perturbation is independent of the adversarial background knowledge and runs in polynomial time, it ignores some inherent characteristics of human mobility which could further diminish perturbation error. Also, unlike query auditing, perturbation is non-truthful, i.e. releases falsified location data.

The second approach can use either a syntactic or a semantic privacy model to anonymize trajectories. Syntactic anonymization techniques providing k-anonymity suffer from the curse of dimensionality and provide inaccurate data in general. k^m-anonymization has smaller error but guarantees weaker privacy and/or has exponential time complexity in m. In addition, all syntactic privacy guarantees can be violated with appropriate background knowledge, which is difficult to model in practice. Semantic anonymization using differential privacy is much more promising, but again, they use perturbation which is non-truthful. In addition, anonymizing trajectories usually provides less accurate density estimation than anonymizing the spatio-temporal density directly. Indeed, density can be modelled accurately with a model which requires less perturbation than the model of complete trajectories. Although some trajectory anonymization techniques have larger time complexity, these are not serious concerns in case of one-shot release.

As the last approach provides the largest accuracy in practice, we detailed the operation of such an anonymization process and showed its performance in a real-world application. This demonstration also shows that differential privacy can be a practical model for the privacy-preserving release of spatio-temporal data, even if it has large dimension. We also showed that, in order to achieve meaningful accuracy, the sanitization process has to be carefully customized to the application and public characteristics of the dataset. The time complexity of this approach is polynomial and also very fast in practice.

As a conclusion, it is unlikely that there is any "universal" anonymization/sanitization solution that fits every application and data, i.e., provides good accuracy in all scenarios. In particular, achieving the best performance requires finding the most faithful model of the data, such that it withstands perturbation. In case of spatio-temporal density, clustering and sampling with Fourier-based perturbation are seemingly the best choices due to the periodic nature and large sensitivity of location counts.

Finally, we emphasize two important properties of semantic anonymization and query perturbation with differential privacy. First, unlike all other schemes, including query auditing and syntactic trajectory anonymization, differential privacy composes and the privacy loss can be quantified and gracefully degrades by multiple releases. This is crucial if the data gets updated and should be "re-anonymized", or, there are other independent releases with overlapping set of individuals (e.g., two CDR datasets about the same city from two different telecom operators). Second, privacy attacks may rely on very diverse background knowledge, which are difficult to capture. For example, not until the appearance of the reconstruction attack in Sect. 12.1.1.1 was it clear that individual trajectories can be recovered merely from spatio-temporal density. Only differential privacy seems to provide adequate defense (with properly adjusted ε and δ) against even such sophisticated attacks.

Nevertheless, there are still many interesting future directions to further improve performance. First, the data generating distribution can be implicitly modeled using generative artificial neural networks (ANNs) such as Recurrent Neural Networks (RNNs) [25]. Generative ANNs have exhibited great progress recently and their representational power has been demonstrated by generating very realistic (but still artificial) sequential data such as texts[10] or music. The intuition is that, as deep ANNs can "automatically" model very complex data generating distributions thanks to their hierarchical structure, they can potentially be used to produce realistic synthetic sequential data such as spatio-temporal densities. Second, current approaches release the spatio-temporal density only for a limited time interval. For example, the solution described in Sect. 12.4 releases the density for only a single week. To release density over multiple weeks, one need to use a the composition property of differential privacy which guarantees $(k\varepsilon, k\delta)$-DP for k-fold adaptive composition based on Theorem 12.1. These are still quite large bounds if we wish to release the density in the whole year with $k = 52$. Fortunately, tighter bound has been derived recently, building on the notions of Concentrated Differential Privacy, which guarantees $\left(O(\varepsilon\sqrt{k}), \delta\right)$-DP after k adaptive releases [1].

Acknowledgements Gergely Acs has been supported by the MTA Premium Post-doctoral Fellowship of the Hungarian Academy of Sciences. Gergely Biczók has been supported by the János Bolyai Research Scholarship of the Hungarian Academy of Sciences.

References

1. M. Abadi, A. Chu, I. Goodfellow, H. B. McMahan, I. Mironov, K. Talwar, and L. Zhang. Deep learning with differential privacy. In *ACM CCS*, 2016.
2. O. Abul, F. Bonchi, and M. Nanni. Never walk alone: Uncertainty for anonymity in moving objects databases. In *ICDE*, pages 376–385, 2008.

---

[10]http://karpathy.github.io/2015/05/21/rnn-effectiveness/.

3. G. Acs, J. P. Achara, and C. Castelluccia. Probabilistic k^m-anonymity (Efficient Anonymization of Large Set-Valued Datasets). In *IEEE International Conference on Big Data (Big Data)*, 2015.
4. G. Acs and C. Castelluccia. A Case Study: Privacy Preserving Release of Spatio-temporal Density in Paris. In *KDD '14 Proceedings of the 20th ACM SIGKDD international conference on Knowledge discovery and data mining*, Aug. 2014.
5. G. Acs, R. Chen, and C. Castelluccia. Differentially private histogram publishing through lossy compression. In *ICDM*, 2012.
6. C. C. Aggarwal. On k-anonymity and the curse of dimensionality. In *VLDB*, 2005.
7. L. Bengtsson, X. Lu, A. Thorson, R. Garfield, and J. Von Schreeb. Improved response to disasters and outbreaks by tracking population movements with mobile phone network data: a post-earthquake geospatial study in haiti. *PLoS Med*, 8(8):e1001083, 2011.
8. R. Chen, G. Acs, and C. Castelluccia. Differentially private sequential data publication via variable-length n-grams. In *ACM Conference on Computer and Communications Security*, pages 638–649, 2012.
9. R. Chen, B. C. M. Fung, and B. C. Desai. Differentially private trajectory data publication. *CoRR*, abs/1112.2020, 2011.
10. A. E. Cicek, M. E. Nergiz, and Y. Saygin. Ensuring location diversity in privacy-preserving spatio-temporal data publishing. *The VLDB Journal*, 23(4):609–625, 2014.
11. G. Cormode. Personal privacy vs population privacy: learning to attack anonymization. In *KDD*, pages 1253–1261, 2011.
12. G. Cormode, C. Procopiuc, D. Srivastava, E. Shen, and T. Yu. Differentially private spatial decompositions. In *ICDE*, pages 20–31, 2012.
13. Y.-A. de Montjoye, C. A. Hidalgo, M. Verleysen, and V. D. Blondel. Unique in the crowd: The privacy bounds of human mobility. *Scientific Reports, Nature*, March 2013.
14. I. Dinur and K. Nissim. Revealing information while preserving privacy. In *PODS*, pages 202–210, 2003.
15. C. Dwork, F. McSherry, K. Nissim, and A. Smith. Calibrating noise to sensitivity in private data analysis. In *TCC*, pages 265–284, 2006.
16. C. Dwork and A. Roth. The Algorithmic Foundations of Differential Privacy. *Foundations and Trends in Theoretical Computer Science*, 9(3–4), 2014.
17. C. Dwork and S. Yekhanin. New efficient attacks on statistical disclosure control mechanisms. In *CRYPTO*, 2008.
18. European Commission. General European Data Protection Regulation (GDPR). http://www.privacy-regulation.eu/en/index.htm, 2016.
19. L. Fan and L. Xiong. Real-time aggregate monitoring with differential privacy. In *ACM CIKM*, pages 2169–2173, 2012.
20. L. Fan, L. Xiong, and V. Sunderam. Differentially private multi-dimensional time series release for traffic monitoring. In *IFIP Annual Conference on Data and Applications Security and Privacy*, pages 33–48. Springer, 2013.
21. B. Fung, K. Wang, R. Chen, and P. S. Yu. Privacy-preserving data publishing: A survey of recent developments. *ACM Computing Surveys (CSUR)*, 42(4):14, 2010.
22. S. R. Ganta, S. P. Kasiviswanathan, and A. Smith. Composition attacks and auxiliary information in data privacy. In *KDD*, pages 265–273, 2008.
23. P. Golle. Revisiting the uniqueness of simple demographics in the US population. In *ACM WPES*, pages 77–80, 2006.
24. M. C. Gonzalez, C. A. Hidalgo, and A.-L. Barabasi. Understanding individual human mobility patterns. *Nature*, 453, 2008.
25. I. Goodfellow, Y. Bengio, and A. Courville. Deep learning. MIT Press, 2016.
26. M. Hardt, K. Ligett, and F. McSherry. A simple and practical algorithm for differentially private data release. In *NIPS*, 2012.
27. M. Hay, A. Machanavajjhala, G. Miklau, Y. Chen, and D. Zhang. Principled evaluation of differentially private algorithms using dpbench. In *Proceedings of the 2016 International Conference on Management of Data*, SIGMOD' 16, pages 139–154, 2016.

28. M. Hay, V. Rastogi, G. Miklau, and D. Suciu. Boosting the accuracy of differentially private histograms through consistency. *PVLDB*, 2010.
29. X. He, G. Cormode, A. Machanavajjhala, C. M. Procopiuc, and D. Srivastava. DPT: differentially private trajectory synthesis using hierarchical reference systems. *PVLDB*, 8(11):1154–1165, 2015.
30. T. Imielinski and W. L. Jr. On th undecidability of equivalence problems for relational expressions. In *Advances in Data Base Theory, Vol. 2, Based on the Proceedings of the Workshop on Logical Data Bases*, pages 393–409, 1982.
31. G. Kellaris and S. Papadopoulos. Practical differential privacy via grouping and smoothing. In *VLDB*, pages 301–312, 2013.
32. G. Kellaris, S. Papadopoulos, X. Xiao, and D. Papadias. Differentially private event sequences over infinite streams. *Proceedings of the VLDB Endowment*, 7(12):1155–1166, 2014.
33. R. Kitchin. The real-time city? big data and smart urbanism. *GeoJournal*, 79(1):1–14, 2014.
34. J. M. Kleinberg, C. H. Papadimitriou, and P. Raghavan. Auditing boolean attributes. In *ACM PODS*, pages 86–91, 2000.
35. C. Li, M. Hay, G. Miklau, and Y. Wang. A data- and workload-aware algorithm for range queries under differential privacy. *Proc. VLDB Endow.*, 7(5):341–352, Jan. 2014.
36. C. Li, M. Hay, V. Rastogi, G. Miklau, and A. McGregor. Optimizing linear counting queries under differential privacy. In *PODS*, pages 123–134, 2010.
37. N. Li, T. Li, and S. Venkatasubramanian. t-closeness: Privacy beyond k-anonymity and l-diversity. In *ICDE*, pages 106–115, 2007.
38. N. Li, W. Yang, and W. Qardaji. Differentially private grids for geospatial data. In *Proceedings of the 2013 IEEE International Conference on Data Engineering (ICDE 2013)*, ICDE'13, pages 757–768. IEEE Computer Society, 2013.
39. A. Machanavajjhala, D. Kifer, J. Gehrke, and M. Venkitasubramaniam. *L*-diversity: Privacy beyond *k*-anonymity. *TKDD*, 1(1), 2007.
40. F. McSherry. Privacy integrated queries: an extensible platform for privacy-preserving data analysis. In *SIGMOD*, pages 19–30, 2009.
41. D. J. Mir, S. Isaacman, R. Cáceres, M. Martonosi, and R. N. Wright. Dp-where: Differentially private modeling of human mobility. In *BigData Conference*, pages 580–588, 2013.
42. N. Mohammed, B. C. M. Fung, P. C. K. Hung, and C. Lee. Anonymizing healthcare data: a case study on the blood transfusion service. In *Proceedings of the 15th ACM SIGKDD International Conference on Knowledge Discovery and Data Mining, Paris, France, June 28 - July 1, 2009*, pages 1285–1294, 2009.
43. A. Monreale, G. Andrienko, N. Andrienko, F. Giannotti, D. Pedreschi, S. Rinzivillo, and S. Wrobel. Movement data anonymity through generalization. *Transactions on Data Privacy*, 3(2):91–121, 2010.
44. S. U. Nabar, K. Kenthapadi, N. Mishra, and R. Motwani. A survey of query auditing techniques for data privacy. In *Privacy-Preserving Data Mining - Models and Algorithms*, pages 415–431. 2008.
45. A. Narayanan and V. Shmatikov. Robust de-anonymization of large sparse datasets. In *IEEE Symposium on Security and Privacy (S&P)*, pages 111–125, 2008.
46. P. Neirotti, A. De Marco, A. C. Cagliano, G. Mangano, and F. Scorrano. Current trends in smart city initiatives: Some stylised facts. *Cities*, 38:25–36, 2014.
47. M. E. Nergiz, M. Atzori, Y. Saygin, and B. Güç. Towards trajectory anonymization: a generalization-based approach. *Trans. Data Privacy*, 2(1):47–75, 2009.
48. G. Poulis, S. Skiadopoulos, G. Loukides, and A. Gkoulalas-Divanis. Distance-based k^m-anonymization of trajectory data. *IEEE MDM*, 2:57–62, 2013.
49. W. Qardaji, W. Yang, and N. Li. Understanding hierarchical methods for differentially private histograms. *Proc. VLDB Endow.*, 6(14):1954–1965, Sept. 2013.
50. A. Rajaraman and J. Ullman. *Mining of Massive Datasets*. Cambridge University Press, New York, NY, USA, 2011.
51. V. Rastogi and S. Nath. Differentially private aggregation of distributed time-series with transformation and encryption. In *SIGMOD*, 2010.

52. L. Sun, D.-H. Lee, A. Erath, and X. Huang. Using smart card data to extract passenger's spatio-temporal density and train's trajectory of mrt system. In *Proceedings of the ACM SIGKDD international workshop on urban computing*, pages 142–148. ACM, 2012.
53. L. Sweeney. *k*-anonymity: A model for protecting privacy. *International Journal on Uncertainty, Fuzziness and Knowledge-based Systems*, 10(5):557–570, 2002.
54. M. Terrovitis, N. Mamoulis, and P. Kalnis. Privacy-preserving anonymization of set-valued data. *VLDB Endow.*, 1(1), 2008.
55. P. M. Vaidya. An algorithm for linear programming which requires o(((m+n)n^2 + (m+n)^1.5 n)l) arithmetic operations. In *ACM STOC*, pages 29–38, 1987.
56. N. Victor, D. Lopez, and J. H. Abawajy. Privacy models for big data: a survey. *International Journal of Big Data Intelligence*, 3(1):61–75, 2016.
57. Q. Wang, Y. Zhang, X. Lu, Z. Wang, Z. Qin, and K. Ren. Rescuedp: Real-time spatio-temporal crowd-sourced data publishing with differential privacy. In *Computer Communications, IEEE INFOCOM 2016-The 35th Annual IEEE International Conference on*, pages 1–9. IEEE, 2016.
58. X. Xiao, G. Bender, M. Hay, and J. Gehrke. iReduct: Differential privacy with reduced relative errors. In *SIGMOD*, pages 229–240, 2011.
59. X. Xiao, G. Wang, and J. Gehrke. Differential privacy via wavelet transforms. In *ICDE*, pages 225–236, 2010.
60. Y. Xiao, L. Xiong, L. Fan, S. Goryczka, and H. Li. Dpcube: Differentially private histogram release through multidimensional partitioning. *Trans. Data Privacy*, 7(3):195–222, Dec. 2014.
61. F. Xu, Z. Tu, Y. Li, P. Zhang, X. Fu, and D. Jin. Trajectory recovery from ash: User privacy is NOT preserved in aggregated mobility data. In *WWW 2017*, pages 1241–1250, 2017.
62. J. Xu, Z. Zhang, X. Xiao, Y. Yang, and G. Yu. Differentially private histogram publication. In *IEEE 28th International Conference on Data Engineering (ICDE 2012), Washington, DC, USA (Arlington, Virginia), 1–5 April, 2012*, pages 32–43, 2012.
63. H. Zang and J. Bolot. Anonymization of location data does not work: a large-scale measurement study. In *MOBICOM*, pages 145–156, 2011.
64. X. Zhang, R. Chen, J. Xu, X. Meng, and Y. Xie. Towards accurate histogram publication under differential privacy. In *Proceedings of the 2014 SIAM International Conference on Data Mining, Philadelphia, Pennsylvania, USA, April 24–26, 2014*, pages 587–595, 2014.

Chapter 13
Context-Adaptive Privacy Mechanisms

Florian Schaub

Abstract Sensing and context awareness are integral features of mobile computing and emerging Internet of Things systems. While context-aware systems enable smarter and more adaptive applications, they also cause privacy concerns due to the extensive collection of detailed information about individuals and their behavior, as well as the difficulties for individuals to understand and manage information flows. However, context awareness also holds significant potential for supporting users in managing their privacy more effectively. Context-adaptive privacy mechanisms can inform users about how changes in context may impact their privacy, recommend privacy-preserving actions tailored to the respective situation, as well as automate certain privacy configuration changes for the user. This chapter provides an overview of research on context-adaptive privacy mechanisms, including an introduction to context-aware computing and the context dependency of personal privacy; a discussion and model for operationalizing context awareness for privacy management, including privacy-relevant context features; as well as an overview of existing context-adaptive privacy mechanisms with various applications. The chapter concludes with a discussion of research challenges for context-adaptive privacy mechanisms.

13.1 Introduction

Context awareness is an essential aspect of mobile computing and the emerging Internet of Things. Today's smartphones, vehicles and other "smart" devices include a multitude of sensors that allow devices and respective applications to determine physical location, orientation, ambient noise, light levels, and many other context features; as well as collect this information periodically or continuously. Such context information can be leveraged to infer user behavior, activities, mobility

F. Schaub (✉)
University of Michigan, Ann Arbor, MI, USA
e-mail: fschaub@umich.edu

© Springer Nature Switzerland AG 2018
A. Gkoulalas-Divanis, C. Bettini (eds.), *Handbook of Mobile Data Privacy*,
https://doi.org/10.1007/978-3-319-98161-1_13

patterns, emotions or mood, as well as learn a user's interests and preferences. Situational awareness gained this way facilitates the adaptation of systems and applications to align with the user's context and appropriately support the user in their activities.

While context-awareness enables smarter and more adaptive technology, the extensive collection of sensor, context and mobility data has implications for personal and information privacy as discussed in prior chapters. The increased sensing, processing, and sharing of detailed information about users and their context further make it inherently difficult for individuals to determine who has access to information, let alone effectively control information flows. This creates inherent user interaction and usability challenges for solutions that aim to provide users with effective information and controls for privacy management.

Sensing and context-aware systems are often collecting data continuously [87, 122]. At the same time, the number of situations and entities potentially requiring privacy decisions and configuration increase constantly. This creates a scaling issue for privacy self-management [141], as it becomes unrealistic to correctly specify privacy settings for each system or situation in advance. Furthermore, the shift towards recording mundane activity and mobility information rather than specific events makes it difficult to grasp potential privacy implications of information collection [87]. Yet, advances in data mining and information retrieval make formerly ephemeral activities more accessible [1, 25, 80, 87, 141] and facilitate profiling through discovery of new patterns and knowledge by combining information from multiple sources [43]. Users may have inconsistent mental models of the capabilities and data practices of systems [117], which hampers their ability to predict what information is actually collected or to whom it is disclosed [2]. Long-term privacy implications of decisions and actions are typically hard to foresee without appropriate support [120], yet, typically "decisions about privacy must be made individually, in isolation, and far in advance" [141]. Users may also not realize that data access once authorized is still active in other situations, or that information collection may occur in unanticipated contexts [10, 19].

However, context awareness and privacy do not have to be mutually exclusive. Prior chapters presented and discussed methods to mitigate privacy issues associated with context and mobility data collection in multiple domains. Furthermore, context awareness can also be leveraged to actively protect privacy and support users in managing their privacy more effectively [128]. Context-adaptive privacy mechanisms can inform users how changes in context may impact their privacy, recommend privacy-preserving actions tailored to the respective situation, as well as automate certain privacy configuration changes for the user.

In this chapter,[1] we first introduce context-aware computing in more detail (Sect. 13.2), before discussing how context information can be leveraged in privacy management (Sect. 13.3). We further provide an overview of different types of

---

[1]Parts of this chapter have appeared in the author's doctoral dissertation [123] and a prior article [128]. This chapter provides an expanded and revised overview of research on context-adaptive privacy mechanisms.

existing or proposed context-adaptive privacy mechanisms (Sect. 13.4) and discuss research challenges in this domain (Sect. 13.5).

13.2   Context-Aware Computing

Context awareness in technology has been studied extensively. Subsequently, we provide a brief introduction to context awareness and context-aware computing from the perspective of ubiquitous computing research.

13.2.1   Defining Context Awareness

In 1987, as part of a critique of the artificial intelligence paradigm of planning, Suchman argued that computer systems should respond to the settings in which they are used [144]. Schilit and Theimer first introduced the term *context-aware computing* in relation to human-computer interaction and ubiquitous computing [129, 130]. They named the current location, other persons in the vicinity, and nearby resources as important aspects of context.

Schmidt [132] distinguishes multiple categories of context-aware computing applications. *Context-adaptive systems* are systems that perform actions when certain context conditions are met. *Adaptive and context-aware user interfaces* dynamically adjust to services and resources available in the current context and make them available to the user. *Context-aware resource management* dynamically maps system functions onto context features and available services. Context-awareness can further facilitate the *management of interruptions based on situations*. Schmidt's *sharing context* category encompasses applications that exchange context information between different systems or users. The category *metadata generation and implicitly user-generated content* reflects the idea that context information can serve as metadata to enrich created information or even as implicitly generated content on its own. For instance, location and mobility data collected via smartphones, connected vehicles or sensing infrastructure with the goal of improving traffic prediction models, maps, or localization accuracy. In context-aware computing, multiple context factors are typically combined to increase the accuracy of collected context information [14] and infer the current situation [1].

The diversity of context-aware applications raises the question what constitutes context. Dey defines context as "any information that can be used to characterize the situation of entities (i.e., whether a person, place, or object) that are considered relevant to the interaction between a user and an application, including the user and the application themselves" [44]. Abowd and Mynatt identified five general dimensions to describe context [1]:

- **Who.** What persons and entities are present in the user's or system's proximity.
- **What.** What are the current activities of the user, present entities, and systems.

- **Where.** What is the current location of the user, the system, or the application.
- **When.** What is the point in time to which the context relates to.
- **Why.** What are the user's reasons and intentions behind an activity.

Context-aware applications might examine the first four dimensions (*who, what, where, when*) in order to determine the user's intentions (*why*) and initiate system actions that support and satisfy these intentions [45, 46]. While the employed terms suggest that context is mainly based on physical aspects, context information can encompass physical, social, emotional, as well as informational aspects [46]. For example, the *who* dimension can also extend to virtually present entities or services. Thus, an entity can be any person, application, service, or object of relevance [44]. Baldauf et al. [13] distinguish between *physical, virtual*, and *logical sensors*. Physical sensors measure real-world phenomena, such as temperature, pressure, light intensity, or radio signal strength. Virtual sensors provide access to digital information, such as calendar data, emails, contacts, or social media posts. Logical sensors combine output from multiple sensors and information sources, also called sensor fusion [133], to obtain higher-level abstractions of the current context.

Consequently, a context-aware application can leverage such context information to provide the user with information or services that are relevant to the user's activities in a specific context [44]. A context instance can be described as a *situation* [134], which is defined as a set of states of relevant context features or entities [44]—a snapshot of a specific context configuration. Different situations can be distinguished by their specific set of values for the relevant context features.

Context information can exist at different levels of abstraction and interpretation. Situations composed of sensor-based cues can be seen as low-level context requiring further interpretation to be useful [21]. High-level context can be obtained through context reasoning and interpretation of available sensor information and, thus, enriching situations with semantic interpretations. An advantage of this view on context is that it becomes possible to distinguish between low-level and high-level context changes. Thus, applications, including context-adaptive privacy mechanisms, can adapt to high-level context changes, while ignoring low-level context changes that do not affect the higher level interpretation.

Situations do not exist in isolation. By viewing context as a process rather than a state, an information space can be modeled as a directed state graph, in which each node represents a different situation and the edges between the nodes are annotated with the conditions of the context change [38]. Modeling such relations can reduce the search space for situation recognition [21], but also requires knowledge of the changing conditions and potential situations.

An important aspect of context-aware computing is the quality of obtained context information. Sensor information can be inaccurate, incomplete, and noisy. Therefore, context-aware applications and systems must take uncertainty into account, e.g., by assigning confidence metrics to context values that represent the estimated likelihood of the value reflecting reality [59, 133]. The assessment of context quality makes it possible to measure improvements in context quality, e.g., with multi-sensor fusion, or to infer high-level context under consideration of potential ambiguities of the low-level context [21, 45]. Many context-aware systems reason over uncertain context by combining different reasoning approaches that support uncertainty [21].

13.2.2 Modeling Context

Context frameworks and middleware have emerged to ease development of context-aware applications and reuse of context information [74]. Most context frameworks support sensor fusion and aggregation of context information, as well as interpretation of raw context data. Many context-aware systems and context middleware follow a similar layered conceptual approach [13, 74], as depicted in Fig. 13.1. Context data is retrieved from physical and virtual sensors. Context processing components interpret the retrieved sensor data. As a first step, logical sensors may combine context information from multiple sources. Reasoning and interpretation of context information provide semantic interpretations or support the transformation of context data into different representations. A context management layer retains historical context information and provides interfaces for applications to access and process context information. Applications either pull context information from the context system or subscribe to specific context updates and events.

Context models are an integral part of context-aware systems. Context models reduce the complexity of context-aware applications by separating representation and evaluation of context information from application logic. They govern preprocessing, interpretation, and representation of context information, improve extensibility of context-aware applications [21] and facilitate exchange and reuse of context information. Formal context models further enable consistency checking

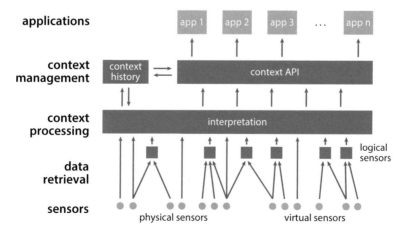

Fig. 13.1 Conceptual view of a context-aware system based on the layer models by Baldauf et al. [13] and Knappmeyer et al. [74]

and support inference of high-level context knowledge [21]. A basic context item can be modeled with five parameters [13]: its *semantic data type* (e.g., location or temperature), the current context *value*, a *time stamp* when the context item was acquired, the *source* it was acquired from (e.g., sensor id), and a *confidence* value that reflects the uncertainty of the measured context item. Bettini et al. [21] further note that a comprehensive context model should be able to represent context information on *different abstraction levels*, model *relations and dependencies between context information*, and support *efficient reasoning with uncertain information*. Furthermore, *context history* should be stored to support adaptation, *modeling formalisms* should ease modeling of real world concepts and support efficient context provisioning. Many models also distinguish between *primary and secondary context* [21]. Primary context information serves to index context situations and enables efficient access, while additional information is considered secondary context. Often location and time are used as primary context [1, 21].

While many different context models have been proposed, existing context models are characterized by a small set of major categories [21, 143]. *Key-value models* are basic context models that represent context as key-value tuples. They provide a flat view on context data. *Hierarchical markup models* can represent hierarchical data structures. Hierarchical markup models are typically represented in an XML dialect, such as RDF [143]. *Object-oriented context models* can be realized in object-oriented programming languages. Such models also enable hierarchical representation of context information, and additionally enable encapsulation [143]. They can combine representation and processing of context information, which can provide advantages for application-specific context models but reduces reusability.

These basic modeling approaches have a number of limitations according to Bettini et al. [21]. Due to pre-defined schemas, such models are limited in the variety of context information they can capture and have limited capabilities for expressing relationships and dependencies between context items. Quality of context information can be included but is often not expressive. Typically, basic models are also limited in terms of consistency checking, reasoning, and inference of high-level context abstractions.

Graphical modeling approaches, such as the Unified Modeling Language (UML) or the more specialized Context Modeling Language (CML) [58, 59], combine expressiveness and formality with ease of specification [143], especially for the modeling of relationships between context items. *Ontological models* leverage knowledge representation methods to describe relationships between context items, as well as the semantics of those relationships and context items. The formalization of context semantics enables consistency checking, the inference of context abstractions, and the derivation of new knowledge from asserted facts with semantic reasoning tools [21]. Ontological models facilitate knowledge sharing and interoperability between different context systems, because semantics of context items are represented in the models. Ontological context models can be described in OWL DL [21].

While ontological context models are most expressive [143], they may not be well suited for the representation of dynamic context information and corresponding adaptation preferences, due to performance limitations of online ontological reasoning and inadequate support for uncertainty [21]. As a consequence, Bettini et al. [21] advocate a *hierarchical hybrid model* that utilizes different context representations with varying levels of formal specification on different layers. The first layer performs low-level sensor fusion of raw sensor data. The second layer provides a basic context representation, e.g., based on markup or RDF, and supports efficient context reasoning to infer context abstractions. A third layer defines context semantics in an ontological model. Applications can choose the context representation layer that is most suitable to their requirements. For instance, Henricksen et al. combine CML with OWL to enable reasoning on uncertain information as supported by CML with OWL's semantic reasoning capabilities [58].

In addition to a context model, context-aware adaptation typically also requires the modeling of user characteristics and preferences to enable personalized adaptation. User characteristics can be integrated into a context model [67] or be maintained in a separate user model [31, 111]. Depending on the application, relevant user characteristics may include personal characteristics, the user's role, user preferences, user tasks and social relations [67, 111].

13.2.3   *Privacy Protection in Context-Aware Systems*

Context information in itself is often privacy sensitive, because it not only supports context-aware adaptation but can also facilitate undesired user profiling [2, 4, 64, 87]. Privacy risks of mobility data have been discussed in detail in the chapters "Privacy risks and inferences with mobility data" by Gambs and "Privacy in location-sensing technologies" by Solti et al. To address these risks of mobility and context data collection, privacy protections for context information have been studied extensively [22, 62]. Especially privacy of location information has received considerable attention as discussed in multiple chapters in Part II of this book. Here, we provide a short introduction on mechanisms and research directions for privacy protection in context-aware systems.

General privacy protection approaches can be applied to and adapted for context-aware systems. This includes strict access control, obfuscation of data through generalization or addition of noise, anonymization and de-identification methods, as well as private information retrieval and privacy-preserving data mining [22]. Privacy engineering and privacy by design principles facilitate the design of context-aware applications and systems that can meet both data quality and privacy requirements [40]. Heiber and Marrón [56] propose a privacy threat modeling framework for context-aware systems, which consists of a *data model*, an *adversary model*, and *inference rules*. The data model describes what context information is available and the adversary model defines what information the adversary could gain access to. The inference rules are a set of rules that the adversary can apply to obtain data, for example, linking or matching of context items. The framework enables evaluation of the amount of information that could be potentially gained by adversaries with varying capabilities against previously defined privacy requirements for the context-aware system or application.

Privacy extensions for context models and systems have been proposed to protect privacy-sensitive context information [13], typically centering on access control. For instance, the CML privacy extensions [60] enable expression of ownership of context facts, object types, fact types, and situations, as well as corresponding usage preferences. Rei is a privacy policy language for the CoBra context middleware [69]. Rei privacy policies govern actions by defining *rights, prohibitions, dispensations,* and *domain-dependent policies*. Available actions are pre-defined in an ontology. For example, location sharing is defined by an action that describes what to share (the location) with whom (a set of recipients). Corresponding privacy policies govern what entity can perform this action. The *info spaces* approach [68] supports access control and privacy management. An info space has a *user*, or a group of users, a *user agent* that handles privacy enforcement for the user, an *owner* that defines permissions for the info space, and a set of *information objects*, which are subject to authorization. Privacy policies are enforced in the info space system when accordingly tagged information crosses info space borders.

The info space approach further introduces support for adapting the granularity of context information. Similarly, Wishart et al. [154] extend CML with granularity support by representing granularity for specific data types as a hierarchical ontology. In their approach, privacy preferences are evaluated first, then granularity preferences are applied to the context information instance before disclosure. They later added dynamic discovery and processing of context sources with declarative rules [155] in order to form a complete privacy-aware context management system based on *context ownership* [60], *privacy and granularity preferences*, and *dynamic handling of disclosure requests*. Pareschi et al. [113] propose semantic aggregation based on local context in order to provide high quality of service while preserving privacy. Information from individual users is aggregated into *stereotypes* in order to enhance privacy by generalizing quasi-identifiers in order to reduce information that could identify the individual. Sheikh et al. [138] draw a connection between the quality of context and its privacy sensitivity. They propose that applications should only receive context information with a granularity that corresponds to the required quality of context.

Bettini and Riboni [22] caution that while hierarchical context models inherently support generalization of context facts—for instance generalizing location data to the city or region level—hierarchical levels of specificity are not necessarily expressions of sensitivity and that continuous data streams pose further challenges for maintaining privacy. Potential solutions are discussed in multiple chapters in this book.

13.3   Leveraging Context in Privacy Management

Privacy expectations and privacy behavior in social interactions have been shown to be subject to dynamic adaptation processes [6, 11, 101]. This dynamism of privacy is often not sufficiently supported in computing systems. Many systems and applications allow a priori configuration of static privacy settings, but do not support dynamic adaptations of those settings to meet the user's privacy expectations in different situations. Context-aware systems, however, have the ability to dynamically adapt to changes in the user's context, environment, and activities. Such context awareness also holds significant potential for dynamically supporting users in managing their privacy [123, 128].

13.3.1   Privacy is Contextual and Dynamic

Throughout their days, individuals constantly adjust their privacy expectations and their sharing behavior based on their activities and surroundings [149]. For example, the amount of information revealed in a conversation depends on who one is talking to, the topic of the conversation, and who else is around. Individual privacy expectations and perceptions of privacy infringement are highly contextualized and shaped by individual, social, and cultural expectations and norms [6, 103, 109].

13.3.1.1   Contextual Integrity

Marx introduced the notion of *personal border crossings* to characterize privacy violations [103]. He argues that privacy expectations are shaped by cultural and individual boundaries. *Natural borders* (e.g., walls and clothes) limit what can be perceived by others. *Social borders* reflect expectations in the roles of persons, e.g., lawyers and doctors keeping client and patient information confidential. *Spatial and temporal borders* separate disjoint events and episodes of life. They reflect the expectation that such events are not linked. *Ephemeral and transitory borders* reflect the expectation that fleeting moments are not recorded. If such borders are breached privacy expectations are being violated and the action that caused the breach is perceived as privacy infringing. For instance, when a user's location traces are used to infer socio-economic status or behavior patterns for targeted advertising.

Nissenbaum expands this perspective by framing privacy as *contextual integrity* [108, 109]. Privacy expectations are shaped by context-relative norms of information flow. The context considered in contextual integrity is elaborate and nuanced, going beyond the primarily sensor-oriented context common in context-aware systems. Context-relative norms of information flow are characterized by contexts, actors, attributes, and transmission principles [109].

Contexts—in the framework of contextual integrity—encompass the general institutional and social circumstances of a situation (e.g., healthcare, education, family, religion, etc.), the activities in which actors engage, as well as the purposes, goals, and *values* associated with those activities. Nissenbaum notes that individuals often engage in multiple such contexts at the same time which can be associated with different, potentially conflicting informational norms. For instance, talking about private matters at work in a specific society and culture [109].

Actors are senders, receivers, and information subjects who participate in activities and contexts. Actors have specific roles and capacities depending on the context. Roles define relationships between various actors, which express themselves through the level of intimacy, expectations of confidentiality, and power dynamics between actors [109]. Informational norms regulate information flow between actors.

Attributes describe the type and nature of the information being collected, transmitted, and processed. Informational norms render certain attributes appropriate or inappropriate in certain contexts. The concept of appropriateness in Nissenbaum's framework serves to describe what are acceptable actions and information practices.

Transmission principles constrain the flow of information between entities. They are associated with specific expectations. Typical transmission principles are confidentiality, reciprocity or fair exchange of information, and whether an actor deserves or is entitled to receive information.

Context-relative norms may be explicitly codified or only implicitly established. Common types of norms are morals, conventions of etiquette, rules, and procedures. Information flows that violate respective norms are perceived as privacy violations by individuals. Furthermore, technology may affect moral and political factors, e.g., power structures, fairness, or social hierarchies; as well as impact goals and values in a specific context.

The aspect of informational norms is also apparent in the concept of *collective information practices* proposed by Dourish and Anderson [48]. In their view, information flows not only transmit information but also serve as social boundaries, which help to define identity, membership, and affiliation in social groups. The acceptance and utilization of the same information practices shapes a group's identity.

13.3.1.2 Privacy Regulation Theory

Contextual integrity provides a framework for understanding privacy expectations in social contexts, and identifying privacy issues of information technology. Because privacy expectations vary with context, privacy regulation in social interactions occurs in a continuous adaptation process in which individuals balance their personal privacy needs with their desire for disclosure [149]. Understanding this process is essential for designing context-adaptive privacy mechanisms to effectively support it.

Altman's *privacy regulation theory* [11] describes privacy as a dynamic, dialectic, and non-monotonic process. In this process, individuals regulate what they disclose (*outputs*) and what level of potential intrusion they are willing to accept (*inputs*) based on internal changes (e.g., changes in personal preference, past experiences, or new knowledge), as well as external changes in the environment and current context. In social interaction, adjustments rely on *verbal*, *paraverbal*, and *nonverbal behavioral mechanisms*, such as revealing or omitting information (verbal), changing intonation and speaking volume (paraverbal), or using posture and gestures to non-verbally express and control personal space and territory.

A critical part of Altman's theory is the distinction between *desired privacy* and *achieved privacy*. Individual privacy preferences and privacy expectations may differ from the level of privacy obtainable in a given situation with the available privacy control means. If achieved privacy is lower than desired privacy, privacy expectations are violated and the individual feels exposed. Achieving more privacy than desired causes *social isolation*. Thus, the privacy regulation process aims for an optimal privacy level in which desired and achieved privacy are aligned.

Validation studies have shown that Altman's theory can be considered a realistic model of individual privacy adaptation behavior [101]. Despite its focus on privacy regulation in social interactions, Altman's theory suggests itself for application to privacy regulation in interactions with information technology, primarily to identify tensions affecting individual dynamic privacy regulation in the presence of technological systems [26, 92, 112].

Lehikoinen et al. extend Altman's theory for privacy in ubiquitous computing [92]. Focused on the bidirectional, dialectic nature of the privacy regulation process, they map Altman's inputs and outputs to different interaction patterns in ubiquitous computing environments. When interacting with an interactive environment or others' personal devices, the inputs are determined by those technical components, own outputs are partially dependent on the sensing capabilities of those components. When the user's personal device interacts with other devices or the environment, inputs and outputs are digital information. Lehikoinen et al. further introduce the concept of *leaking* to describe situations where the actual outputs exceed desired privacy [92]—a case of importance in information systems where individuals may not be fully aware of their outputs, i.e., what information about them is being sensed or communicated. Romero et al. focus on the dialectic aspect of the privacy regulation process [118]. They propose additional phases (*collaboration*, *signaling*, *joint understanding*) before the actual boundary regulation in order to better capture the influence of technology support in mediated communication in contrast to Altman's verbal, paraverbal, and nonverbal regulation mechanisms.

13.3.2 Operationalizing Context Awareness for Privacy

Altman's privacy regulation theory significantly influenced reasoning about privacy and has been found to be a realistic model to describe privacy regulation from an individual's perspective [101]. Altman's theory, as well as Nissenbaum's framework of contextual integrity, recognize that privacy regulation is a dynamic and dialectic process. Perception and awareness of the given situation shape a person's privacy concerns and expectations, together with personal privacy preferences, individual knowledge and experiences, as well as cultural and social background and constraints. This privacy decision making process results in a consciously or subconsciously desired privacy level, which is put into practice with available means of privacy control. Subsequently, the individual may receive feedback on the effectiveness of the exercised control, i.e., what level of privacy was actually achieved. Such feedback in turn leads to internal adjustments of individual privacy concerns, expectations, and preferences.

Context-adaptive privacy mechanisms can mirror aspects of a user's cognitive privacy regulation processes in order to provide privacy adaptation and privacy decision making support specific to the user's situation. For the sake of operationalizing privacy regulation theory for context-adaptive privacy mechanisms, the dynamic regulation process can be coarsely divided into three inter-related phases [128]:

- **Awareness.** Awareness of privacy-relevant processes and information flow shape individual privacy concerns [16]. An individual becomes aware of a contextual aspect or change in her environment that potentially necessitates regulative action to maintain a desired level of privacy (e.g., another person appears that could overhear a private conversation). The recognition of context changes as potential privacy risks depends on the individual's perception. However, with modern sensing technologies, a user's awareness and privacy perception is likely incomplete [87, 92], because sensors and information flows may not be apparent. Potential consequences are wrongly formed mental models and misconceptions about afforded privacy in a given context.

- **Decision.** Based on contextual and situational awareness, personal preferences and experiences, as well as cultural background and social motivations, an individual decides whether to decrease or increase exposure in the changed situation (e.g., including or excluding the new person from the conversation). Privacy decisions often need to be made based on incomplete information and are subject to cognitive biases and decision heuristics [6, 7], as well as susceptible to framing and manipulation [5, 6].

- **Control.** Once the individual formed a privacy decision, the decision needs to be mapped onto controls available in the current context. Available controls are determined by the means at disposal for asserting control (e.g., a door that can be closed to prevent eavesdropping) as well as the prevalent socio-cultural expectations and norms, which may restrict available controls (e.g., closing a door may be considered inappropriate in some cultures [135]). Although deciding on a regulation action and acting upon it are closely related, we argue that decision and control should be considered separately. Consciously or subconsciously forming an intention for a desired level of privacy (*desired privacy*) is an internal process, while the ability to implement the desired privacy is subject to external contextual constraints in a given situation (*achieved*

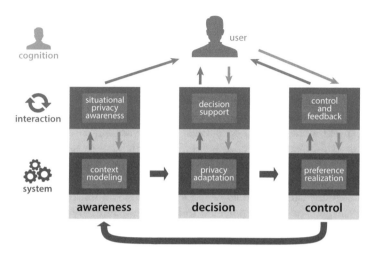

Fig. 13.2 Context-adaptive privacy mechanisms align context-aware system capabilities with the cognitive privacy regulation process in order to support privacy decision making and regulation

privacy). The results of performed control actions can be potentially perceived and verified by the user or the system and thus influence future awareness and subsequent regulation decisions [11].

In social interactions, these phases may overlap and influence each other. For instance, control actions may lead to awareness about their effectiveness, which may potentially require re-evaluation and re-adaptation. The regulation process runs continuously, resulting in micro-level privacy adaptations, such as adjusting what degree of information is revealed in a conversation, as well as macro-level adaptations, such as moving to a different, more private location.

Context-adaptive privacy mechanisms can support privacy regulation by supporting the different phases of this cognitive process on the system level, as shown in Fig. 13.2. Context-aware systems for privacy align well with the cognitive privacy regulation process. An individual combines situational awareness with individual preferences and experiences to make informed decisions on how to regulate privacy boundaries. Such decisions are implemented through actions, and personal preferences are adapted by learning from positive and negative experiences. Similarly, a context-adaptive privacy mechanism can leverage context awareness, elicited privacy preferences, and previous decisions to predict the user's privacy preferences and desired level of privacy for a given situation. Context-triggered changes in privacy expectation can either be addressed by automatically reconfiguring privacy controls and settings to the changed needs, or by suggesting privacy actions suitable for the current context to the user. This may include suggestions for more restrictive privacy configuration, but could also lead to a more permissive configuration [128]. Oppermann and Zimmermann [111] similarly distinguish three components of context-adaptive systems: the *sensory function* for obtaining relevant information, the *inference function* guiding adaptation, and the *effector function* implementing adaptations in the system.

Context-adaptive privacy mechanisms can further tailor what information and recommendations they provide to the user's current situation and information needs in order to optimally support privacy decisions without getting in the way of the user's activity or overburdening them with irrelevant privacy information or settings [124]. Interaction strategies may be constrained by application requirements that determine the appropriateness and opportunities for user interactions in order to avoid disrupting the user's primary activities. Awareness of the user's context and preferences also offers the potential for integrating privacy regulation tasks into the user's primary activities. Output modalities and the presentation of recommendations can be tailored to the user's primary context and activity. Thus privacy management has the potential to become a natural by-product of using a system rather than a burdensome configuration task [91].

The level of automation should further align with the expectations of individual users. For instance, some users may be content with largely automated adaptations of their privacy settings, while others may prefer explicit awareness and control. Automation preferences may also be different for different situations. For instance, most people would likely not object to their location being automatically shared with emergency services in the case of a severe car crash, but may have diverse preferences for everyday situations.

In behavioral privacy regulation, individuals leverage combinations of different privacy mechanisms to achieve a desired privacy level, depending on the environment and context [11]. Similarly, technical privacy control mechanisms often function as systems, which are combined and configured according to application needs and privacy requirements. Rather than merely preventing information exchange, control mechanisms should enable users to form and maintain realms of exclusion within a certain socio-technological context [142], which would facilitate interaction with specific desired people, devices, or systems without interference.

13.3.3   Privacy-Relevant Context Features

Context-adaptive privacy mechanisms aim to identify and adapt to privacy-relevant changes in context. This requires maintaining a context model composed of privacy-relevant context features. Much research has been conducted to gain a deeper understanding of what contextual factors affect privacy perceptions, concerns, and behavior.

In contrast to general-purpose context models and context systems, a privacy context model constitutes a high-level abstraction of context that focuses on privacy-relevant context features only, in order to most effectively support dynamic adaptation of privacy mechanisms. In practice, context-adaptive privacy mechanisms can act as a consumer of more detailed context information provided by a context middleware.

Adams and Sasse identified *information sensitivity*, *information receivers*, and *information usage* as key factors for privacy perceptions [8]. The framework of contextual integrity provides a more generalized perspective on those factors as *actors*, *attributes*, and *transmission principles* [109]. In order to support privacy decision making of individual users, a privacy context model should provide a user-centric perspective on context [126]. Therefore, we can distinguish between the user, the environment (including other entities), as well as activities that link the user to other entities.

13.3.3.1 User Features

The *user* is characterized by *privacy-sensitive items*, which are potentially exposed in the current situation. Privacy-sensitive items loosely correspond to attributes defined by the contextual integrity framework [109]. Privacy-sensitive items can either be information sources or disturbance endpoints, corresponding to Altman's outputs and inputs [11].

Information sources potentially reveal information about the user, this could be the user's behavior, presence or activity that can be observed, e.g., by sensors, or digital information created by or about the user. Information sources can reveal time-variant or static information [116]. For time-variant information, an observer's scope is limited by the observation window. While an observer may be able to predict past or future values of an information source, the prediction scope is bounded in relation to the observation scope. Other information is static, such as a name, social security number or fingerprint. Once disclosed, they are known to the observer, and can only be changed with substantial effort.

Some information sources can further provide information at different levels of *granularity* and abstraction [8, 57, 68, 154]. For example, location can be provided as exact geo coordinates, as an address, on a street level, city level, or region level. Changes in information granularity are privacy relevant, because coarser information increases the difficulty for an observer to derive the exact information and thus potentially affords higher anonymity or privacy. In the contextual integrity framework, granularity adjustments are considered as a transmission principle for restricting information flow [109]. How granularity is specified depends on the semantic type of an information source. If granularity is expressed numerically, the scale must be mapped to a generalization function for the specific information source, which then transforms the original information into a version with the respective specificity. If specificity is expressed by class identifiers (e.g., *street* or *city* for location), a partial order of semantics between classes, e.g., provided by an ontology, is required in order to be able to determine if granularity is increased or decreased.

Disturbance endpoints are an individual's physical aspects that constitute potential targets for physical disturbances [28, 75, 126]. Any action occurring in the user's physical proximity can be seen as a potential disturbance. For instance, a household robot may have no direct means to observe the user, but the device's activity or presence could still be perceived as an intrusion by the individual. Similarly, notifications in the user's environment can disrupt the user's solitude. The endpoint of a disturbance can be seen as a privacy-sensitive item, as it constitutes the connection point where the actual intrusion on an individual's privacy occurs. Therefore, disturbance endpoints should be included in privacy context models for smart environments and physical spaces [126]. Intuitively, the user's body is a potential endpoint for physical disturbances, as the user's senses perceive physical privacy intrusions. The user can be disturbed by touch, sound, smell, taste, and visual aspects [75]. In addition to the user's body, devices that are closely associated with the user may also be considered disturbance endpoints of the user. This may include the user's smartphone and its notifications but also wearable or implanted devices. Including such additional disturbance endpoints in privacy context models when relevant, allows to model changes of such endpoints and resulting privacy implications. In contrast to information sources, disturbance endpoints have typically no granularity. A disturbance endpoint is either exposed or not. Yet, certain disturbance endpoints can be perceived more invasive than others. For example, a vibrating phone is generally perceived less disturbing than a ringing phone.

13.3.3.2   Environment Features

The user's physical and virtual surroundings can be described as the user's *environment*. The environment contains context features that have an extrinsic effect on the user's privacy, compared to the intrinsic effects of user features. If one takes a user-centric perspective on modeling privacy context [126], the environment and its context features change based on the user's actions, e.g., when the user changes location. A different perspective can be to model physical and virtual aspects in multiple environments of which the user can be a part [34]. In this approach a user can participate in multiple environments at once. Chang et al. give the example of a video conference at the office, in which the user is in the virtual environment video conference and the physical environment office [34]. With a user-centric modeling perspective, the environment is implicitly defined by the user's location (the office), but includes any virtual entities that are able to participate in the user's physical environment, including any communication partners in a video call [126].

While environment models for context awareness can be highly detailed, a small number of environmental context features is most relevant for privacy. Primary privacy-relevant context features are other *entities* that participate physically or virtually in the user's environment, as well as the *observation or disturbance channels* [77], with which they are connected to the user's information sources and disturbance endpoints [126]. This corresponds to findings identifying receivers as a salient factor for privacy decisions [8, 109, 114]. Modeling of other environmental aspects, such as room layout or inanimate objects is typically not required, because any privacy-relevant change to such environment features would be reflected by changes to the set of present entities and whether and how they can access information sources and disturbance endpoints. For example, closing an office door should remove people in the hallway from the model.

Entities determine the relevant actors [109] in relation to privacy. For privacy preferences pertaining to the sharing of information, the receiving entity has been shown to be a key factor [8, 41, 63, 66, 95, 110, 115, 139]. Patil et al. identified it as the highest ranked factor for location sharing preferences [114]. Thus, present entities are often considered in the modeling of privacy-relevant context [53, 126]. An entity could be a *person, device, software agent,* or *service* that participates in the user's environment. Both *physical and virtual entities* can be represented in the same model [75, 126]. Entities in the user's physical environment potentially forward information to virtual entities. Virtual entities rely on physical entities in order to participate in the user's physical environment. For example, the user's location can only be observed by users of a location sharing service if some physical entity senses the information and relays it, e.g., a smartphone with a GPS sensor.

Channels model how entities participate in the user's environment. They define the underlying transmission principles considered in the contextual integrity framework [109]. *Observation channels* originate at an information source and are connected to one or more observing entities. *Disturbance channels* originate from one or more entities and end at one of the user's disturbance endpoints. This perspective corresponds to Altman's inputs and outputs in the sense that inputs, i.e., disturbances, depend on the capabilities of entities in the environments and that the user's own outputs also partially depend on the sensing capabilities of physically present entities [92]. A channel can consist of multiple *links* between a set of entities. Such a multi-link channel defines a directed graph between the user and multiple hierarchically organized entities [75].

In addition to entities and channels, *location* [53, 66, 77, 136] and *time* [77, 121] are often considered as environmental context factors when modeling privacy-relevant context. Benisch et al. find that location, the time of the day, and the day of the week influence privacy preferences of users [20]. Tsai et al. also note the importances of time [146], while Massimi et al. find that location is an important aspect in determining privacy sensitivity [104]. Location should be considered on different levels of abstraction, including the user's physical location (e.g. geo coordinates), a semantic interpretation of the location (e.g., a specific room in a building), as well as the type of the location or environment. Semantic location information can be derived from the user's position or other location-specific environment cues, such as nearby WiFi access points. Kargl et al. propose a semantic retrieval process for geographic locations [71]. The environment type gives the location further semantic meaning [106] and describes the prevailing social context [109]. Human association of location is based on actions rather than coordinates [29]. Massimi et al. found that the type of environment strongly influenced the expectations and perceptions of being recorded [104]. *Home, work,* and *other* are common types to categorize environments [55, 95]. The environment type can be derived to some extent from the semantic location and the user's activities. Krumm et al. propose a method for deriving general semantic labels for environments from geographic positions [81].

13.3.3.3 Activities

Dourish frames context as an interactional rather than a representational problem [47]. Grounded in an analysis of the sociological origins of context, he argues that context arises from the user's activity rather than being purely representational. Thus, the relevance of context features is dynamically defined for individual users by their activities. The same intuitively holds true for privacy in context. Depending on activity, certain personal aspects that are shared as part of an activity, may be considered sensitive in other activities or situations. Therefore, user's *activities* have been considered as a privacy-relevant context feature [77, 106, 121, 126]. For privacy in ambient assisted living systems, Shankar et al. note that activity can be a crucial discriminant [137], because such systems typically pertain to the same location—the home—and different activities at that location are likely associated with different preferences.

Purpose plays an important role in privacy decision making [19, 37, 65, 104]. The notion of activity can describe or be associated with purposes. An activity is an abstract description of what the user is doing. To a certain extent, the user's activities reflect the user's intentions and goals in a specific situation [109].

For context-adaptive privacy mechanisms, activities can further describe which entities must have access to certain privacy-sensitive items of the user so that the user can actually pursue the activity. However, activities should not be confused with privacy preferences or privacy settings. An activity describes what the user is doing in a situation and with whom, and is therefore part of context information; privacy preferences describe which entities are allowed to observe or disturb that situation.

Activity recognition is a well-researched topic of ubiquitous and pervasive computing research. Recognition of complex human activities is challenging because individuals may engage in concurrent or interleaved activities, which results in ambiguity for interpretation [72]. Common approaches for inferring activity from sensor data rely on machine learning or rule-based inference from different context cues and sensors [99], including wearable sensors [30, 89]. Activity recognition typically requires the collection of training data, which is used for relevant feature extraction and training and validation of a recognition model for specific activities or classes of activities [89]. Activity recognition in context-aware systems requires the real-time processing and analysis of continuous sensor data streams [80], such as location, body motion, or interaction with a system. Activities can be organized hierarchically, by decomposing them into smaller actions that are easier to detect than complex activities [66, 100]. Higher-level activities composed from a set of actions can then be used in reasoning [100]. Knowledge about previously recognized activity and past activities can improve activity recognition [80].

13.4   Context-Adaptive Privacy Mechanisms

Privacy-relevant context factors can be utilized by context-adaptive privacy mechanisms to actively support privacy decision making or automatically configure privacy settings in reaction to context changes. Context-adaptive privacy mechanisms take context information and knowledge about the user's privacy preferences, expectations or concerns as inputs and determine whether privacy adaptation is required, as well as how privacy aspects should be adapted to align with the user's privacy needs or which actions should be recommended to the user. While privacy preferences may be used in the reasoning process, reasoning results need to be translated into privacy specifications that map privacy decisions—made by the user or the system on behalf of the user—onto configurations for privacy controls and tools available in the current situation. The differentiation between higher level privacy preferences for reasoning and lower level policies for enacting preferences allows to decouple privacy reasoning from technical realization and enforcement aspects, in order to better align with the user's cognitive privacy regulation process [123].

In this section, we first discuss a number of requirements for context-adaptive privacy mechanisms before providing an overview of common approaches and applications for context-adaptive privacy mechanisms (see Table 13.1).

13.4.1   Requirements

The process of privacy reasoning is not only subject to the *internal constraints* posed by the preferences of the individual user and the *external constraints* posed

Table 13.1 Overview of context-adaptive privacy mechanisms

Category	Approaches
Privacy context acquisition and dissemination	Privacy labeling of context features
	Automated analysis of privacy implications
	Collaborative identification of sensors
	Machine-readable privacy specification
	Proactive communication of privacy information
Context-aware authentication	Adapt required level of authentication to context
	Context as authentication secret
Context-adaptive information disclosure	Context-based disclosure policies
	Limit access to context of origin
	Create and enforce contextual privacy borders
Context-aware content adaptation	Content hiding based on context
	Content adaptation based on present entities
	Privacy-friendly output modality selection
Context-adaptive privacy automation	Predict desired level of user involvement
	Automated blocking of disclosure and access requests

by the given context, but must also respect additional *systemic constraints* posed by system characteristics and available technology. Each of these aspects may introduce uncertainty into the privacy reasoning process that must be taken into account by context-adaptive privacy mechanisms. Context features represented in the privacy context model are derived from sensor data, which can be noisy, inaccurate, or incomplete [12]. This uncertainty needs to be reflected in privacy context models [126]. The user model, which captures the privacy preferences of an individual user in reference to contextual factors, is also subject to uncertainty. Any elicited or inferred privacy preferences can only be a discrete approximation of the user's true privacy preferences. Even if the user explicitly states a preference, uncertainty remains, because the user may not have been able to properly express the desired preference with the available interaction methods, or may not even be able to articulate a privacy preference consistently [6]. Furthermore, users' preferences may change over time or may only apply to specific situations. Hence, context-adaptive privacy mechanisms need to consider uncertainty in their reasoning processes. The confidence in the outcome of the reasoning process should reflect the uncertainty of the considered inputs.

In order to be trusted by users, context-aware systems must be perceived as reliable, which can be achieved with predictable and consistent behavior [35]. Following the principle of least astonishment, context-aware systems should aim to consistently match the user's expectations and preferences. At the same time, reasoning results should be explainable to the user [12]. The reasoning process must be intelligible and understandable to ensure that reasoning results are perceived as credible by the user [49]. Therefore, context-adaptive privacy mechanisms should align with the user's privacy decision making, for instance by modeling it after Altman's privacy regulation process [11] and its three phases [128], as described in Sect. 13.3. Furthermore, reasoning processes and their outcomes should be accompanied by intelligible explanations that can help the user understand actions taken by the system or recommendations provided to them [90, 93], e.g., by reducing the complexity of rules constituting the user model [35].

In order to be able to provide meaningful decision support in previously unknown situations and dynamically adapt to such new situations, context-adaptive privacy mechanisms should operate under the open world assumption. User models should be extensible in order to accommodate new situations and adapt to the individual user's privacy preferences over time by integrating explicit and implicit user feedback. Users should also be enabled to inspect and adjust inferred privacy preferences. Learning of privacy preferences and new contexts needs to occur online during normal operation. Considering the subjective and fluid nature of privacy preferences [6], the user model requires continuous maintenance [12] to account for changes in preferences, to add new preferences, and to correct erroneously learned preferences.

Furthermore, context-adaptive systems typically have to operate continuously which has interesting consequences for context-adaptive privacy mechanisms. For instance given a context change, a context-adaptive privacy mechanism may have to anticipate future context changes in order to prevent infringing situations before they occur, as well as perform adaptations immediately when certain context features change, e.g., remove sensitive information from a display before someone entering a space can see it [127]. Similarly, context-adaptive privacy mechanisms need to consider carefully when to prompt users for input. On the one hand, privacy mechanisms should not alienate the user with unexpected autonomous actions; on the other hand, the system should not annoy the user with copious messages and notifications. The crucial issue is to develop a system that provides user support for privacy management without being obtrusive or overwhelming.

13.4.2 Privacy Context Acquisition and Dissemination

A crucial aspect of any context system is the acquisition and dissemination of context information. In the case of privacy mechanisms, sensing of context information has been complemented with proactive approaches that communicate privacy-relevant information. For instance, sending out wireless privacy beacons [76, 87] enables devices to communicate their sensing and actuation capabilities, as well as their data practices in machine-readable formats.

Multiple machine-readable privacy specification formats and policy languages have been proposed [42, 82]. A well-known example is P3P [148]—the platform for privacy preferences—which was designed to enable website operators to express a legal privacy policy in machine-readable form. When users also specify their privacy preferences in a machine-readable format (e.g., with APPEL [88] or XPref [9]), a website's data practices can be matched against the user's personal privacy preferences with privacy practices of visited websites and detect conflicts. Machine-readable privacy policies, such as P3P, can be seen as *labeling protocols* [3] that enhance context features, namely present services and entities, with privacy-relevant information. Such machine-readable privacy specifications can be either integrated with service discovery protocols or actively announced—either by the respective entity [76, 85] (e.g., a localization system, a surveillance camera, a smart thermostat, or a vacuuming robot) or a separate infrastructure component or third party that gathers, aggregates and disseminates machine-readable privacy information about devices and systems in an environment to user devices (e.g., the user's smartphone). *Privacy proxies* [85]—dedicated entities trusted by the user and other stakeholders—can manage privacy and match user preferences with a system's data practices in the case of sensing systems and sensors with limited resources.

Langheinrich implemented these approaches in pawS [85]. Privacy beacons communicate data collection and processing practices of nearby systems to the user's trusted device. The user's privacy proxy obtains the respective machine-readable privacy policies and matches them with the user's pre-specified privacy preferences. Yee proposes a similar architecture [156], in which a smart environment has a dedicated privacy controller that performs policy matching between a user's privacy preferences and present system's privacy policies. In case of mismatch, the privacy controller initiates policy negotiation among the involved parties. Similarly,

P4P [83] is a context-based negotiation framework that arbitrates between a service provider's P3P privacy policy and the user's privacy preferences selected according to the user's context.

A challenge for approaches that rely on machine-readable privacy specifications is that they require cooperation by the entity collecting information and translating their natural language privacy policies into machine-readable formats. This process is often met by resistance due to fears and uncertainty regarding the legal nature of machine-readable formats and associated liability. As a result, even widely popularized and standardized formats, such as P3P, lacked adoption and have been largely abandoned [39]. However, recent advances in natural language processing, machine learning, crowdsourcing, and static code analysis, provide the opportunity for service operators, third parties as well as regulators to automate the analysis of systems' natural language privacy policies and program code to infer their data practices [23, 27, 119, 125, 151, 152, 157]. Such analysis results could then be encoded in machine-readable formats to support context-adaptive privacy mechanisms.

Similar approaches can be employed to detect and identify sensors in physical environments. Winkler and Rinner propose collaborative tagging of cameras to gain privacy awareness with respect to video surveillance [153], i.e., individuals mark locations and characteristics of spotted cameras in a mapping system. Korayem et al. propose an automated method to identify computer screens from camera images [78]. Information about identified sensors or devices and their locations can be leveraged by context-adaptive privacy mechanisms to determine privacy risks.

How and what privacy-relevant context information is gathered ultimately depends on the purpose and requirements of the respective context-adaptive privacy mechanism. Furthermore, environment constraints, such as the level of trust in the present infrastructure or the willingness and capability of other entities to cooperate with privacy mechanisms. Next, we discuss different kinds of context-adaptive privacy mechanisms.

13.4.3 Context-Aware Authentication

Context awareness has been proposed as an improvement for user authentication. Sigg notes that context awareness could enable password-less authentication and unobtrusive adaptive security [140]. He also suggests the use of ambient audio as a location- or situation-based secret key. Bardram et al. leveraged the user's location in an environment as an additional factor in user authentication [15]. Langheinrich proposed a password-free authentication scheme for RFID tags [86]. Instead of revealing the complete tag ID, an RFID tag releases secret shares of the ID over time, thus requiring an attacker to be in close proximity for a longer time, while legitimate users can identify their tags efficiently by relying on simple caching strategies. Mayrhofer and Gellersen leverage parallel motion as a context factor in device paring [105]. They propose a method that relies on shaking two mobile devices together in order to derive a shared key from accelerometer data.

Gupta et al. propose what they call intuitive and sensible access control (ISAC) as an approach to autonomously select adequate authentication methods for smartphones depending on the user's context [52, 53]. For instance, password or PIN entry may be required in public but not when the user is alone. Their approach uses location, present Bluetooth devices, and WiFi access points as features to define *contexts of interest*. How frequently users encounter a context of interest is used to determine *familiarity of context* in order to identify likely "safe" locations.

Hayashi et al. propose context-aware scalable authentication (CASA) [55], which employs a similar approach. They uses passively collected context features to determine whether a user is at home, work or another place (other). Depending on the user's context, the user is either authenticated implicitly without interaction based on the passive context features, or active authentication is required (e.g., with PIN or password) when not at home or work.

Context-aware authentication approaches can be leveraged in context-aware privacy mechanisms to identify and authenticate the primary user as well as other entities detected in the environment.

13.4.4 Context-Adaptive Information Disclosure

The integration of context awareness into information disclosure and access control mechanisms has received considerable attention. Particularly location is a frequently used context feature to manage information sharing and access control. For instance, Behrooz and Devlic propose the context-aware privacy policy language (CPPL) [17]. With CPPL, machine-readable privacy specifications are scoped to certain contexts, e.g., based on a where a specific system is active. CPPL facilitates filtering of relevant privacy policies based on current context.

Jagtap et al. propose a privacy system that constrains information flow from mobile devices with dynamic semantic reasoning over context and pre-specified privacy preferences [66]. Their context features include user location, surroundings, other present entities, and inferred user activity, which are associated with entity roles. The Android-based implementation of their approach [50] supports context-aware privacy preferences encoded as privacy policies. These policies can be used for instance to specify under what circumstances smartphone apps should receive correct or fake location information and other sensor data.

Context and proximity have further been used to limit access to information to the context in which it has been collected originally [70, 73, 84]. For example, Kriplean et al. developed an approach that makes RFID readouts only accessible to devices that were physically present when the readout occurred [79].

The *info spaces* concept [61, 68] is a technical realization of Marx's personal border crossings theory [103]. Information spaces are defined by physical, social or activity-based borders, which are supported by location, entity and activity detection as context features. Privacy preferences (encoded as privacy policies) are enforced when accordingly-tagged information crosses such borders. This may include granting or denying read or write access to the information, adjusting the granularity and accuracy with which information is sensed or shared, as well as aggregating information to obtain higher-level interpretations.

Moncrieff et al. use a context model to manage privacy in a smart home equipped for ambient assisted living [106]. Their context features include a user's location (room in the house), social interactions (other present persons), hazardous activity (e.g., leaving the kitchen stove on), and unusual periods of inactivity. Context are matched against pre-defined privacy disclosure rules for care givers. Their policies support different granularity levels to regulate sensor access. An active feedback display provides occupants with information about current and past observers on demand.

13.4.5 Context-Aware Content Adaptation

Approaches in the previous category primarily provide context-aware adaptation of what information is disclosed or forwarded to outside parties. However, particularly in the context of smart environments and media spaces, context awareness has further been used to dynamically adapt content and information within the environment in order to provide privacy protections.

An early example of context-aware content adaptation is Schmidt's Context NotePad, which is a PDA application that dynamically hides its content when the user is not alone in the room and is not actively using the NotePad [131]. The TreasurePhone system [136] uses location to specify and activate preference spheres. The user-specified preference spheres limit which information on a mobile phone can be accessed in a certain location.

Presence and position of individual users in smart spaces can been used to implicitly regulate privacy and content visibility. Neustaedter and Greenberg propose a system in which the video stream in a media space automatically stops when the user leaves the chair or other persons enter the room [107]. The ProD system uses access control lists to define privacy preferences for content adaptation [36]. Similarly, the Angel system [51] poses privacy restrictions on displayed content based on the user's activity, which is associated with user-defined privacy rules. Marquardt and Greenberg suggest the use of proxemic interaction, which leverages context information about present persons and devices to guide device adaptation, for privacy management [102]. The PriCal system [127] enables context-aware privacy adaptation on calendar wall displays. More specifically, displayed calendar views are dynamically adapted to present persons and their privacy preferences for each other. In addition to known persons, users can also specify privacy preferences for unknown persons, i.e., people who are not registered system users. The system uses case-based reasoning to learn a user's nuanced privacy preferences based on users adjusting the visibility of individual calendar entries (full, busy, hidden) in deviation from their pre-specified rules.

The ATRACO system [77] considers both physically present entities as well as virtual entities connected via communication channels to the user's environment in context-adaptive privacy management. Privacy preferences describe what information items or spaces can be accessed by whom and how, in relation to a specific context. Changes in context trigger dynamic privacy evaluation. Privacy preferences are pre-specified as ontology instances, which are part of a larger user profile for dynamic adaptation of the smart environment.

When multiple output modalities are available in the user's physical environment, information display and means of interaction can be adapted to block observations by undesired entities. For example, private notifications could be displayed only on the user's personal device, such as a smartphone, rather than on a wall display to reduce the opportunity of visual observation channels by other entities [32]. Similarly, auditive output can be moved from speakers to earphones or translated into a visual representation. Furthermore, observation granularity can be tied to the physical arrangement of entities. For instance, Vogel and Balakrishnan propose a calendar application for public displays that only displays a user's calendar entries if the user is close enough to the display so that the display's content is shielded by the user's body [147] .

13.4.6 Context-Adaptive Privacy Automation

A major promise of context-adaptive privacy mechanisms is that they can reduce privacy management and configuration effort for users. However, not all privacy or information sharing decisions can or should be automated—automation and user autonomy need to be balanced carefully. Furthermore, systems need to enable users to correct decisions [54]. Bellotti and Edwards discuss how intelligibility of context-aware systems can be enhanced through user involvement [18]. Depending on the level of uncertainty about an inferred decision, a system should either provide means for correction, require confirmation from the user, or offer the user the available choices for selection. Multiple approaches have been proposed to leverage context awareness in determining if and when users should be involved in privacy decisions and when certain decisions can be automated.

The Super-Ego framework dynamically determines whether the user should be involved in decisions about location disclosure requests [145]. A decision engine uses a set of previous disclosure decisions from the larger user base to decide about user involvement based on two thresholds for manual decision and automatic decision. Location requests where the average of previous disclosure decisions is below the manual decision threshold are denied, requests between the manual and automatic decision threshold require user intervention, and requests above the automatic decision threshold are granted automatically. By adjusting these thresholds different decision strategies can be supported. Toch finds that mixed strategies provide the best tradeoff between accuracy and automation [145]. Similarly, SPISM [24] semi-automatically determines if for a given location request the user's location should be shared and at what granularity. Context information is used in classifying the request and assigning a sharing class (no; yes with low/medium/high granularity). Depending on the level of confidence in the classification result, the system prompts the user to make the decision.

Wijesekera et al. leverage context awareness to explore opportunities for automatically blocking permission requests from smartphone apps [150]. They conducted a field study to determine how context of an app's permission access affects users' privacy concerns. They identified the app's visibility (is the display on, is the user interacting with the app or not) and the request frequency as important factors in user's privacy preferences. In further experiments, they find that automatically blocking requests when the screen is off is unlikely to interfere with the user experience but enhances privacy. If an app that the user is currently not using requests resource access a prompt should be shown to obtain the user's consent. Apps that are running in the background should use passive indicators, such a GPS icon, when resources are being accessed.

13.5   Research Challenges

The variety of context-adaptive privacy mechanisms in different domains demonstrates the benefits of leveraging context awareness to actively support and partially automate privacy management. Context-adaptive privacy mechanisms are subject of active research, with multiple research challenges requiring further investigation:

- **Secure and trustworthy context acquisition.** Context-adaptive privacy mechanisms and other context systems rely on the integrity and trustworthiness of context information. Some sensors and context information can potentially be spoofed, which could trick context-adaptive privacy mechanisms into revealing personal information in the wrong situations. Therefore, context-adaptive privacy mechanisms should be designed to be resilient against spoofing attacks. For instance, by triangulating context information with different types of sensors and context acquisition methods, sensor fingerprinting, or distributed trust management in sensing and context infrastructures. In reasoning, confidence in context information and uncertainty about undetected context features should be considered.
- **Protection of contextual privacy information.** As discussed in Sect. 13.2.3, context information itself can often be privacy-sensitive and needs to be adequately secured, especially if systems retain context and privacy decision history. Similarly, context-specific privacy preferences and rules should be treated as privacy-sensitive information in their own right, as they detail what entities are given access to what kind of information in different situations. Some approaches rely on a trusted privacy assistant or privacy proxy [85] to aggregate

and protect a user's privacy information. Many of the approaches for protecting context information, such as authentication, obfuscation and other cryptographic techniques, can also provide opportunities for protecting privacy preference information against information leakage.

- **Privacy adaptation in heterogeneous environments.** A particular challenge for privacy mechanisms is to provide individual users with control over the collection, dissemination, and use of their information in collaborative and shared information ecosystems [1]. Different systems, infrastructure, sensors, services, or devices may be controlled by different stakeholders [2]. Thus, how a specific privacy decision can be translated into a privacy adaptation largely depends on the control capabilities available in the given context, i.e., the level of control and trust concerning other devices, infrastructure, and entities [75]. Thus, context-adaptive privacy mechanisms may have to consider different adaptation strategies within the same context as well as across contexts and systems. Furthermore, privacy controls available within a specific environment should be discoverable in order to facilitate privacy management and adaptation in previously unknown environments and contexts.

- **Privacy adaptation in physical environments.** Context-aware adaptation of content and information flows faces a particular challenge in physical spaces: other persons may be able to observe content and information adaptation, which could potentially be interpreted negatively and reveal the user's privacy preferences, potentially resulting in awkward situations. Thus, context-adaptive privacy mechanisms should be designed with plausible deniability in mind. Resulting adaptations should either be difficult to observe by others or others should not be able to determine whether the adaptation occurred because of them. For instance, the PriCal system [127] implements a hide-then-reveal paradigm: Whenever a person enters a room, the calendar display is cleared instantly when the new entity is detected and subsequently populated with content that has been adapted to the user's privacy preferences for the changed context. Because the display updates every time someone enters the room, study participants did not perceive those adaptations as specifically related to them [127]. Similarly, ATRACO [77] adapts a photo slideshow dynamically to the user's privacy preferences for present persons by seamlessly filtering out photos they should not see.

- **Privacy adaptation in multi-user environments.** Depending on the application and system context, context-adaptive privacy mechanisms may have to take privacy preferences from multiple users into account. In such situations, diverging privacy preferences have to be resolved by the system while respecting individuals' privacy preferences. Privacy preference negotiation and resolution could occur automatically or delegate the final decision whether some information should be disclosed to each user.

- **User trust in adaptation capabilities.** Context-adaptive privacy mechanisms must be perceived as trustworthy and reliable by users in order for users to trust the mechanisms with dynamically regulating privacy for them. This is particularly relevant for autonomous privacy adaptation on the user's behalf.

Trust evaluation in inter-personal relations as well as in technical systems has to rely on external trust signals in order to obtain information about internal trust facets [33], thus context-adaptive privacy mechanisms need to provide indications to users that they are functioning as configured or expected.

- **Personalized Privacy Adaptation.** Context-adaptive privacy mechanisms leverage context information to support users with privacy management. Privacy mechanisms can further learn a user's privacy preferences over time to not only contextualize but also personalize privacy management. Existing personalized privacy approaches learn from user feedback [63, 120] or derive privacy preference profiles from many users [94, 96, 98]. These approaches can further be combined to bootstrap privacy preferences by matching a user to one of a small set of privacy profiles [97, 98] and then leveraging user feedback and behavior to further refine and extend the user model to account for individual nuances in privacy preferences [123].

13.6 Summary

While context aware systems and the collection of context information pose challenges for personal privacy, we outlined the potential for supporting privacy management with context awareness in this chapter. Interpersonal privacy regulation has been shown to be highly dynamic and dependent on context. We presented a model for operationalizing context awareness by developing privacy mechanisms that align in their way of operation with individuals' privacy decision making processes. Context-adaptive privacy mechanisms can leverage context awareness to detect and determine privacy-relevant context changes in a user's environment and either provide context-specific privacy recommendations to the user or automatically adjust and adapt privacy configurations to ensure that the user's privacy preferences are respected in the changed context. We further provided an overview of privacy-relevant context features that have been shown to play a role in individual privacy decision making, as well as an overview of existing context-adaptive privacy mechanisms in various domains and associated research challenges. Context-adaptive privacy mechanisms are a promising approach for reducing the user effort in privacy management, in particular in sensor-rich environments.

References

1. G. D. Abowd and E. D. Mynatt. Charting past, present, and future research in ubiquitous computing. *ACM Transactions on Computer-Human Interaction (TOCHI)*, 7(1):29 –58, 2000.
2. M. Ackerman and T. Darrell. Privacy in context. *Human-Computer Interaction*, 16(2–4):167–176, 2001.
3. M. S. Ackerman. Privacy in pervasive environments: next generation labeling protocols. *Personal and Ubiquitous Computing*, 8(6):430–439, Sept. 2004.

4. M. S. Ackerman and S. D. Mainwaring. Privacy Issues and Human-Computer Interaction. In L. F. Cranor and S. Garfinkel, editors, *Security and Usability*, chapter 19, pages 381–400. O'Reilly, 2005.

5. A. Acquisti, I. Adjerid, R. Balebako, L. Brandimarte, L. F. Cranor, S. Komanduri, P. G. Leon, N. Sadeh, F. Schaub, M. Sleeper, Y. Wang, and S. Wilson. Nudges for privacy and security: Understanding and assisting users choices online. *ACM Computing Surveys*, 50(3), 2017.

6. A. Acquisti, L. Brandimarte, and G. Loewenstein. Privacy and human behavior in the age of information. *Science*, 347(6221):509–514, 2015.

7. A. Acquisti and J. Grossklags. What Can Behavioral Economics Teach Us About Privacy? In *Digital Privacy: Theory, Technologies, and Practices*, chapter 18, pages 363–377. Auerbach Publications, 2008.

8. A. Adams and M. Sasse. Privacy in multimedia communications: Protecting users, not just data. In *Human-Computer Interaction/Interaction d'Homme-Machine (IMH-HCI '01)*, pages 49–64, 2001.

9. R. Agrawal, J. Kiernan, R. Srikant, and Y. Xu. An XPath-based preference language for P3P. In *12th international conference on World Wide Web (WWW '03)*. ACM, 2003.

10. H. Almuhimedi, F. Schaub, N. Sadeh, I. Adjerid, A. Acquisti, J. Gluck, L. F. Cranor, and Y. Agarwal. Your location has been shared 5,398 times! a field study on mobile app privacy nudging. In *Proc. CHI '15*. ACM, 2015.

11. I. Altman. *The Environment and Social Behavior: Privacy, Personal Space, Territory, Crowding*. Brooks/Cole Publishing company, Monterey, California, 1975.

12. A. Aztiria, A. Izaguirre, and J. C. Augusto. Learning patterns in ambient intelligence environments: a survey. *Artificial Intelligence Review*, 34(1):35–51, May 2010.

13. M. Baldauf, S. Dustdar, and F. Rosenberg. A survey on context-aware systems. *International Journal of Ad Hoc and Ubiquitous Computing*, 2(4):263–277, 2007.

14. J. Bardram and A. Friday. Ubiquitous Computing Systems. In J. Krumm, editor, *Ubiquitous Computing Fundamentals*, chapter 2, pages 37–94. CRC Press, 2009.

15. J. E. Bardram, R. E. Kjær, and M. Ø. Pedersen. *Context-Aware User Authentication – Supporting Proximity-Based Login in Pervasive Computing*, pages 107–123. Springer Berlin Heidelberg, Berlin, Heidelberg, 2003.

16. L. Barkhuus. The Mismeasurement of Privacy: Using Contextual Integrity to Reconsider Privacy in HCI. In *ACM annual conference on Human Factors in Computing Systems (CHI '12)*, page 367, New York, New York, USA, 2012. ACM.

17. A. Behrooz and A. Devlic. A Context-aware Privacy Policy Language for controlling access to context information of mobile users. In *Conference on Security and Privacy in Mobile Information and Communication Systems (MobiSec '11)*. Springer, 2011.

18. V. Bellotti and K. Edwards. Intelligibility and Accountability: Human Considerations in Context-Aware Systems. *Human-Computer Interaction*, 16(2):193–212, Dec. 2001.

19. V. Bellotti and A. Sellen. Design for Privacy in Ubiquitous Computing Environments. In *Third European Conference on Computer-Supported Cooperative Work (ECSCW '93)*. Springer, 1993.

20. M. Benisch, P. G. Kelley, N. Sadeh, and L. F. Cranor. Capturing location-privacy preferences: quantifying accuracy and user-burden tradeoffs. *Personal and Ubiquitous Computing*, 15(7):679–694, Dec. 2010.

21. C. Bettini, O. Brdiczka, K. Henricksen, J. Indulska, D. Nicklas, A. Ranganathan, and D. Riboni. A survey of context modelling and reasoning techniques. *Pervasive and Mobile Computing*, 6(2):161–180, Apr. 2010.

22. C. Bettini and D. Riboni. Privacy protection in pervasive systems: State of the art and technical challenges. *Pervasive and Mobile Computing*, 17:159–174, 2015.

23. J. Bhatia, T. D. Breaux, and F. Schaub. Mining privacy goals from privacy policies using hybridized task recomposition. *ACM Trans. Softw. Eng. Methodol.*, 25(3):22:1–22:24, May 2016.

24. I. Bilogrevic, K. Huguenin, B. Agir, M. Jadliwala, and J.-P. Hubaux. Adaptive information-sharing for privacy-aware mobile social networks. In *ACM international joint conference on Pervasive and ubiquitous computing (UbiComp '13)*, page 657, New York, New York, USA, 2013. ACM.

25. J. Bohn, V. Coroam, M. Langheinrich, F. Mattern, and M. Rohs. Social, Economic, and Ethical Implications of Ambient Intelligence and Ubiquitous Computing. In W. Weber, J. M. Rabaey, and E. Aarts, editors, *Ambient Intelligence*, chapter 1, pages 5–29. Springer, 2005.

26. M. Boyle and S. Greenberg. The language of privacy: Learning from video media space analysis and design. *ACM Transactions on Computer-Human Interaction*, 12(2):328–370, June 2005.

27. T. D. Breaux and F. Schaub. Scaling requirements extraction to the crowd: Experiments with privacy policies. In *2014 IEEE 22nd International Requirements Engineering Conference (RE)*, pages 163–172, Aug 2014.

28. P. Brey. The Importance of Privacy in the Workplace. In S. O. Hansson and E. Palm, editors, *Privacy in the Workplace*, chapter 5, pages 97–118. Fritz Lang, 2005.

29. B. Brown, A. S. Taylor, S. Izadi, A. Sellen, J. J. Kaye, and R. Eardley. Locating Family Values: A Field Trial of the Whereabouts Clock. In *9th international conference on Ubiquitous computing (UbiComp '07)*, pages 354–371. Springer, 2007.

30. A. Bulling, U. Blanke, and B. Schiele. A tutorial on human activity recognition using body-worn inertial sensors. *ACM Comput. Surv.*, 46(3):33:1–33:33, Jan. 2014.

31. H. Byun and K. Cheverst. Exploiting user models and context-awareness to support personal daily activities. In *Personal Daily Activities, Workshop in UM2001 on User Modelling for Context-Aware Applications*, 2001.

32. H. Cao, P. Olivier, and D. Jackson. Enhancing Privacy in Public Spaces Through Crossmodal Displays. *Social Science Computer Review*, 26(1):87–102, Feb. 2008.

33. C. Castelfranchi and R. Falcone. *Trust Theory: A Socio-Cognitive and Computational Model*. John Wiley & Sons, 1 edition, 2010.

34. Y.-L. Chang, E. Barrenechea, and P. Alencar. Dynamic user-centric mobile context model. In *Fifth International Conference on Digital Information Management (ICDIM '10)*, pages 442–447. IEEE, July 2010.

35. K. Cheverst, N. Davies, K. Mitchell, and C. Efstratiou. Using Context as a Crystal Ball: Rewards and Pitfalls. *Personal and Ubiquitous Computing*, 5(1):8–11, Feb. 2001.

36. B. Congleton, M. S. Ackerman, and M. W. Newman. The ProD framework for proactive displays. In *21st annual ACM symposium on User interface software and technology (UIST '08)*, page 221, New York, New York, USA, 2008. ACM.

37. S. Consolvo, I. Smith, T. Matthews, A. LaMarca, J. Tabert, and P. Powledge. Location disclosure to social relations: why, when, & what people want to share. In *SIGCHI conference on Human factors in computing systems (CHI '05)*, pages 81–90. ACM, 2005.

38. J. Coutaz, J. L. Crowley, S. Dobson, and D. Garlan. Context is key. *Communications of the ACM*, 48(3):49–53, Mar. 2005.

39. L. F. Cranor. Necessary but not sufficient: Standardized mechanisms for privacy notice and choice. *Journal on Telecommunications and High Technology Law*, 10:273, 2012.

40. G. Danezis, J. Domingo-Ferrer, M. Hansen, J.-H. Hoepman, D. L. M tayer, R. Tirtea, and S. Schiffner. Privacy and data protection by design – from policy to engineering. resreport, European Union Agency for Network and Information Security (ENiSA), 2015.

41. S. Davis. Using relationship to control disclosure in Awareness servers. In *Graphics Interface (GI '05)*, pages 145–152, 2005.

42. J. De Coi and D. Olmedilla. A review of trust management, security and privacy policy languages. In *International Conference on Security and Cryptography (SECRYPT '08)*. INSTICC Press, 2008.

43. P. De Hert, S. Gutwirth, A. Moscibroda, D. Wright, and G. González Fuster. Legal safeguards for privacy and data protection in ambient intelligence. *Personal and Ubiquitous Computing*, 13(6):435–444, Oct. 2009.

44. A. K. Dey. Understanding and Using Context. *Personal and Ubiquitous Computing*, 5(1):4–7, Feb. 2001.
45. A. K. Dey. Context-Aware Computing. In J. Krumm, editor, *Ubiquitous Computing Fundamentals*, chapter 8, pages 321–352. CRC Press, 2009.
46. A. K. Dey and G. D. Abowd. Towards a Better Understanding of Context and Context-Awareness. In *CHI 2000 Workshop on The What, Who, Where, When, and How of Context-Awareness*, 2000.
47. P. Dourish. What we talk about when we talk about context. *Personal and Ubiquitous Computing*, 8(1):19–30, Feb. 2004.
48. P. Dourish and K. Anderson. Collective Information Practice: Exploring Privacy and Security as Social and Cultural Phenomena. *Human-Computer Interaction*, 21(3):319–342, Sept. 2006.
49. B. J. Fogg and H. Tseng. The elements of computer credibility. In *SIGCHI Conference on Human Factors in Computing Systems (CHI '99)*. ACM, 1999.
50. D. Ghosh, A. Joshi, T. Finin, and P. Jagtap. Privacy Control in Smart Phones Using Semantically Rich Reasoning and Context Modeling. In *IEEE Symposium on Security and Privacy Workshops*, pages 82–85. IEEE, May 2012.
51. G. Go aszewski and J. Górski. Context sensitive privacy management in a distributed environment. In *On the Move to Meaningful Internet Systems (OTM '10)*, pages 639–655, 2010.
52. A. Gupta, M. Miettinen, and N. Asokan. Using context-profiling to aid access control decisions in mobile devices. In *International Conference on Pervasive Computing and Communications (PerCom '11) Workshops*, pages 310–312. IEEE, Mar. 2011.
53. A. Gupta, M. Miettinen, N. Asokan, and M. Nagy. Intuitive security policy configuration in mobile devices using context profiling. In *2012 ASE/IEEE International Conference on Privacy, Security, Risk and Trust (PASSAT '12)*. IEEE, 2012.
54. B. Hardian, J. Indulska, and K. Henricksen. Balancing Autonomy and User Control in Context-Aware Systems - a Survey. In *3rd Workshop on Context Modeling and Reasoning (CoMoRea'06)*, pages 51–56. IEEE, 2006.
55. E. Hayashi, S. Das, S. Amini, J. Hong, and I. Oakley. CASA: context-aware scalable authentication. In *Ninth Symposium on Usable Privacy and Security (SOUPS '13)*, New York, New York, USA, 2013. ACM.
56. T. Heiber and P. Marrón. Exploring the Relationship between Context and Privacy. In P. Robinson, H. Vogt, and W. Wagealla, editors, *Privacy, Security and Trust within the Context of Pervasive Computing*. Springer, 2005.
57. K. Henricksen and J. Indulska. A software engineering framework for context-aware pervasive computing. In *Second IEEE Annual Conference on Pervasive Computing and Communications (PerCom '04)*, pages 77–86. IEEE, 2004.
58. K. Henricksen and J. Indulska. Developing context-aware pervasive computing applications: Models and approach. *Pervasive and Mobile Computing*, 2(1):37–64, Feb. 2006.
59. K. Henricksen, J. Indulska, and A. Rakotonirainy. Modeling context information in pervasive computing systems. In *1st International Conference on Pervasive Computing (Pervasive '02)*, pages 167–180, 2002.
60. K. Henricksen, R. Wishart, T. McFadden, and J. Indulska. Extending Context Models for Privacy in Pervasive Computing Environments. In *2nd Workshop on Context Modeling and Reasoning (CoMoRea '05), PerCom '05 workshops*, pages 20–24. IEEE, 2005.
61. J. I. Hong and J. A. Landay. An architecture for privacy-sensitive ubiquitous computing. In *2nd International Conference on Mobile systems, applications, and services (MobiSys '04)*. ACM, 2004.
62. J.-Y. Hong, E.-H. Suh, and S.-J. Kim. Context-aware systems: A literature review and classification. *Expert Systems with Applications*, 36(4):8509–8522, May 2009.
63. G. Hsieh, K. P. Tang, W. Y. Low, and J. I. Hong. Field Deployment of IMBuddy: A Study of Privacy Control and Feedback Mechanisms for Contextual IM. In *9th International Conference on Ubiquitous Computing (UbiComp '07)*, pages 91–108. Springer, 2007.

64. S. E. Hudson and I. Smith. Techniques for addressing fundamental privacy and disruption tradeoffs in awareness support systems. In *ACM conference on Computer supported cooperative work (CSCW '96)*, pages 248–257, New York, New York, USA, 1996. ACM.
65. G. Iachello, K. N. Truong, G. D. Abowd, G. R. Hayes, and M. Stevens. Prototyping and sampling experience to evaluate ubiquitous computing privacy in the real world. In *SIGCHI conference on Human Factors in computing systems (CHI '06)*, page 1009, New York, New York, USA, 2006. ACM.
66. P. Jagtap, A. Joshi, T. Finin, and L. Zavala. Preserving Privacy in Context-Aware Systems. In *Fifth International Conference on Semantic Computing*, pages 149–153. IEEE, Sept. 2011.
67. A. Jameson. Modelling both the Context and the User. *Personal and Ubiquitous Computing*, 5(1):29–33, Feb. 2001.
68. X. Jiang and J. A. Landay. Modeling privacy control in context-aware systems. *IEEE Pervasive Computing*, 1(3):59–63, July 2002.
69. L. Kagal, T. Finin, and A. Joshi. A policy language for a pervasive computing environment. In *4th International Workshop on Policies for Distributed Systems and Networks (POLICY '03)*, pages 63–74. IEEE, 2003.
70. A. Kapadia, T. Henderson, J. J. Fielding, and D. Kotz. Virtual Walls: Protecting Digital Privacy in Pervasive Environments. In *5th International Conference on Pervasive Computing (PERVASIVE '07)*, pages 162–179. Springer, 2007.
71. F. Kargl, G. Dannhäuser, S. Schlott, and J. Nagler-Ihlein. Semantic information retrieval in the COMPASS location system. In *Third International Symposium on Ubiquitous Computing Systems (UCS '06)*. Springer, Oct. 2006.
72. E. Kim, S. Helal, and D. Cook. Human activity recognition and pattern discovery. *IEEE Pervasive Computing*, 9(1):48–53, Jan 2010.
73. T. Kindberg and A. Fox. System software for ubiquitous computing. *IEEE Pervasive Computing*, 1(1):70–81, Jan. 2002.
74. M. Knappmeyer, S. L. Kiani, E. S. Reetz, N. Baker, and R. Tonjes. Survey of Context Provisioning Middleware. *IEEE Communications Surveys & Tutorials*, 15(3):1492–1519, 2013.
75. B. Könings and F. Schaub. Territorial Privacy in Ubiquitous Computing. In *Eighth International Conference on Wireless On-Demand Network Systems and Services (WONS '11)*, pages 104–108, Bardoneccia, Jan. 2011. IEEE.
76. B. Könings, F. Schaub, and M. Weber. PriFi Beacons: Piggybacking Privacy Implications on WiFi Beacons. In *UbiComp '13 Adjunct Proceedings*. ACM, 2013.
77. B. Könings, F. Schaub, and M. Weber. Privacy and Trust in Ambient Intelligence Environments. In *Next Generation Intelligent Environments*, chapter 7, pages 133–164. Springer, second edition, 2016.
78. M. Korayem, R. Templeman, D. Chen, D. Crandall, and A. Kapadia. Enhancing lifelogging privacy by detecting screens. In *Proceedings of the 2016 CHI Conference on Human Factors in Computing Systems*, CHI '16, pages 4309–4314, New York, NY, USA, 2016. ACM.
79. T. Kriplean, E. Welbourne, N. Khoussainova, V. Rastogi, M. Balazinska, G. Borriello, T. Kohno, and D. Suciu. Physical Access Control for Captured RFID Data. *IEEE Pervasive Computing*, 6(4):48–55, 2007.
80. N. C. Krishnan and D. J. Cook. Activity recognition on streaming sensor data. *Pervasive and Mobile Computing*, 10(Part B):138–154, 2014.
81. J. Krumm and D. Rouhana. Placer: semantic place labels from diary data. In *ACM international joint conference on Pervasive and ubiquitous computing (UbiComp '13)*, page 163, New York, New York, USA, 2013. ACM.
82. P. Kumaraguru, L. Cranor, J. Lobo, and S. Calo. A survey of privacy policy languages. In *Third Symposium on Usable Privacy and Security (SOUPS '07) Workshops*, 2007.
83. O. Kwon. A pervasive P3P-based negotiation mechanism for privacy-aware pervasive e-commerce. *Decision Support Systems*, 50(1):213–221, Dec. 2010.
84. M. Langheinrich. *Privacy by Design — Principles of Privacy-Aware Ubiquitous Systems*, pages 273–291. Springer Berlin Heidelberg, Berlin, Heidelberg, 2001.

85. M. Langheinrich. A Privacy Awareness System for Ubiquitous Computing Environments. In *4th International Conference on Ubiquitous Computing (UbiComp '02)*, pages 237–245. Springer, 2002.

86. M. Langheinrich. RFID privacy using spatially distributed shared secrets. In *4th International Symposium on Ubiquitous Computing Systems (UCS '07)*, pages 1–16. Springer, 2007.

87. M. Langheinrich. Privacy in Ubiquitous Computing. In J. Krumm, editor, *Ubiquitious Computing Fundamentals*, chapter 3, pages 95–160. CRC Press, 2009.

88. M. Langheinrich, L. F. Cranor, and M. Marchiori. A P3P Preference Exchange Language 1.0 (APPEL1.0). W3c working draft, W3C, 2002.

89. O. D. Lara and M. A. Labrador. A survey on human activity recognition using wearable sensors. *IEEE Communications Surveys and Tutorials*, 15(3):1192–1209, 2013.

90. D. Leake, A. Maguitman, and T. Reichherzer. Cases, Context, and Comfort: Opportunities for Case-Based Reasoning in Smart Homes. In *Designing Smart Homes: The Role of Artificial Intelligence*, pages 109–131. Springer, 2006.

91. S. Lederer, J. I. Hong, A. K. Dey, and J. A. Landay. Five Pitfalls in the Design of Privacy. In L. F. Cranor and S. Garfinkel, editors, *Security and Usability*, chapter 21, pages 421–446. O'Reilly, 2005.

92. J. T. Lehikoinen, J. Lehikoinen, and P. Huuskonen. Understanding privacy regulation in ubicomp interactions. *Personal and Ubiquitous Computing*, 12(8):543–553, Mar. 2008.

93. B. Y. Lim, A. K. Dey, and D. Avrahami. Why and why not explanations improve the intelligibility of context-aware intelligent systems. In *27th international conference on Human factors in computing systems (CHI '09)*, page 2119, New York, New York, USA, 2009. ACM.

94. J. Lin, S. Amini, J. I. Hong, N. Sadeh, J. Lindqvist, and J. Zhang. Expectation and Purpose: Understanding Users Mental Models of Mobile App Privacy through Crowdsourcing. In *ACM Conference on Ubiquitous Computing (Ubicomp '12)*. ACM, 2012.

95. J. Lin, M. Benisch, N. Sadeh, J. Niu, J. Hong, B. Lu, and S. Guo. A comparative study of location-sharing privacy preferences in the United States and China. *Personal and Ubiquitous Computing*, 17(4):697–711, 2013.

96. J. Lin, B. Liu, N. Sadeh, and J. I. Hong. Modeling users' mobile app privacy preferences: Restoring usability in a sea of permission settings. In *Symposium On Usable Privacy and Security (SOUPS 2014)*, pages 199–212, Menlo Park, CA, 2014. USENIX Association.

97. B. Liu, M. S. Andersen, F. Schaub, H. Almuhimedi, S. A. Zhang, N. Sadeh, Y. Agarwal, and A. Acquisti. Follow my recommendations: A personalized privacy assistant for mobile app permissions. In *Twelfth Symposium on Usable Privacy and Security (SOUPS 2016)*, pages 27–41, Denver, CO, 2016. USENIX Association.

98. B. Liu, J. Lin, and N. Sadeh. Reconciling mobile app privacy and usability on smartphones: Could user privacy profiles help? In *Proceedings of the 23rd International Conference on World Wide Web*, WWW '14, pages 201–212, New York, NY, USA, 2014. ACM.

99. S. Loke. On representing situations for context-aware pervasive computing: six ways to tell if you are in a meeting. In *3rd Workshop on Context Modeling and Reasoning (CoMoRea '06)*, pages 35–39. IEEE, 2006.

100. P. Lukowicz, S. Pentland, and A. Ferscha. From Context Awareness to Socially Aware Computing. *IEEE Pervasive Computing*, 11(1):32–41, 2012.

101. S. T. Margulis. On the Status and Contribution of Westin's and Altman's Theories of Privacy. *Journal of Social Issues*, 59(2):411–429, June 2003.

102. N. Marquardt and S. Greenberg. Informing the Design of Proxemic Interactions. *IEEE Pervasive Computing*, 11(2):14–23, Feb. 2012.

103. G. Marx. Murky conceptual waters: The public and the private. *Ethics and Information technology*, pages 157–169, 2001.

104. M. Massimi, K. N. Truong, D. Dearman, and G. R. Hayes. Understanding Recording Technologies in Everyday Life. *IEEE Pervasive Computing*, 9(3):64–71, July 2010.

105. R. Mayrhofer and H. Gellersen. Shake Well Before Use: Intuitive and Secure Pairing of Mobile Devices. *IEEE Transactions on Mobile Computing*, 8(6):792–806, June 2009.

106. S. Moncrieff, S. Venkatesh, and G. West. Dynamic privacy assessment in a smart house environment using multimodal sensing. *ACM Transactions on Multimedia Computing, Communications, and Applications*, 5(2), 2008.

107. C. Neustaedter and S. Greenberg. The design of a context-aware home media space for balancing privacy and awareness. In *International Conference on Ubiquitous Computing (UbiComp '03)*. Springer, Mar. 2003.

108. H. Nissenbaum. Privacy as Contextual Integrity. *Washington Law Review*, 79(1):119–159, 2004.

109. H. Nissenbaum. *Privacy in Context - Technology, Policy, and the Integrity of Social Life*. Stanford University Press, 2009.

110. J. S. Olson, J. Grudin, and E. Horvitz. A study of preferences for sharing and privacy. In *CHI '05 extended abstracts on Human factors in computing systems*, New York, New York, USA, 2005. ACM.

111. R. Oppermann and A. Zimmermann. Context Adaptive Systems. *i-com*, 10(1):18–25, May 2011.

112. L. Palen and P. Dourish. Unpacking "privacy" for a networked world. In *Conference on Human factors in computing systems (CHI '03)*, pages 129–136, New York, New York, USA, 2003. ACM.

113. L. Pareschi, D. Riboni, A. Agostini, and C. Bettini. Composition and Generalization of Context Data for Privacy Preservation. In *Sixth International Conference on Pervasive Computing and Communications (PerCom '08)*, pages 429–433. IEEE, Mar. 2008.

114. S. Patil, Y. L. Gall, and A. Lee. My Privacy Policy: Exploring End-user Specification of Free-form Location Access Rules. In *Workshop on Usable Security (USEC '12)*, 2012.

115. S. Patil and J. Lai. Who gets to know what when: configuring privacy permissions in an awareness application. In *SIGCHI conference on Human factors in computing systems (CHI '05)*, page 101, New York, New York, USA, 2005. ACM.

116. B. A. Price, K. Adam, and B. Nuseibeh. Keeping ubiquitous computing to yourself: A practical model for user control of privacy. *International Journal of Human-Computer Studies*, 63(1-2):228–253, July 2005.

117. A. Rao, F. Schaub, N. Sadeh, A. Acquisti, and R. Kang. Expecting the unexpected: Understanding mismatched privacy expectations online. In *Twelfth Symposium on Usable Privacy and Security (SOUPS 2016)*, pages 77–96, Denver, CO, June 2016. USENIX Association.

118. N. A. Romero, P. Markopoulos, and S. Greenberg. Grounding Privacy in Mediated Communication. *Computer Supported Cooperative Work (CSCW)*, 22(1):1–32, 2013.

119. N. Sadeh, A. Acquisti, T. D. Breaux, L. F. Cranor, A. M. McDonald, J. R. Reidenberg, N. A. Smith, F. Liu, N. C. Russell, F. Schaub, and S. Wilson. The usable privacy policy project: Combining crowdsourcing, machine learning and natural language processing to semi-automatically answer those privacy questions users care about. techreport CMU-ISR-13-119, Carnegie Mellon University, 2013.

120. N. Sadeh, J. Hong, L. Cranor, I. Fette, P. Kelley, M. Prabaker, and J. Rao. Understanding and capturing people's privacy policies in a mobile social networking application. *Personal and Ubiquitous Computing*, 13(6):401–412, Aug. 2009.

121. R. Saleh, D. Jutla, and P. Bodorik. Management of Users' Privacy Preferences in Context. In *International Conference on Information Reuse and Integration*, pages 91–97. IEEE, Aug. 2007.

122. M. Satyanarayanan. Pervasive computing: vision and challenges. *IEEE Personal Communications*, 8(4):10–17, 2001.

123. F. Schaub. *Dynamic Privacy Adaptation in Ubiquitous Computing*. Phd dissertation, University of ulm, 2014.

124. F. Schaub, R. Balebako, A. L. Durity, and L. F. Cranor. A design space for effective privacy notices. In *Proc. SOUPS'15*, 2015.

125. F. Schaub, T. D. Breaux, and N. Sadeh. Crowdsourcing privacy policy analysis: Potential, challenges and best practices. *it-Information Technology*, 58(5):229–236, 2016.

126. F. Schaub, B. Könings, S. Dietzel, M. Weber, and F. Kargl. Privacy Context Model for Dynamic Privacy Adaptation in Ubiquitous Computing. In *6th International Workshop on Context-Awareness for Self-Managing Systems (CASEMANS '12), ACM UbiComp 2012 workshops*, pages 752–757, New York, New York, USA, 2012. ACM.

127. F. Schaub, B. Könings, P. Lang, B. Wiedersheim, C. Winkler, and M. Weber. Prical: Context-adaptive privacy in ambient calendar displays. In *Proceedings of the 2014 ACM International Joint Conference on Pervasive and Ubiquitous Computing*, UbiComp '14, pages 499–510, New York, NY, USA, 2014. ACM.

128. F. Schaub, B. Könings, and M. Weber. Context-adaptive privacy: Leveraging context awareness to support privacy decision making. *IEEE Pervasive Computing*, 14(1):34–43, 2015.

129. B. Schilit, N. Adams, and R. Want. Context-aware computing applications. In *Workshop on Mobile Computing Systems and Applications*, pages 85–90. IEEE, 1994.

130. B. Schilit and M. Theimer. Disseminating active map information to mobile hosts. *IEEE Network*, 8(5):22–32, Sept. 1994.

131. A. Schmidt. Implicit human computer interaction through context. *Personal Technologies*, 4(2-3):191–199, June 2000.

132. A. Schmidt. Context-Aware Computing: Context-Awareness, Context-Aware User Interfaces, and Implicit Interaction. In M. Soegaard and R. F. Dam, editors, *Encyclopedia of Human-Computer Interaction*, chapter 14, pages 1–28. The Interaction-Design.org Foundation, Aarhus, Denmark, 2012.

133. A. Schmidt, K. Aidoo, A. Takaluoma, U. Tuomela, K. Van Laerhoven, and W. Van de Velde. Advanced interaction in context. In *First International Symposium on Handheld and ubiquitous computing (HUC '99)*, pages 89–101. Springer, 1999.

134. A. Schmidt, M. Beigl, and H.-W. Gellersen. There is more to context than location. *Computers & Graphics*, 23(6):893–901, Dec. 1999.

135. B. Schwartz. The Social Psychology of Privacy. *American Journal of Sociology*, 73(6):741–752, May 1968.

136. J. Seifert, A. De Luca, B. Conradi, and H. Hussmann. TreasurePhone : Context-Sensitive User Data Protection on Mobile Phones. In *8th International Conference on Pervasive Computing (Pervasive '10)*, pages 130–137. Springer, 2010.

137. K. Shankar, L. J. Camp, K. Connelly, and L. Huber. Aging, Privacy, and Home-Based Computing: Developing a Design Framework. *IEEE Pervasive Computing*, 11(4):46–54, Oct. 2012.

138. K. Sheikh, M. Wegdam, and M. V. Sinderen. Quality-of-Context and its use for Protecting Privacy in Context Aware Systems. *Journal of Software*, 3(3):83–93, Mar. 2008.

139. P. Shi, H. Xu, and Y. Chen. Using contextual integrity to examine interpersonal information boundary on social network sites. In *SIGCHI Conference on Human Factors in Computing Systems (CHI '13)*, page 35, New York, New York, USA, 2013. ACM.

140. S. Sigg. Context-based security: state of the art, open research topics and a case study. In *5th ACM International Workshop on Context-Awareness for Self-Managing Systems (CASEMANS '11)*, pages 17–23, New York, New York, USA, 2011. ACM.

141. D. Solove. Privacy Self-Management and the Consent Dilemma. *Harvard Law Review*, 126:1880–1903, 2013.

142. D. J. Solove. *Understanding Privacy*. Harvard University Press, 2008.

143. T. Strang and C. Linnhoff-Popien. A context modeling survey. In *First International Workshop on Advanced Context Modelling, Reasoning And Management*, UbiComp '04, 2004.

144. L. A. Suchman. *Plans and situated actions: the problem of human-machine communication*. Cambridge University Press, 1987.

145. E. Toch. Super-Ego: a framework for privacy-sensitive bounded context-awareness. In *5th ACM International Workshop on Context-Awareness for Self-Managing Systems (CASEMANS '11)*, pages 24–32, New York, New York, USA, 2011. ACM.

146. J. Y. Tsai, P. Kelley, P. Drielsma, L. F. Cranor, J. Hong, and N. Sadeh. Who's viewed you? the impact of feedback in a mobile location-sharing application. In *27th international conference on Human factors in computing systems (CHI '09)*, New York, New York, USA, 2009. ACM.

147. D. Vogel and R. Balakrishnan. Interactive public ambient displays: transitioning from implicit to explicit, public to personal, interaction with multiple users. In *17th annual ACM symposium on User interface software and technology (UIST '04)*, pages 137–146, New York, New York, USA, 2004. ACM.

148. R. Wenning, M. Schunter, L. Cranor, B. Dobbs, S. Egelman, G. Hogben, J. Humphrey, M. Langheinrich, M. Marchiori, M. Presler-Marshall, J. Reagle, and D. A. Stampley. The Platform for Privacy Preferences 1.1 (P3P1.1) Specification. W3c working group note, W3C, 2006.

149. A. F. Westin. *Privacy and Freedom*. Atheneum, New York, 1967.

150. P. Wijesekera, A. Baokar, A. Hosseini, S. Egelman, D. Wagner, and K. Beznosov. Android permissions remystified: A field study on contextual integrity. In *24th USENIX Security Symposium (USENIX Security 15)*, pages 499–514, Washington, D.C., 2015. USENIX Association.

151. S. Wilson, F. Schaub, A. A. Dara, F. Liu, S. Cherivirala, P. G. Leon, M. S. Andersen, S. Zimmeck, K. M. Sathyendra, N. C. Russell, et al. The creation and analysis of a website privacy policy corpus. In *Proceedings of the 54th Annual Meeting of the Association for Computational Linguistics (ACL)*, 2016.

152. S. Wilson, F. Schaub, R. Ramanath, N. Sadeh, F. Liu, N. A. Smith, and F. Liu. Crowdsourcing annotations for websites' privacy policies: Can it really work? In *Proceedings of the 25th International Conference on World Wide Web*, WWW '16, pages 133–143, Republic and Canton of Geneva, Switzerland, 2016. International World Wide Web Conferences Steering Committee.

153. T. Winkler and B. Rinner. User-centric privacy awareness in video surveillance. *Multimedia Systems*, 18(2):99–121, July 2012.

154. R. Wishart and K. Henricksen. Context obfuscation for privacy via ontological descriptions. In *First International Workshop on Location-and Context-Awareness (LoCa '05)*. Springer, 2005.

155. R. Wishart, K. Henricksen, and J. Indulska. Context Privacy and Obfuscation Supported by Dynamic Context Source Discovery and Processing in a Context Management System. In *4th International Conference on Ubiquitous Intelligence and Computing (UIC)*, number 1, pages 929–940, 2007.

156. G. Yee. A privacy-preserving UBICOMP architecture. In *International Conference on Privacy, Security and Trust (PST '06)*, New York, New York, USA, 2006. ACM.

157. S. Zimmeck, Z. Wang, L. Zou, R. Iyengar, B. Liu, F. Schaub, S. Wilson, N. Sadeh, S. Bellovin, and J. Reidenberg. Automated analysis of privacy requirements for mobile apps. In *NDSS'17: Network and Distributed System Security Symposium*, 2017.

Chapter 14
Location Privacy-Preserving Applications and Services

Ioannis Boutsis and Vana Kalogeraki

Abstract Mobile location-based applications have recently prevailed due to the massive growth of the mobile devices and the mobile network. Such applications give the opportunity to the users to share content with the community which is coupled with their current geographical location. However, sharing such information might have serious privacy implications as an adversary might monitor the system and use such information to expose sensitive user information including user mobility traces and sensitive locations. This problem has led both the research community and the commercial mobile applications to develop several solutions to handle these privacy implications so as to enable users to disclose content without compromising their privacy. This chapter provides a survey of the state-of-the-art location-based mobile applications, describes the privacy implications that arise from contributing information in such applications and the respective existing countermeasures to deal with the privacy preservation issues. Furthermore, we describe our experiences from deploying a real-world location-based application that aims to allow the user contribute content and protect the user's privacy.

14.1 Introduction

Mobile applications have recently become a core part of commercial systems targeted to end-users, as they enable users to access their services from everywhere. Mobile applications typically provide the same features with the respective desktop applications, but they are developed for mobile devices such as smartphones and tablets. However, the vast majority of the mobile applications also take advantage of the ability to acquire the user location from the mobile devices' sensors, a feature which is not available in traditional desktop applications, in order to generate

I. Boutsis · V. Kalogeraki (✉)
Athens University of Economics and Business, Athens, Greece
e-mail: mpoutsis@aueb.gr; vana@aueb.gr

© Springer Nature Switzerland AG 2018
A. Gkoulalas-Divanis, C. Bettini (eds.), *Handbook of Mobile Data Privacy*,
https://doi.org/10.1007/978-3-319-98161-1_14

personalized content, depending on the respective location, and provide a richer experience compared to the desktop version of the service.

From the user perspective, it seems that more and more users appreciate mobile applications that exploit the geo-location features and are willing to share their location information with the service provider and potentially with other users, despite their privacy concerns. This is due to the fact that location-based applications provide additional features as they enable them to share their location with their acquaintances, including their friends and family, and to explore content which is related to their current location, such as weather, news, nearby places or nearby friends.

Although users are tempted to share their location information to enjoy the location-based services, sharing such data makes them vulnerable to several types of privacy attacks. This is because location data constitutes sensitive user information and the nature of the location information enables an adversary to expose a great amount of information per user with only a small set of geo-located information. Hence, if this data is exploited from potentially untrusted parties it can lead to severe consequences that range from user profiling for advertising purposes to real-world crimes such as stalking, robberies, etc.

This fact has led both the research community as well as the commercial mobile applications to develop several privacy preservation approaches to cope with this problem. However, as we explain in this chapter the commercial mobile applications typically use simple approaches which are prone to attacks rather than exploiting the complex but effective approaches that have been proposed in the research literature.

The goal of this chapter is to provide a survey of the most popular location-based applications and state-of-the-art privacy mechanisms. The rest of this chapter is organized as follows. Section 14.2 presents the most popular location-based applications and explains the benefits of sharing location-based data. Section 14.3 describes the characteristics of the shared data and the respective privacy implications that derive from these characteristics. Section 14.4 summarizes the existing privacy mechanisms that have been developed as countermeasures. Section 14.5 gives a discussion of our experiences from employing a real-world location based application and the privacy mechanism that we used. Finally, Sect. 14.6 concludes the chapter.

14.2   Popular Location Sharing Applications

The first section of this chapter introduces the most popular location sharing applications. In this chapter we focus on the following categories:

1. Social Applications
2. Transportation Applications
3. Travel Applications

4. Fitness Applications
5. Image Sharing Applications
6. Location Sharing Applications initiated by the research community and focus on preserving the user privacy

All the location sharing applications that are presented in this chapter have been selected due to their popularity and constitute the most prominent location sharing applications at the time of writing this chapter.

14.2.1 Social Applications

Social services have recently become extremely popular as they enable users to connect and interact with their friends and family. Social services give the opportunity to the users to create social profiles and share content such as text posts and photos or videos with their social connections. They also facilitate the development of online social networks by connecting a user's profile with those of other individuals and/or groups.

The prevalence of mobile technologies has enabled the social services to develop mobile apps that exploit the embedded sensors in order to provide a richer experience for the users. Hence, they allow the users to include location-embedded information or content. Introducing location capabilities, such as geotagging, provides additional features to the users such as letting their acquaintances know where they are or where a specific image was captured. However, revealing the physical location of the user can lead to significant privacy implications.

14.2.1.1 Facebook

Facebook (http://www.facebook.com) is one of the prominent location-based social networks with 1.66 billion mobile monthly active users (as of September 30, 2016).[1] It allows users to share various types of content with their designated set of users, including posts, photos and videos and to chat through Messenger with the rest of the Facebook users.

Each type of content which is shared by a user in Facebook from the mobile app may be coupled with the user geo-location. Hence, except from social posts, the user can also share her geo-location even during a chat. However, Facebook includes a consistent indicator as a reminder when the users share their location.

In Facebook, the content published by the users can be visible by all the users who are authenticated to access the shared content. Thus, the user has the

---

[1]http://newsroom.fb.com/company-info/.

responsibility to determine her audience and manage her privacy by assigning the users and groups that should be available to view shared content.

14.2.1.2 Twitter

Twitter (http://www.twitter.com) is another well-known social service used by 313 million monthly active users, from which 82% are mobile.[2] The idea behind Twitter is that it allows users to send short 140-character messages called "tweets" to interact with the community. Registered users can post and read tweets, but those who are unregistered can only read them. Mobile users are also able to embed their geographical location in these tweets to share the location where the tweet was produced.

Twitter provides two levels of privacy for the tweets that can be selected by the users: (1) Public Tweets and (2) Protected Tweets. Public Tweets, which is the default setting, makes the tweets visible to anyone, even users that do not own a Twitter account. On the other hand, Protected Tweets can only be visible by users that have been authenticated from the producer of the tweet so as to protect user privacy.

14.2.1.3 Foursquare

Foursquare (http://www.foursquare.com) is a leading location-based social networking website for mobile devices with more than 50 million people using Foursquare each month, through the web service and the mobile app.[3] Foursquare allows registered users to post their attendance at a venue (referred as "check-in") that can also be shared to other social networks such as Facebook or Twitter.

In Foursquare, users are encountered to be very specific with their check-ins indicating their precise location or activity while at a venue, and they receive awards as incentives for checking in. This enables Foursquare to collect important information from the users that can be used to provide personalized recommendations and business deals. Although the real-time location of the users is not shared on the Foursquare app, all the user interaction with a venue, such as writing a tip is time-stamped and publicly available to the community. This allows other users to infer when the user was at a specific place. Moreover, information like checking-in at a place might not be public, but it can be accessed by the followers of the user and allows them to know when the user visited the venue.

Foursquare assumes that the users are aware regarding the location privacy issues and responsible for the visibility of their geo-located posts. Hence, the users can only protect their privacy by defining the visibility of their shared posts.

---

[2]https://about.twitter.com/company.

[3]https://foursquare.com/about.

14.2.2   Transportation Applications

Another type of apps that require the location of the users are transportation apps. These applications take advantage of the mobile sensors in order to provide real-time navigation to the users. Hence, transportation applications need to acquire the user real-time location frequently to propose the optimal route and to constantly provide directions to a user to reach her destination.

14.2.2.1   Google Maps

Google Maps (http://maps.google.com) is an online service that offers satellite imagery, street maps, panoramic views of streets, real-time traffic conditions and route planning for traveling by foot, car, bicycle or public transportation. Moreover, Google Maps provides an online service for the users to navigate to places that requires sharing the real-time location of the user, captured through the sensors of her mobile device.

Besides navigation purposes, Google Maps also collects the speed and location information of the users anonymously. This is used to calculate traffic conditions in real-time so as to provide better estimations for the travel times. Hence, the users need to accept this fact in order to be able to take advantage of the Google Maps.

Google has mentioned that it permanently deletes the start and end points of every user trip it monitors so that information about where each user came from and went to remains private. Moreover, users can set locations that user activity will not be captured such as their home or work.

14.2.2.2   Moovit

Moovit (http://www.moovitapp.com) is a public transit app and mapping service that features trip planning, real-time arrival and departure times, line schedules, alerts, and advisories that may affect the trip of the user. Moovit uses several transit modes including buses, ferries, metro, trains, trams, and trolleybuses.

Moovit allows users to send reports actively including reasons for delays, overcrowding, satisfaction with their bus driver, and wifi availability which is shared with the community. Moreover, the users can also share data passively. By riding with the 'Live Directions' feature active allows Moovit to collect passively and anonymously the user speed and location data. This data is used in addition to the public transit schedules to improve trip plan results based on current conditions.

Moovit utilizes the information gathered from the application, including information regarding the user location and public transport preferences. They also make anonymous, statistical use of this information in order to estimate the arrival times of various bus lines, analyze their frequency and convey the information to third parties for whom this information is likely to be relevant.

14.2.2.3 Waze

Waze (http://www.waze.com) is a navigation application that provides turn-by-turn information to the users. The Waze app captures real-time information that translates into traffic conditions and road structure from its users. Moreover, it enables them to actively report to the community traffic, accidents, police traps, blocked roads, weather conditions and much more. Waze collects and analyzes this information to provide other Wazers with the most optimal route to their destination.

Similar to the rest of the Transportation Applications, Waze may use anonymous, statistical or aggregated information, including anonymous location information. Any other content that the users submit manually, such as geo-located reports, as well as their current location during a route is publicly available to all users of Waze. However, Waze users are able to share information as anonymous users to preserve their location privacy, that prevents them from collecting the rewards for their contributions. We also note, that, during navigation Waze does not share data within 500 m from the user home to preserve user privacy.

14.2.2.4 Uber

Uber (http://www.uber.com) provides a service for hiring a private driver. It allows users with smartphones to submit a trip request, which is automatically sent to the nearest Uber driver, alerting the driver regarding the location of the customer. Uber drivers typically use their own personal cars and their payment is calculated by the Uber app.

When someone uses Uber her precise location data about the trip is collected from the Uber app used by the Driver. Moreover, if the user permits the Uber app to access location services of the mobile device Uber also collects the precise location of the device when the app is running in the foreground or background.

14.2.3 Applications for Traveling

Traveling applications is another category of apps that share interactive travel-related content including ratings and experiences for specific points of interest. Although users may opt-out from sharing their location with the app, sharing their experiences implicitly validates their presence.

14.2.3.1 TripAdvisor

TripAdvisor (http://www.tripadvisor.com) is one of the early adopters of user-generated content that allows users to contribute content based on places that they visit to advise the rest of the community based on their experience. TripAdvisor offers advice from millions of travelers for a large variety of places including accommodations, restaurants and attractions. Nevertheless, all these data are implicitly geo-located and shared publicly with the community.

TripAdvisor may collect information about the user location if the user has instructed the device to send such information to the application or if the user has uploaded photos tagged with location information. This location information is used to provide relevant content and contextual advertising to the user.

14.2.3.2 Yelp

Yelp (http://www.yelp.com) uses a similar model with TripAdvisor where users provide reviews and ratings for points of interest that can be exploited by the community when they make decisions. In addition to reviews, Yelp can also be used to find events, lists and to talk with other users. In Yelp every business owner can setup a free account to post photos and messages from their customers. Nevertheless, similar to TripAdvisor, these reviews can reveal the users' spatiotemporal presence.

14.2.4 Fitness Applications

Several fitness applications have recently emerged due to the massive prevalence of smartphones and wearable devices such as smartwatches and smartbands. These devices are equipped with sensors such as GPS, accelerometer, gyrometer, pedometer and heart rate sensor that give the opportunity to the users to capture and share several aspects of their training including distance travelled, steps, heart rate, calories burned, etc. Nevertheless, these applications capture massive amounts of data from the users, especially when the users are active.

14.2.4.1 Strava

Strava (http://www.strava.com) is one of the most prominent fitness apps especially for cyclists. It allows them to record their bike rides and runs, compare their performance over time and share them with the community so that they compare their performance with other users. The users can either share these data with the community to compete with other users or preserve them in their account. However, inevitably, users that share these data allow potentially untrusted users to access their trajectories, including their commutes. In order to preserve user privacy for users that share their data, Strava hides all user traces within a predefined radius from the user's home.

14.2.4.2 Endomondo

Another similar social fitness app is Endomondo (http://www.endomondo.com), which allows users to track their fitness and health statistics with a mobile app. Endomondo is focused on running and walking and encourages users to track their workouts so as to reach their fitness goals. Similar to Strava it allows its users to share their workouts with the community and, thus, share their mobility traces.

14.2.5 Image Sharing Applications

Another type of social location-sharing applications are Image Sharing Applications. These apps allow users to share images with their social interactions or publicly. In addition they also give the opportunity to the user to tag the location where an image is captured.

14.2.5.1 Instagram

Instagram (http://www.instagram.com) is one of the most prominent location sharing applications with more than 500 million users.[4] It allows users to share images with geo-location with the community and interact with the shared images. In Instagram anyone can view the images shared by the users by default. However, each individual user can choose to make them private so that only approved followers can see them.

14.2.5.2 Flickr

Flickr (http://www.flickr.com) provides a similar service for the users to upload their photos and share them with the community and has been widely used by photo researchers and by bloggers. In Flickr, each photo can be geo-located either by the mobile device or manually by the user and the photo's location can then be shown on a map. In Flickr each user can set her geo-privacy to determine the users who are allowed to access the location of the published photos.

14.2.6 Friend/Family Finder Applications

Another type of location sharing applications are the Friend/Family Finder Applications. These applications have been initiated due to the simplicity of acquiring

---

[4]https://www.instagram.com/about/us/.

and sharing user location in mobile devices. Hence, they track the users in order to notify them when there is a nearby friend and to allow their friends and family to know where they are.

14.2.6.1 GPS Phone Tracker

One of the most popular Friend Finder application is the GPS Phone Tracking Pro App (http://gpsphonetracker.org). It has two main features: (1) it allows users to find their friends and navigate towards them and (2) it allows the user to find her mobile device from the app's website.

GPS Phone Tracker uses the geographical coordinates of the user to report her real-time whereabouts to her friends. Once registered to the system, each friend appears as a unique icon on the map to let the user know where each of her friends is. However, this implies that the users are constantly tracked by the operator but also from all their relatives.

14.2.6.2 Family Locator: GPS Tracker

Another popular app for location sharing is the Family Locator app (https://www. life360.com/family-locator/). It allows users to create their own groups of people such as friends or family, called "Circles". This enables them to view the real-time location of Circle Members and receive real-time alerts when circle members arrive at or leave destinations. Moreover, similar to the GPS Phone tracker, it enables users to locate their mobile devices. Again, the issue is that the users are constantly tracked and their location is shared, providing no privacy among users in a particular circle.

14.2.7 University Initiatives

The following apps are differentiated from the above categorization because they have been initiated by Universities. The main difference of these applications is that they have tried to incorporate innovative techniques to provide users with features that require their real-time location, while preserving user privacy at the same time.

14.2.7.1 PCube

PCube (http://www.everywaretechnologies.com/apps/pcube) is a location-aware social networking app that aims to alert users when their friends are in close proximity. Hence, it allows users to observe which of their friends happen to be in the area.

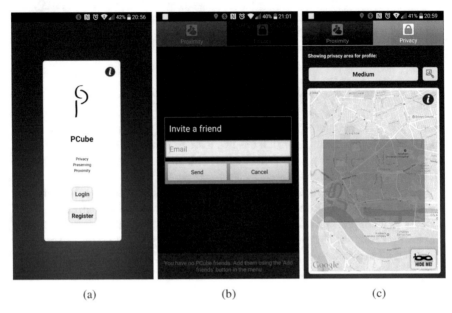

Fig. 14.1 The PCube app. (**a**) Login. (**b**) Add friends. (**c**) Set proximity

As we mentioned above, such Friend/Family Finder apps need to constantly share the user location to provide this functionality. However, PCube aims to preserve user privacy for the location-based data shared by the users using a novel approach, further described in [18]. Hence it exploits encryption techniques to encrypt the data before leaving the device to prevent user exposure even from their servers. Moreover, it takes advantage of the privacy preferences, set by the users, to alert friends in proximity, but the alert received does not reveal the actual user location with higher precision than the one she has decided.

Figure 14.1a illustrates the login screen of PCube that enables users to login and register. After login, the user is presented with a screen which is divided into two tabs. The first tab (Fig. 14.1b) allows the user to add friends and create social communities while the second one (Fig. 14.1c) provides the proximity functionality to set the radius within which her friends will be alerted that they are in close proximity.

14.2.7.2 Locaccino

Another Friend/Family Finder app initiated by the academic community, is Locaccino (http://locaccino.org), that allows users to share their location with their friends.

Similar to PCube, Locaccino focuses on preserving user privacy but it uses a different approach. Locaccino exploits a rule-based approach that allows users to create rules rules regarding their visibility from the privacy settings. The rules can be as simple or complex as the user wants them to be, such as setting a different

Fig. 14.2   The Locaccino app. (**a**) Map. (**b**) Privacy settings. (**c**) History

proximity for each particular user or disabling the visibility of the user for specific hours or places.

Figure 14.2a shows the map presented to the user that presents the geo-location shared by her acquaintances. Moreover, the user can modify the rules that will be used to enable others to access her location, as shown in Fig. 14.2b, and the view the history of the users that could access her location, as illustrated in Fig. 14.2c.

14.2.7.3   CrowdAlert

CrowdAlert (http://crowdalert.aueb.gr) is a location-based app, designed for Android users that enables them to report real-world events of interest and receive real-time alerts for nearby events, by exploiting real-time data from human users, road sensors, bus sensors and web sources. Similar to Social and Transportation apps it enables users to provide sparse geo-located data such as geo-located reports for real-world events (e.g., Traffic, Floods, etc). Such data can reveal the location of the user to the community. However, CrowdAlert is built with privacy in mind and focuses on effectively preserving user privacy for her shared data as we describe in detail in Sect. 14.5.

14.3   Privacy Characteristics of Location Sharing Applications

In this section we present the types of data shared by the users in location-based applications. Then we explain the information that can be inferred by the data and finally we describe thoroughly the potential privacy risks that arise from exposing such data per service.

14.3.1 Type of Data Shared

Location-based applications share different types of content but in all applications the content is coupled with the user location that can lead to privacy implications. The data shared in each application is illustrated in Tables 14.1 and 14.2. In the following we enumerate the most important categorization among the different types of data sharing in terms of privacy.

14.3.1.1 Data Density

One important categorization depends on the density of the sharing location-based data. Applications such as Transportation applications, Fitness applications and Friend/Family Locator applications typically share dense data that include the user mobility, such as user trajectories. On the other hand, Social applications, Travel applications and Image Sharing applications typically share sparse data since every post is coupled with a specific geographical location.

Obviously, providing dense data makes the privacy preservation task even more difficult as an adversary acquires richer information from the system. However, it has been proved that even a small number of spatiotemporal points is enough for user identification [7]. Hence, even sparse data might be enough for an adversary to expose user privacy.

14.3.1.2 Visibility

We can also discriminate the applications depending on whether they allow sharing the location-based data publicly. For instance, Social, Image Sharing and Fitness Applications give the opportunity to the users to share their data publicly if they want to. On the contrary, the majority of the Transportation apps allow users only to share data privately with the service operator or anonymously. Based on this characteristic we argue that public data are prone to exposing user privacy since all the data are linked with a specific user identifier. Furthermore, even anonymized data can be de-anonymized through complex data analysis [10].

14.3.1.3 Data Source

Finally, another categorization of the data is whether they are generated from the users explicitly or they are generated by the mobile sensors. Hence, Social, Image Sharing and Travel applications require from the users to generate the content which is coupled with the geographical location. On the other hand, Transportation applications and Friend/Family Locator applications produce dense data which is extracted from the mobile sensors without requiring user interaction.

Table 14.1 Location-based applications, data sharing and privacy mechanisms

App	Type	Google play downloads	Data shared	Privacy mechanism
Facebook	Social	1,000,000,000– 5,000,000,000	Share posts with the community Posts may include geo-location	Visibility scopes
Twitter	Social	500,000,000– 1,000,000,000	Share tweets with users Tweets may include geo-location	Visibility Scopes
Foursquare	Social	10,000,000– 50,000,000	Share user attendance with friends (check-in) Share public content like public user profile information, tips, likes, saves, public photos, badges/stickers, mayorships, and lists of friends	Visibility scopes
Yelp	Travel	10,000,000– 50,000,000	Users can share reviews regarding places that include implicit location	Disabling location sharing
TripAdvisor	Travel	100,000,000– 500,000,000	Users can share reviews and ratings regarding places that include implicit location Users can share their location with TridAdvisor to receive relevant content and contextual advertising	Disabling location sharing
Moovit	Transportation	10,000,000– 50,000,000	Users can passively and anonymously transmit their speed and location Users can actively provide reports including reasons for delays, overcrowding, etc	Anonymous data sharing
Waze	Transportation	100,000,000– 500,000,000	Users can share geo-located traffic reports that may be viewed by all users in the community Waze may use anonymous, statistical or aggregated information	Privacy zones Anonymous data sharing
Google traffic maps	Transportation	1,000,000,000– 5,000,000,000	User accurate location is shared during navigation Users share live traffic conditions anonymously (speed and location information)	Privacy zones Anonymous data sharing
Uber	Transportation	100,000,000– 500,000,000	Share user location with drivers, other riders, general public, third parties	Disabling location sharing
Strava	Fitness	5,000,000– 10,000,000	Users shares bike rides, runs Users can also share their achievements	Privacy zones Visibility scopes

(continued)

Table 14.1 (continued)

App	Type	Google play downloads	Data shared	Privacy mechanism
Endomondo	Fitness	10,000,000–50,000,000	Users can shares their workouts Users can also share their achievements	Visibility scopes
Instagram	Image sharing	1,000,000,000–5,000,000,000	Share images with geo-location	Visibility scopes
Flickr	Image sharing	10,000,000–50,000,000	Share images with geo-location	Visibility scopes
GPS phone tracker	Friend finder	10,000,000–50,000,000	User shares her real-time location with her friends	Disabling location sharing
Family locator—GPS tracker	Friend finder	10,000,000–50,000,000	User shares her real-time location with her friends	Disabling location sharing

Table 14.2 University initiative location based app applications, data sharing and privacy mechanisms

Apps	Type	Data shared	Privacy mechanism
PCube	Friend finder	Alerts the users about friends who are in proximity	Rule-based scheme encryption
Locaccino	Friend finder	Share user location with friends	Rule-based scheme
CrowdAlert	Event alerting	Share public geolocated reports with the community	Data suppression

14.3.2   Inferring Information from the Data

The problem of exposing all this data is that they reveal the user's spatiotemporal instance when the data are published. Assume a user that passively observes and records the location information published by an individual user. Although the information shared in a particular post may not reveal a lot of information when considered individually, it can expose a large fraction of the user mobility as the user shares more and more data with the community. The problem can become worse when the user identifier can be linked across different applications and, thus, all geo-located information shared across these networks can be linked to the specific user.

The issue with sharing location-based data is that the majority of the data typically reside at locations or along routes that are mostly visited by the individuals. Hence, as the amount of content contributed by the user increases, the shared data will gradually expose the user's most frequent trajectories, important locations and even the user's physical identity, which could place a user in physical danger [15] and lead to crimes such as stalking the user or robbing the user when she is absent from home.

We also note that due to the nature of the data, an adversary might be able to generate the user mobility even using sparse data. For instance assume that a user produces a couple of geo-located posts in a social app as she commutes from her home to work. Although these data can be very distant, the intermediate points can easily be inferred using navigational tools. Furthermore, such information can also be linked with posts shared previously at the same time window where the user commutes to her work, in case that they overlap spatially.

14.3.3 Privacy Threats

The data shared in such applications enable an adversary to pose the following three types of privacy threats [9]: (1) Tracking Threat, (2) Identification Threat and (3) Profiling Threat. These threats originate from sharing location-based data with potentially untrusted entities and need to be taken into consideration when developing an application that shares the user location to be able to preserve user privacy.

- **Tracking Threat:** In a location-based application an untrusted party might extract continuous location updates that enable him to track the user in real-time. Hence, an adversary should not be able to determine the user mobility and predict her future location with high accuracy when leveraging the location-based data shared by the user.
- **Identification Threat:** Even if the untrusted party is able to sporadically accesses a user's location, the untrusted party should not be able to identify the user's most frequently visited places, such as the user's home and work location. This is because an adversary can exploit such information to reveal the identity of the user even from anonymous mobility traces.
- **Profiling Threat:** The user mobility traces, shared by the mobile application, might not reveal only places that can help to identity the user but also places that can be used by an adversary to profile the user. For instance, an adversary will be able to profile the user when he acquires location data showing that the user has visited religious places or attended political meetings.

Note, that, the user might explicitly share information that reveal sensitive information (e.g., posting a geo-located tweet that he just arrived at home). Although a privacy-concerned app will not prevent the user from sharing this information, it should consider these privacy threats when this information can be inferred from the shared data.

14.4 Existing Approaches for Privacy Preservation

14.4.1 Privacy Mechanisms in Popular Location-Based Applications

Subsequently, we present the mechanisms which are used in each popular location-based service to prevent unauthorized parties from learning the user's current or past locations. The main schemes can be grouped into the following categories:

1. **Visibility Scopes** allow the user to select the groups of users which are allowed to access the user's posts and respective location.
2. **Privacy Zones** essentially hide the user location in specific areas to anonymize the user.
3. **Rule Based approaches** where the users develop rules per user group in order to determine the visibility of their data depending on multiple factors.
4. **Anonymous Data Sharing** allow the users to share location based data but the service consider them as anonymous data, without coupling them with the user.

14.4.1.1 Visibility Scopes

The most prominent approach used in popular location-based services is the use of visibility scopes. This approach enables the users to develop groups of users such as friends and family manually. Then they can select for every piece of information that they share if the information will be publicly available or determine the groups of users that will be able to access this information. The idea behind visibility scopes is that the users are capable of managing the visibility of their posts and evaluate their privacy exposure.

Visibility scopes have been widely used as a privacy preservation paradigm in multiple social location-based services such as Facebook, Twitter and Foursquare. This is because users in these networks tend to share content with multiple users and thus they do not focus on their privacy. Similarly, Image Sharing applications like Flickr and Instagram depend on the users to decide who can view their photos as well as the location of their photos.

However, this approach can lead to serious privacy issues especially when the shared data embed the users' geographical location. The problem with visibility scopes is that they require constant effort from the user to preserve her privacy. Hence, users tend to share information publicly or with users that they have accepted as their friends, that they might have never met. Thus, as we mentioned above sharing this information can be exploited by the community to track the user, extract sensitive user information or profile the user.

14.4.1.2 Privacy Zones

Another well-known approach that also requires user interaction are Privacy Zones. This approach allows the user to select one or more sensitive locations that will be considered by the app in order to hide information produced near these locations. This is typically achieved using a radius around sensitive locations within which no information will be shared with the community.

Privacy Zones have been used particularly for apps that share frequent user locations such as the Strava fitness app and Waze. Hence, in Strava when a user shares a bike route the data near the sensitive locations will be hidden from the rest of the users. Similarly, Waze hides all the data within 500 m from the user's home.

Although Privacy Zones are more sophisticated than Visibility Scopes as they require less interaction with the user, they have one important drawback. Since the radius is typically fixed in these apps, an adversary can easily identify the sensitive locations with high accuracy with triangulation. However, even if the radius can be modified, the sensitive locations will start to be exposed as the users contribute more data.

14.4.1.3 Rule-Based Scheme

A different technique to preserve user privacy is the Rule-Based Scheme. This approach also requires user interaction in order to determine several rules regarding her visibility. However, these rules can be very complex, such as hiding the user location for specific places, hours or people (e.g., hiding the user location from her employer during work hours).

Rule-Based approaches have been used from Friend Finder applications such as PCuble and Locaccino, that produce massive location-based data from the users as they constantly track them. Hence, it allow them to filter the portion of the data that each individual will be able to observe.

Rule-Based approaches constitute an improvement compared to Visibility Scopes and Privacy Zones as they use their concepts but extend them to make more complex decisions regarding the data that can be shared among users.

14.4.1.4 Anonymous Data Sharing

Another technique employed by the commercial location-based applications to protect privacy for privacy-concerned users is Anonymous Data Sharing. This approach allows users to share data with the service provider or the community without revealing their identifiers.

Anonymous Data Sharing has been extensively used for traffic and transportation related applications such as Moovit, Google Maps, HERE Maps and Waze. These applications collect user data from mobile sensors, typically using a background service, to compute real-time road conditions. Additionally, user-identified services can also be provided in an anonymous way. For instance, Waze facilitates anonymous contributions for privacy-concerned users. However, such alternatives are not always attractive to end users, as they do not receive any reward for their contributions, and from the system's point of view, anonymous data has shown to increase spamming.

14.4.1.5 Disabling Location Sharing

Finally, another option for the users to preserve their privacy is to completely turn off the "Location Services" from their devices. This action prevents the mobile operating system to give access to the location-based readings of the mobile sensors to the applications. Similarly a user can typically reject permission to access the location data for a particular app. This approach can be used by the users for applications like Yelp, TripAdvisor, Uber, GPS Phone Tracker and Family Locator - GPS Tracker, as they do not provide any other way to prevent tracking the mobility traces of the user. However, this option is out of the scope of this chapter since it removes all the functionality that depends on the user location, making some of these applications ineffectual.

14.4.2 Other Approaches

Finally, several approaches have been proposed in the literature to deal with privacy threats and preserve user privacy for location-based application. We summarize them here for completeness but they are discussed in detail in Chap. 5.

The first set of privacy preservation approaches focuses on modifying location-based information using a variety of mechanisms: (a) **Path Confusion** [8, 14] that aim at confusing the paths of users that reside in the same region by connecting their traces, (b) **Mix Zones** [11], which are generated when there are enough users located in the same place at the same time, making it hard for an adversary to distinguish an individual from others that reside in the same zone at the same time, (c) **Fake Data Injection** [12, 20], where fake data are injected along with real data to confuse an attacker, (d) **Data perturbation** [2, 25] where the goal is to modify the original data set with some noise drawn from a selected statistical distribution to preserve user privacy, (e) **Data generalization** [9, 17] that generalizes the user location information by reducing the spatial accuracy of the user, and (f) **K-anonymity approaches** [6, 16, 19] that release data that hold the k-anonymity property.

The second set of approaches to protect user privacy focuses on maintaining the original location-based data in a way that an adversary will not be able to extract sensitive user information. Here we have two main categories: (a) **Encryption techniques** [13, 23] that encrypt the user data before sharing them so that they disable untrusted parties from accessing the data, and (b) **Data Suppression techniques** [5, 24] where the user location is suppressed when the set of data shared by the user can expose her privacy. That way they prevent adversaries from exposing sensitive user information although they are capable of publishing the original information shared by the users.

14.5   CrowdAlert: Experiences from Deploying a Location-Based App in a Smartcity

In this section, we discuss our experience from deploying CrowdAlert, a location-based application, in a Smart City environment [4]. We explain our design objectives for our privacy preservation approach and we discuss the trade-offs between privacy and utility when publicly sharing data. Moreover, we give an overview of how our privacy approach works and we point out the effort needed to develop an interface that can satisfy both novice and expert users. Finally, we discuss our experience and the knowledge that we obtained from deploying CrowdAlert.

CrowdAlert is a location-based mobile application that enables users to report and receive information regarding unusual events in SmartCities that include Accidents, Constructions, Traffic, Natural Disasters, etc. CrowdAlert provides great benefits to both citizens and authorities in a SmartCity. In particular, it allows citizens to be alerted about local unusual events in real-time, and gives the opportunity to city authorities to identify, supervise and react if necessary, to these events in a cost-effective manner.

14.5.1   Data Sharing in CrowdAlert

Users connect with CrowdAlert using the mobile application, which is freely available at Google Play.[5] All registered users receive real-time information for a wide variety of events such as Accidents, Constructions, Hazards and more. This information arrives both in the form of notifications as well as on a map-based interface as can be seen in Fig. 14.3a. Moreover, registered users are able to report such information through CrowdAlert, as can be observed in Fig. 14.3b and receive questions for local information from the authorities when needed.

CrowdAlert app is designed to consider user privacy preservation, since users share information, coupled with their personal geographical location, that can lead to privacy exposure. Hence, we have incorporated our privacy preserving mechanisms to allow users share data with the community without compromising their privacy, as we explain in the following.

14.5.2   Design Objectives for Data Sharing

Before developing our privacy preservation approach we set the following design objectives for our solution:

---

[5]CrowdAlert—http://crowdalert.aueb.gr/.

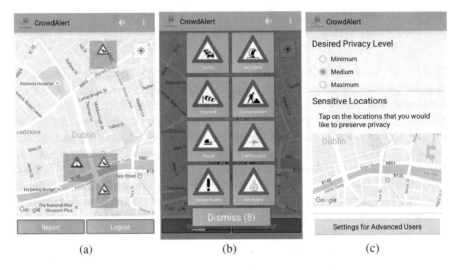

Fig. 14.3 The CrowdAlert app. (**a**) Login. (**b**) Report. (**c**) Settings

1. **Users should be able to post data coupled with their identifier, location and timestamp.** In CrowdAlert we allow the users to share data publicly with their identifiers, location and timestamp as happens in other real-world applications like Social and Transportation applications. However, users are not capable of estimating the effect of the sharing such data on their privacy. The goal of our approach is to alert the user before sharing location-based data that may expose her privacy to decide whether to share it.

2. **Users should be able to tune their privacy levels.** Every user has a different preference in terms of privacy levels and desired utility. However, existing approaches are basically limited to only two options. Users can either allow the location-based application to have access their data or accept generic, potentially low quality levels of service such as anonymous contributions that does not allow the users to receive any reward for their contributions. Our goal is to allow the users to tune the privacy levels they aim to achieve when contributing data.

3. **Users should keep their mobility traces locally.** The majority of the privacy protection approaches that exist in the literature [3], use a centralized approach to analyze user mobility traces. However, such centralized approaches have important shortcomings: (1) they are prone to different types of attacks like eavesdropping and the possibility that the server might become malicious, (2) they introduce additional communication costs that may degrade the user experience that needs to wait for a response before sharing her data. Our goal is to be able to achieve privacy preservation by evaluating the privacy exposure of the user locally on her mobile device and share the data after deciding that they pose no privacy risks.

14.5.3   Trade Offs Among Privacy and Sharing

Due to our first design objective, it is clear that existing solutions to preserve user privacy, that modify the original user data, could not be used. Since we focus on data which is publicly available we used an approach that proactively determines whether sharing information can expose user privacy and suppress such data.

The challenge is that users share content publicly coupled with their spatiotemporal instance as happens in Social Apps or Transportation Apps. Hence, users share more data in different locations, they expose larger parts of their mobility. In CrowdAlert we deal with the privacy preservation problem using a novel data suppression approach that aims to balance the trade off among user privacy and amount of shared data, based on the user preferences.

14.5.4   Privacy Problem

The problem that we deal with is how to prevent honest-but-curious adversaries that monitor the shared geo-located, user identified data from extracting sensitive information about the users. We define that the set of shared data effectively preserves user privacy only if:

- An attacker cannot approximate the user trajectories, and
- The user sensitive locations such as home and work cannot be determined by an adversary from the shared data based on their amount and frequency.

Essentially, our goal is to answer effectively the following questions:

1. How to evaluate the privacy exposure of the user trajectories from the set of the publicly shared reports on resource-constrained mobile devices in real-time?
2. How to preserve user privacy in terms of the user's sensitive locations when sharing geo-located data publicly?
3. Should the user share a newly produced report without the risk of compromising her privacy?

14.5.5   Privacy Preservation Approach

To preserve user privacy for CrowdAlert we developed our privacy mechanism called PROMPT [5]. The core idea of PROMPT is to employ the novel geometric approximation approach of ϵ-Coresets. We selected the ϵ-Coreset approach since ϵ-Coresets can effectively reduce the computation overhead for several complex geometric and graph problems [1]. Thus, they can be used to effectively process user mobility in resource constrained devices. The idea of ϵ-Coresets is to select a small subset that approximately represents the original data that is able to process a given query with up to $(1+\epsilon)$-multiplicative error. This reduction allows us to perform queries on the coreset and greatly reducing the computation time.

PROMPT is executed whenever a user desires to publish content with the community. It exploits the user mobility traces, which are compressed and preserved locally in the mobile device, in order to compute whether user privacy can be exposed. Hence, we extract all user trajectories that reside near the current user location and then we develop a coreset for each of these trajectories. Each coreset is developed by the current user location and the set of locations that the user has previously shared which are spatiotemporally close to the trajectory. This allow us to compute the fraction of the trajectories that can be approximated by the shared user data and prevent exceeding a predefined threshold, which can be modified by the user, in order to tune her privacy levels. Hence, whenever this threshold will be exceeded we alert the user, before sharing the data, to prevent her from exposing her privacy.

Moreover, in PROMPT we aim to preserve privacy near the user sensitive location; sensitive locations can be determined automatically as the most frequent locations based on the user mobility which is available on the mobile device and manually from the user. However, as mentioned above Privacy Zones are prone to expose sensitive locations through triangulation. Hence, in PROMPT we allow users to share data near sensitive locations but we aim to prevent the user from sharing a large percentage of location-based data near her sensitive locations, compared to the rest of the locations, to confuse an adversary regarding their importance. This can be achieved using the entropy metric [22]. Thus, in PROMPT, we only allow users to share data near sensitive locations when the entropy of the shared data increases. Since entropy increases when the frequency of the locations becomes more similar, we discourage sharing a report near a sensitive location when there already exists a large number of location-based data contributed from that location, compared to other locations.

14.5.6 Privacy Settings Interface

PROMPT has been integrated in a beta version of CrowdAlert to preserve user privacy. The user interface to adjust the privacy levels of PROMPT is illustrated in Fig. 14.3c. Since PROMPT depends on several variables, in order to allow the users to tune their privacy levels, we allow them to select their desired privacy level with three privacy profiles: Minimum, Medium and Maximum. This is due to several recent surveys which state that the users often find it difficult to adjust their privacy preferences [21]. Hence, profiles are able to simplify the privacy choices for the users [26].

Nevertheless, we give the opportunity to advanced users to access the parameters of our approach and fine tune their privacy. An advanced users can adjust: (1) the maximum percentage approximation can be achieve for her trajectories before preventing the user to share data, (2) the spatial radius to consider from the user sensitive locations to preserve privacy, and (3) the spatiotemporal thresholds that are used to correlate similar user presences.

Moreover, as can be observed from Fig. 14.3c the users can also define their sensitive locations using a map-based interface. These locations are considered from PROMPT in addition to the most frequent locations extracted from the local mobility traces.

14.5.7   Experiences from Our Deployment and Lessons Learnt

The development of CrowdAlert enabled us to interact with best-testers and with the end-users that gave us important feedback which led to the re-design of several aspects of our approach in terms of privacy preservation. There were three significant lessons that we learned from deploying our CrowdAlert app in the real-world that we discuss in the following.

14.5.7.1   Privacy Preservation

Our original version of CrowdAlert was developed without any privacy mechanism as we expected users would be sharing location-based data with multiple services. However, when we shared the app with our beta-testers this became an issue because the app required the users to share their location as a permission upon installation. During our discussions we understood that, although, the app acquired the user location when the user wanted to share a public report, it was not clear to the users whether their privacy could be exposed. Moreover, we felt that there were some users that were extremely concerned regarding their privacy.

In order to deal with this problem we developed our PROMPT privacy preservation approach that was briefly presented above. PROMPT enables users to quantify their privacy exposure in terms of location before sharing their data. Employing such a mechanism made the users were more willing to share their data with the community.

Furthermore, as mentioned above, our PROMPT approach enables the users to tune their privacy levels. This is essential to the privacy preservation mechanism since we realized that different users had different needs. That way we can attract both privacy concerned users as well as users that aim to share information with the community without prioritizing their privacy.

14.5.7.2   Data Sharing

Another important lesson was that the privacy preservation should not degrade the utility of the application. We explored several existing solutions before developing PROMPT, but none of them was able to preserve user privacy effectively without degrading system utility. The majority of the existing approaches modify user information and, thus, were rejected since our app focuses on real-world events that require accurate location. On the other hand since we focus on public data reports, similar to Waze, an encryption-based approach would not provide any benefit to the user privacy.

To deal with this problem we developed PROMPT, which is based on data suppression to preserve user privacy. One significant benefit of PROMPT, as we show in [5] is that it can preserve high levels of utility in terms of data sharing even with strict privacy settings. This enables even privacy concerned users to be able to share a lot of data without exposing their sensitive information.

14.5.7.3   User Interface

Finally, the most important lesson was that the user interface can play a fundamental role to the users' understanding of their privacy exposure. Our initial interface allowed the users to modify several parameters of our PROMPT approach. Nevertheless, this freedom was also a bottleneck especially for novice users did not have the technical knowledge to understand and set these parameters. Hence, several beta testers complained regarding the privacy settings as they were not able to tune them properly.

In order to deal with the above problem, we developed a new user interface where the privacy setting was selected among three privacy profiles, as explained above. This feature enabled the users to select their desired privacy levels even if they did not tune every aspect of the algorithm. Nevertheless, we kept the parameter tuning as a separate functionality that can be accessed by advanced users.

14.6   Conclusion

In this chapter we provided a survey of the most popular location-based applications and we explained the benefits that they provide to the users. Then, we discussed the respective privacy threats that may arise in such settings and presented the existing privacy mechanisms. Finally, we presented our own experiences upon developing a real-world location-based application, the privacy mechanism that we developed and the lessons that we learnt based on the issues that we encountered.

One important message that can be extracted from this chapter is that location-based applications can expose a great deal of sensitive user information and that the most popular applications rely on the user to handle such issues. Although, several approaches have been proposed from the research community, the commercial applications use simple solutions and expect that the users are capable of quantifying their exposure and preserve their privacy. On the other hand, we argue that popular real-world applications should benefit from the solutions that have already been proposed in the research literature so as to allow users to continue contributing information without being concerned about their privacy.

References

1. P. K. Agarwal, S. Har-Peled, and K. R. Varadarajan. Geometric approximation via coresets. *Combinatorial and computational geometry*, 52:1–30, 2005.
2. R. Agrawal and R. Srikant. Privacy-preserving data mining. In *SIGMOD*, Dallas, Texas, United States, May 2000.
3. C. Bettini and D. Riboni. Privacy protection in pervasive systems: State of the art and technical challenges. *Pervasive and Mobile Computing*, 17:159–174, 2015.
4. I. Boutsis and V. Kalogeraki. Crowdalert: a mobile app for event reporting and user alerting in real-time. In *UbiComp*, Heidelberg, Germany, 2016.
5. I. Boutsis and V. Kalogeraki. Location privacy for crowdsourcing applications. In *UbiComp*, Heidelberg, Germany, 2016.
6. C.-Y. Chow, M. F. Mokbel, and W. G. Aref. Casper*: Query processing for location services without compromising privacy. *ACM Transactions on Database Systems (TODS)*, 34(4):24, 2009.
7. Y.-A. de Montjoye, C. A. Hidalgo, M. Verleysen, and V. D. Blondel. Unique in the crowd: The privacy bounds of human mobility. *Scientific reports*, 3, 2013.
8. K. Dong, T. Gu, X. Tao, and J. Lu. Complete bipartite anonymity: Confusing anonymous mobility traces for location privacy. In *ICPADS*, pages 205–212, Singapore, December 2014.
9. K. Fawaz and K. G. Shin. Location privacy protection for smartphone users. In *CCS*, pages 239–250, Scottsdale, Arizona,, november 2014.
10. S. Gambs, M.-O. Killijian, and M. Núñez del Prado Cortez. De-anonymization attack on geolocated data. *Journal of Computer and System Sciences*, 80(8):1597–1614, 2014.
11. S. Gao, J. Ma, W. Shi, G. Zhan, and C. Sun. Trpf: A trajectory privacy-preserving framework for participatory sensing. *Information Forensics and Security, IEEE Transactions on*, 8(6):874–887, 2013.
12. A. Gkoulalas-Divanis and V. S. Verykios. A privacy-aware trajectory tracking query engine. *SIGKDD Explorations Newsletter*, 10(1):40–49, May 2008.
13. T. Higuchi, P. Martin, S. Chakraborty, and M. Srivastava. AnonyCast: privacy-preserving location distribution for anonymous crowd tracking systems. In *UbiComp*, pages 1119–1130, Osaka, Japan, Sep 2015.
14. B. Hoh and M. Gruteser. Protecting location privacy through path confusion. In *SECURECOMM*, Athens, Greece, September 2005.
15. C.-C. Hung, W.-C. Peng, and W.-C. Lee. Clustering and aggregating clues of trajectories for mining trajectory patterns and routes. *The VLDB Journal*, pages 1–24, 2011.
16. A. Kapadia, N. Triandopoulos, C. Cornelius, D. Peebles, and D. Kotz. Anonysense: Opportunistic and privacy-preserving context collection. In *Pervasive Computing*, pages 280–297. Sydney, Australia, May 2008.
17. S. Mascetti, L. Bertolaja, and C. Bettini. Safebox: adaptable spatio-temporal generalization for location privacy protection. *Transactions on Data Privacy*, 7(2):131–163, 2014.
18. S. Mascetti, D. Freni, C. Bettini, X. S. Wang, and S. Jajodia. Privacy in geo-social networks: proximity notification with untrusted service providers and curious buddies. *The VLDB Journal*, 20(4):541–566, 2011.
19. B. Niu, Q. Li, X. Zhu, G. Cao, and H. Li. Achieving k-anonymity in privacy-aware location-based services. In *INFOCOM*, pages 754–762, Toronto, CA, April 2014.
20. N. Pelekis, A. Gkoulalas-Divanis, M. Vodas, D. Kopanaki, and Y. Theodoridis. Privacy-aware querying over sensitive trajectory data. In *CIKM*, Glasgow, Scotland, October 2011.
21. N. Sadeh, J. Hong, L. Cranor, I. Fette, P. Kelley, M. Prabaker, and J. Rao. Understanding and capturing people's privacy policies in a mobile social networking application. *Personal and Ubiquitous Computing*, 13(6):401–412, 2009.
22. C. E. Shannon. A mathematical theory of communication. *SIGMOBILE Mob. Comput. Commun. Rev.*, 5(1):3–55, 2001.

23. J. Shao, R. Lu, and X. Lin. Fine: A fine-grained privacy-preserving location-based service framework for mobile devices. In *INFOCOM*, pages 244–252, Toronto, CA, April 2014.
24. M. Terrovitis and N. Mamoulis. Privacy preservation in the publication of trajectories. In *MDM*, Beijing, China, April 2008.
25. I. J. Vergara-Laurens, D. Mendez, and M. A. Labrador. Privacy, quality of information, and energy consumption in participatory sensing systems. In *PerCom*, pages 199–207, Budapest, Hungary, March 2014.
26. S. Wilson, J. Cranshaw, N. Sadeh, A. Acquisti, L. F. Cranor, J. Springfield, S. Y. Jeong, and A. Balasubramanian. Privacy manipulation and acclimation in a location sharing application. In *UbiComp*, pages 549–558, Zurich, CH, September 2013.

Index

A

Absence inference attack, 217
Abstraction level, 342
Accountability, 207, 236, 237, 239, 241, 242
Accuracy, 3, 6, 7, 24, 25, 40, 41, 44–45, 57, 64,
    85, 92, 102, 106–108, 111, 115, 117,
    120, 121, 132, 136, 151, 198, 200, 201,
    216–220, 245, 275, 281, 302, 311, 315,
    322, 323, 331, 339, 360, 362, 387, 389,
    390
Activity privacy, 215
Adaptive grid (AG), 186, 188–191
Adversary knowledge, 259, 260, 265
Aggregate mobility, 24–27, 308
Analytics quality/utility, 262–263, 267–268
Anonymization/anonymity, 4, 6, 58, 61–63,
    81–87, 90, 91, 93, 99, 143, 185, 204,
    208, 212, 213, 215, 239, 241, 245, 257,
    262–268, 274, 282, 284–286, 298–300,
    312–315, 317–326, 330–332, 343, 351
Anonymous data sharing, 385, 388, 389
Artificial neural network (ANN), 332
Assignment link, 172, 173, 175, 177
(location-based) Attack, 4, 171, 180
Attack model, 257–260, 265
Authentication, 78, 85, 237–239, 355,
    358–359, 363

B

Blowfish privacy, 99, 111, 117–123
Bluetooth, 3, 16, 41–45, 55, 131, 132, 135,
    137, 147, 200, 202, 359
Budget absorption, 108
Budget distribution, 108

C

Call detail records (CDR), 15–18, 20–23, 26,
    28, 321–325, 332
Car-to-car (C2C) communication, 230, 239,
    245
Cellular service provider (CSP), 186–188,
    190–192
Cellular tower, 3, 43
Center for Disease Control and Prevention
    (CDC), 208
Citizen sensing, 129
(spatial) Cloaking technique/algorithm, 82, 83,
    179–180, 185
Clustering, 5, 219, 255, 258, 263, 275, 276,
    294, 297, 299, 320, 324, 327–331
Collective information practices, 219, 257,
    268, 270, 272, 273, 346
Co-location privacy threat, 215, 218
Community sensing, 129
Confidence value, 340, 342
Connected car ecosystem/vehicle, 5, 17, 27,
    97, 207, 230–248, 310, 311
Consent, 45, 57–62, 64, 76, 79, 80, 172, 179,
    187, 202, 301, 302, 362
Constraints, 73, 85–87, 107, 110, 115, 119,
    120, 144, 148, 180–185, 236, 312, 320,
    346, 348, 355, 356, 358
Content adaptation, 355, 360–361
Context-adaptive privacy mechanism, 6,
    337–364
Context-aware browsing, 38, 51, 53, 55
Context-aware computing system/context-
    adaptive, 6, 248, 337–364
Context-aware resource management, 339
Context-aware user interface, 339

© Springer Nature Switzerland AG 2018
A. Gkoulalas-Divanis, C. Bettini (eds.), *Handbook of Mobile Data Privacy*,
https://doi.org/10.1007/978-3-319-98161-1

Context-based service, 37
Context framework, 341
Context model, 341–344, 350–352, 356, 360
Context modeling language (CML), 342–344
Context ownership, 344
Contextual integrity, 345, 346, 348, 351, 353
Contributed geographic information (CGI),
        199–202, 209
Cooperative awareness message (CAM), 230,
        238
Coordinate transformation, 86, 88–89
Counting queries, 105–109, 111, 115, 123,
        317, 318
Coverage, 2, 75, 92, 130, 133, 136–139, 151,
        183–185, 198, 200, 201, 235, 237, 238,
        267, 268
CrowdAlert, 383, 386, 391–396
Crowdsensing application, 4, 130–134, 136,
        138–142, 145–149, 151, 153
Crowdsensing backend/infrastructure, 132–139
Crowdsensing frontend, 131
(spatial) Crowdsourcing, 4, 129, 167–192
Cryptographic hashing, 89, 218
Cryptographic method, 81, 89
Curse of dimensionality, 320, 331

D
Data generalization, 6, 390
Data minimisation, 57, 61, 63
Data perturbation, 6, 144, 145, 274, 284, 285,
        390
Data quality, 130, 135–138, 148, 151, 200,
        201, 343
Data quantity, 135
Data suppression, 386, 390
De-identification, 254, 258, 363
Differential privacy, 4, 5, 58, 81, 83, 90–93,
        99–105, 110–112, 114–124, 143, 185,
        186, 191, 192, 254, 255, 269–271, 274,
        275, 314–316, 319, 321–323, 327, 331,
        332
Distance estimation, 183
Distributed anonymization, 84–85
Disturbance endpoints, 351–353
Diversity (problem), 84
Dummy based techniques, 86–88
Dynamic privacy, 347, 361

E
Efficiency applications, 233, 239
Electronic Privacy Information Center (EPIC),
        208

Elliptic-curve digital signature algorithm
        (ECDSA), 239
Encryption (techniques), 191–192, 382
Endomondo, 134, 380, 386
Entropy, 21, 178, 179, 219, 394
Environment features, 352–354
EU Data Protection Directive (DPD), 57
EU General Data Protection Regulation
        (GDPR), 57, 59, 60, 75, 79, 254, 257
EU privacy framework, 256
Event-differential privacy, 103
ω-Event privacy, 104–106, 108, 323
Exchange-based techniques, 178–179

F
Facebook, 106, 199, 209, 212, 375–376, 385,
        388
Fake data injection, 6, 390
Fake trajectory, 287–297, 304
Fingerprinting, 3, 40–43, 54, 64, 74, 131, 141,
        238, 364
Flickr, 380, 386, 388
Floating car data (FCD), 137, 230, 231, 237,
        238, 243, 247, 248
Foursquare, 16, 20, 209, 212, 217, 376, 385,
        388
Frequent pattern mining, 298
Full disclosure, 317

G
Generalized trajectory, 258, 270
Geocast region (GR), 186–190
Geofencing, 38, 55
Geographic information system (GIS), 17, 203
Geo-indistinguishable mechanism, 121
Geo-surveillance, 107, 202–208
Global positioning system (GPS), 3, 15, 16,
        42, 74, 80, 91, 97, 131, 132, 135, 139,
        146, 200, 202, 211, 212, 219, 244, 254,
        258, 260, 261, 272, 324, 325, 353, 362,
        379, 381, 386, 390
Global sensitivity, 269, 315, 322, 323
Google maps, 17, 377, 389
Granularity preference, 344
Graphical modeling approach, 342
Gravity model, 24–26, 28

H
Health Insurance Portability and
        Accountability Act (HIPAA),
        207, 208

Hermes++, 5, 283, 284, 286–298, 304
Hierarchical hybrid model, 343
Hippocratic principles, 6, 284, 301, 302, 304
HipStream, 6, 284, 301–304
Human behavior, 15, 18, 20, 28–30, 98, 266,
     274
Human mobility, 4, 7, 13–31, 258, 309, 311,
     319, 331

I
Identification threat, 387
Information disclosure, 210, 3259–360
Instagram, 212, 380, 386, 388
Institutional review board (IRB), 208
Internet of Things (IoT) system, 98, 337

K
k-anonymity, 4, 6, 58, 81–86, 98, 102, 179,
     181, 185, 245, 259, 261, 268, 274, 284,
     311–314, 331, 390
Knowledge discovery, 5

L
Laplace mechanism/noise, 101, 102, 105, 106,
     116, 117
l-diversity, 84, 185
Linkability, 63, 64, 236, 239
Locaccino, 80, 382–383, 386, 389
Location-based application, 6, 97, 112, 374,
     384–387, 389, 390, 392, 396
Location-based service (LBS), 3, 7, 36, 37,
     73–93, 120, 196, 197, 259, 275, 374,
     388
Location-based service classification, 74–75,
     77, 78, 92
Location-based social network (LBSN), 3,
     73–93, 195–221
   adversaries/threats, 213–216
   goals, 212
Location indicative words (LIW), 217
Location privacy protection, 2, 78, 93, 202,
     247
Location sensing technology, 36, 39, 40, 97
Location sharing application, 6, 7, 374–375,
     380
l_1-sensitivity, 101–102, 105, 106, 109,
     115–117, 121, 122

M
Markov model/process, 21, 22, 109, 110, 113,
     115, 119, 121, 220, 322

Minimum bounding box (MBB), 289, 290,
     292, 293, 295, 296
Mix zone, 6, 242, 390
Mobile device, 1, 2, 4, 19, 38, 74, 82, 130–133,
     135, 137, 138, 168, 170, 192, 196, 198,
     200, 202, 209, 212, 218, 281, 359, 373,
     376–378, 380, 381, 392–394
(human) Mobility model, 13–31
Mobility trace, 2, 23, 81, 133, 246, 323, 380,
     387, 392, 394, 395
Moovit, 377, 385, 389
Move frequency, 270
Multiple data releases, 77

N
Nearest neighbor (NN), 82, 87–89, 283, 286,
     288, 297, 298
Neighboring databases, 100–105, 111, 112,
     114, 116, 118, 119, 122

O
(location) Obfuscation, 81, 85–88
Offline data release, 76, 77
Online data release, 76, 77, 91
Ontological model, 342, 343
Optimization, 39, 75, 106–107, 110, 176,
     183–185, 235, 310, 324, 330

P
Participatory sensing, 4, 7, 129, 169–171, 178,
     179
Participatory urbanism, 129
Path confusion, 6, 390
Pay-as-you-drive (PAYD), 231, 233, 243–244,
     247
PCube, 80, 90, 381–382, 386
People-centric sensing, 129
Personal data, 57–59, 61, 62, 79, 90, 92, 139,
     200, 214, 254, 257, 301, 309, 314
Personal LBS, 74, 75, 345
Personally identifiable information (PII), 57,
     62
(location) Perturbation, 85–89, 191, 323, 331
Place of interest (POI), 81, 82, 84, 199
Planar isotropic mechanism, 121
Positioning technologies, 1, 7, 281
Precision, 74, 77–80, 82, 85, 90, 132, 136, 263,
     274, 382
Preferences, 3, 46, 47, 52, 55, 78–80, 90, 92,
     93, 113, 124, 149, 168, 218, 248, 302,
     338, 343, 344, 347–350, 353–364, 377,
     382, 392–394

Privacy automation privacy-aware analytical
    process, 257
Privacy-aware querying, 298
Privacy budget, 101, 106–109, 116, 186, 327
Privacy by design, 5, 7, 79, 243, 253–276, 343
Privacy challenges, 4, 5, 17, 169, 197
Privacy controls, 36, 57–64, 78, 248, 302, 347,
    350, 357, 363
Privacy-enhancing technology (PET), 197, 210
Privacy model, 2, 6, 186, 187, 196, 259–260,
    264–265, 268–269, 274, 291, 311–316,
    321, 330, 331
Privacy preference, 78–80, 90, 124, 218, 302,
    344, 347–349, 353–364, 382, 2148
Privacy-preserving application, 373–396
Privacy-preserving charging, 245–247
Privacy-preserving (mobility data)
    management, 7, 281–304
Privacy-preserving (mobility data) publication,
    5
Privacy-preserving service, 5, 8, 88, 308
Privacy-preserving (data) sharing, 244–245
Privacy-preserving technique, 63, 81, 215,
    260–261, 265–266, 268, 270–271
Privacy regulation theory, 346–348
Privacy service provider (PSP), 63, 81, 215,
    260–261, 265–266, 268, 270–271
Privacy settings interface, 394–395
Privacy threat/privacy harm, 3, 4, 6, 36, 40,
    45–51, 53–57, 61, 62, 64, 74–77, 92,
    169, 171–175, 180, 210, 212–221, 247,
    274, 295, 297, 308, 330, 343, 387, 396
Privacy zone, 385, 388–389, 394
Private-Hermes, 6, 283, 284, 297–301, 304
Private information retrieval (PIR), 89–90,
    216, 343
Private spatial decomposition (PSD), 186–191
Profiling (threat), 387
Progressive retrieval technique, 88, 215
Protected health information (PHI), 207, 208
Pseudonym change, 238, 240–242, 248
Pseudonym lifecycle, 239–241
Pseudonym/pseudonymity, 4, 61, 176–178,
    192, 215, 230, 238–242, 248
Pufferfish privacy, 99, 112–117
Pull mode, 4, 74, 168–177, 179, 192
Push mode, 4, 74, 168–176, 179–192

Q
Quasi-identifier (QI), 82, 259, 265, 266, 274,
    312, 344
Query auditing, 291–298, 317–319, 331, 332

Query engine, 5, 283–287, 291, 304
Query enlargement, 86–87, 215

R
Radiation model, 25, 26
Radio-frequency identification (RFID), 3, 7,
    36, 39, 44–58, 61–63, 203, 358, 359
Real-time locating system (RTLS), 44, 52–56,
    63, 64
Reasoning, 264, 308, 340–343, 347, 354–356,
    359, 362
Received signal strength indicator (RSSI),
    41–44
Reciprocity, 83, 84, 86, 87, 346
(trajectory) Reconstruction, 332
Recurrent neural networks (RNN), 332
Redundancy, 136
Re-identification (attack), 2, 259
Reporting (link), 172, 173, 175, 178
Resource investment, 145–147
Retail, 3, 36–39, 42–52, 55, 56, 64, 310
Revocation, 237, 241–242
Rule-based scheme, 386, 389

S
Safety applications, 5, 231, 232, 234–236, 242,
    248
Sampling, 18, 110, 121, 135, 276, 290, 298,
    323–325, 328, 329, 331
Sanitization, 185–190, 267, 315, 323, 331
Scaling, 21, 25, 30, 327–329, 338
Secret sharing, 89, 90, 242, 245
Secure multi-party computation (SMC), 89,
    90, 144, 218
Semantic anonymization, 319, 321–322, 331
Semantic data type, 342
Semantic privacy model, 311, 313–316, 331
Semantic trajectory, 264, 266
Semi-honest third party, 192, 214, 216
Sensing accuracy, 44–45
Sensitive attribute (SA), 259, 274, 312, 313
Sensitive location tracking attack, 286, 291,
    292
Sensor, 2, 4, 15, 16, 27, 130–153, 168, 196,
    201, 202, 208, 211, 212, 231–233, 337,
    338, 340–343, 345, 348–351, 353–360,
    362–364, 373, 375, 377, 379, 383, 384,
    389, 390
Sequential composition, 102, 104, 120
Sequential tracking attack, 283, 286, 292, 295
Single data release, 77

Sliding window, 107, 108
Smart device, 45
Smoothing, 289, 290, 323, 325, 326, 328–329
Social mining, 254, 274
Spatial index, 188, 275, 323
Spatiotemporal annotations, 144, 145, 152
Spatiotemporal density, 6, 307–332
Spatio-temporal resolution, 261
Strava, 379–380, 385, 389
Stream, 4, 6, 36, 37, 104, 105, 108–109, 124,
        137, 153, 233, 268, 284, 301–304, 344,
        354, 360
Syntactic anonymization, 313, 317–319, 331
Syntactic privacy model, 311–314

T

Tasking, 4, 132, 133, 141–143, 168–177, 179,
        180, 191
TaskRabbit, 168–172, 174–175
Thematic maps, 4, 196, 197, 204, 205, 221
Threat mitigation, 142–145
Threat modalities, 142
Tracking threat, 387
Trajectory, 4–6, 21, 73, 77, 98, 104, 105,
        109–111, 115, 116, 121, 124, 136, 144,
        178, 179, 203, 212, 230, 236, 257–261,
        264–266, 270, 271, 274–276, 283–298,
        304, 309, 312, 313, 316, 320, 321, 331,
        332, 394
    anonymization, 6, 298, 331, 332
    linkage attack, 259
    representative, 297
Transmission principles, 345, 346, 351, 353
Triangulation attack/trilateration, 3, 16, 40–44,
        54, 64, 175, 187, 200, 389, 394
Trip advisor, 378–379, 385, 390

Trusted anonymizer, 83–85
Twitter, 16, 20, 199, 209, 211, 212, 216, 217,
        376, 385, 388

U

Uber, 17, 378, 385, 390
Ultrasonic waves, 43–44
Uniqueness, 23, 42, 47, 58, 93, 98, 141, 308,
        309
Urban sensing, 129
User identification attack, 286, 294
User incentives, 140
User profiling, 343, 374
User tailing, 282

V

Vehicle-to-vehicle (V2V) communication, 230
Vehicle-to-X communication, 231
Vehicular services, 231–233, 243–247
Velocity, 137, 138, 153, 233, 235
Visibility scope, 385, 386, 388, 389
Volunteered geographic information (VGI),
        107, 199–202, 209

W

Waze, 140, 168, 171, 378, 385, 389, 396
Wifi, 3, 15, 16, 36, 41–48, 54–56, 58, 63–64,
        74, 84, 141, 146, 200, 202, 212, 235,
        353, 359, 377
Window-differential privacy, 104

Y

Yelp, 379, 385, 390

Printed in the United States
By Bookmasters